Global Agricultural and Food Marketing in a Global Context:

Advancing Policy, Management, and Innovation

Aluwani Maiwashe-Tagwi
University of South Africa, South Africa

Ailweli Solomon Mawela
University of South Africa, South Africa

Phineas Khazamula Chauke
University of South Africa, South Africa

IGI Global
PUBLISHER of TIMELY KNOWLEDGE

A volume in the Advances in Marketing, Customer Relationship Management, and E-Services (AMCRMES) Book Series

Published in the United States of America by
IGI Global
Business Science Reference (an imprint of IGI Global)
701 E. Chocolate Avenue
Hershey PA, USA 17033
Tel: 717-533-8845
Fax: 717-533-8661
E-mail: cust@igi-global.com
Web site: http://www.igi-global.com

Library of Congress Cataloging-in-Publication Data

Names: Tagwi, Aluwani Maiwashe, 1984- editor. | Mawela, Ailwei Solomon, editor. | Chauke, Phineas K., 1955- editor.
Title: Global agricultural and food marketing in a global context : advancing policy, management, and innovation / Aluwani Maiwashe Tagwi, Ailwei Mawela, and Phineus Chauke, editor.
Description: Hershey, PA : Business Science Reference, [2023] | Includes bibliographical references and index.
Identifiers: LCCN 2022021437 (print) | LCCN 2022021438 (ebook) | ISBN 9781668447826 (ebook) | ISBN 9781668447802 (hardcover) | ISBN 9781668447819 (paperback)
Subjects: LCSH: Food--Africa--Marketing. | Farm produce--Africa--Marketing. | Food industry and trade--Africa. | Agricultural industries--Economic aspects--Africa.
Classification: LCC HD9000.5 (ebook) | LCC HD9000.5 .G573 2023 (print) | DDC 338.196068/8 23/eng/20220--dc18
LC record available at https://lccn.loc.gov/2022021437

This book is published in the IGI Global book series Advances in Marketing, Customer Relationship Management, and E-Services (AMCRMES) (ISSN: 2327-5502; eISSN: 2327-5529)

British Cataloguing in Publication Data
A Cataloguing in Publication record for this book is available from the British Library.

For electronic access to this publication, please contact: eresources@igi-global.com.

Advances in Marketing, Customer Relationship Management, and E-Services (AMCRMES) Book Series

Eldon Y. Li
National Chengchi University, Taiwan & California Polytechnic State University, USA

ISSN:2327-5502
EISSN:2327-5529

Mission

Business processes, services, and communications are important factors in the management of good customer relationship, which is the foundation of any well organized business. Technology continues to play a vital role in the organization and automation of business processes for marketing, sales, and customer service. These features aid in the attraction of new clients and maintaining existing relationships.

The Advances in Marketing, Customer Relationship Management, and E-Services (AMCRMES) Book Series addresses success factors for customer relationship management, marketing, and electronic services and its performance outcomes. This collection of reference source covers aspects of consumer behavior and marketing business strategies aiming towards researchers, scholars, and practitioners in the fields of marketing management.

Coverage

- CRM strategies
- Customer Relationship Management
- Legal Considerations in E-Marketing
- CRM and customer trust
- Social Networking and Marketing
- Cases on Electronic Services
- Mobile Services
- Electronic Services
- CRM in financial services
- Ethical Considerations in E-Marketing

IGI Global is currently accepting manuscripts for publication within this series. To submit a proposal for a volume in this series, please contact our Acquisition Editors at Acquisitions@igi-global.com or visit: http://www.igi-global.com/publish/.

The Advances in Marketing, Customer Relationship Management, and E-Services (AMCRMES) Book Series (ISSN 2327-5502) is published by IGI Global, 701 E. Chocolate Avenue, Hershey, PA 17033-1240, USA, www.igi-global.com. This series is composed of titles available for purchase individually; each title is edited to be contextually exclusive from any other title within the series. For pricing and ordering information please visit http://www.igi-global.com/book-series/advances-marketing-customer-relationship-management/37150. Postmaster: Send all address changes to above address.

Titles in this Series

For a list of additional titles in this series, please visit: www.igi-global.com/book-series

Enhancing Customer Engagement Through Location-Based Marketing
Amandeep Singh (Chitkara Business School, Chitkara University, India) Amit Mittal (Chitkara Business School, Chitkara University, India) and Murat Unanoglu (İstanbul Aydin University, Turkey)
Business Science Reference • © 2023 • 300pp • H/C (ISBN: 9781668481776) • US $250.00

Big Data Marketing Strategies for Superior Customer Experience
Jose Ramon Saura (Rey Juan Carlos University, Spain)
Business Science Reference • © 2023 • 305pp • H/C (ISBN: 9781668464540) • US $240.00

Managing Festivals for Destination Marketing and Branding
Sharad Kumar Kulshreshtha (North-Eastern Hill University, India)
Business Science Reference • © 2023 • 300pp • H/C (ISBN: 9781668463567) • US $250.00

Handbook of Research on the Interplay Between Service Quality and Customer Delight
Sarmistha Sarma (Institute of Innovation in Technology and Management, India) and Neha Gupta (IBCS, SOA University (Deemed), India)
Business Science Reference • © 2023 • 438pp • H/C (ISBN: 9781668458532) • US $295.00

Cases on Social Justice in China and Perspectives on Chinese Brands
Youssef Elhaoussine (Beijing Normal University-Hong Kong Baptist University United International College, China) and Lulu Wang (Sinotrend Consulting, China)
Business Science Reference • © 2023 • 276pp • H/C (ISBN: 9781668449554) • US $215.00

Handbook of Research on the Future of Advertising and Brands in the New Entertainment Landscape
Blanca Miguélez-Juan (University of the Basque Country, Spain) and Gema Bonales-Daimiel (Complutense University of Madrid, Spain)
Business Science Reference • © 2023 • 441pp • H/C (ISBN: 9781668439715) • US $295.00

Promoting Consumer Engagement Through Emotional Branding and Sensory Marketing
Monika Gupta (Chitkara Business School, Chitkara University, India) Priya Jindal (Chitkara Business School, Chitkara University, India) and Shubhi Bansal (Indian Institute of Technology, Indore, India)
Business Science Reference • © 2023 • 298pp • H/C (ISBN: 9781668458976) • US $250.00

IGI Global
PUBLISHER of TIMELY KNOWLEDGE

701 East Chocolate Avenue, Hershey, PA 17033, USA
Tel: 717-533-8845 x100 • Fax: 717-533-8661
E-Mail: cust@igi-global.com • www.igi-global.com

Table of Contents

Detailed Table of Contents

Over a period of time, the South African agricultural sector has been divided into two sub-sectors, i.e., commercial and subsistence farming for whites and black farmers respectively. It also went through three marketing environmental phases, commencing pre 1913 to date. Use is made of the systemic literature review methodology, considered superior to its more unstructured narrative review counterpart. The findings of the review process are that the early 1900s reflected what a competitive non-regulated market can achieve even for those with minimal productive capacities, while the second phase that commenced with intensification of discriminatory legislation reflected the negative outcomes of the system. The last phase, covering the period after 1994, reflected development of an inclusive policy environment with minimal incorporation of the emerging black farming community. The chapter recommends more concerted efforts at graduating emerging farmers, implementation of well-intended and crafted policies and improved monitoring of CPAs.

Trade in agricultural products plays a particularly crucial role in addressing food security and reduction of poverty. By making the ability of the continent to feed itself a priority goal in the Malabo Declaration, African leaders shared renewed faith in agriculture, and showed how stepped-up investments in production and distribution can make a strong contribution towards the attainments of goals of the African Continental Free Trade Area agreement (AfCFTA). Food security goes beyond quantity of food available to inclusion of nutritious diets in households' menu. These commodities are vital to the attainment of food and nutritional security, soil conservation, and, if commercially produced in efficient systems, can boost

ECOWAS participation in African trade. This chapter therefore examines the prospects for promoting trade in pulses and legumes within the ECOWAS region in the context of the AfCFTA protocol, paying particular attention to trends, prospects and barriers and where there are flexibilities that allow for beneficial and productive policy intervention.

The chapter focused on assessing the nature of voluntary sustainability standards (VSSs) that affect South Africa's producers of horticultural produce (fruits in particular) destined for markets within Europe and Africa. Eight VSSs affecting South Africa's fruit trade with Europe and within Africa under food safety and traceability themes were identified. The outcomes of global certification as a requirement to participate in global international markets are outlined, with a specific focus on the implication of these standards for smallholder farmers. Given the independent nature of the end goals from the institutions responsible for these VSSs, it becomes problematic and impractical to argue that VSSs that focus on similar thematic areas, e.g., GLOBALG.A.P. Crops and IFOAM, be merged into one standard. Thus, it is prudent that smallholder producers outrightly identify a specific target market and the corresponding standard requirements such that activities are undertaken in accordance with what the market needs.

Public/social behaviour is one of the main elements that determines the success or failure of any commercial product launched in the market, and food product marketing is no exception. There have been several studies conducted to assess attitude, awareness, or acceptance of genetically modified (GM) food products, giving a good insight about the perception towards this product. However, limited attempts have been made to determine the antecedents that influence the attitude towards GM food products in southeast Asian context. This chapter attempts to investigate the influence of perceived benefits, perceived risks, knowledge, general attitude, and trust on attitude towards GM food products in Malaysia, a fast-developing economy. To effectively develop the GM food industry, it is important for the government, policy makers, non-governmental organisations and the industry in general to seriously consider the market's attitude towards GM food products, while formulating commercialisation strategies.

Agricultural marketing determines the extent of smallholder farmer commercialization. However, this needs effective agricultural markets. Evaluating these markets and their interaction with small farmers is critical in harnessing the potential from the rural farming sector. The chapter aims to evaluate formal and informal agricultural markets, highlighting opportunities for smallholder agriculture. The findings reveal that smallholder producers' interaction with formal markets is limited by a lack of knowledge and resources, poor safety and quality standards, and inconsistency in supply. However, they can benefit from opportunities in the formal markets through effective collective action arrangements. Informal markets are critical to smallholders' access to markets, and with legislation that seeks to address infrastructure, safety, and quality standards, they can be a conduit for small producers' access to formal markets. Some informality still exists in the supply chain of formal markets. Thus, informal markets remain important in agriculture marketing in Africa.

Compared to the other fields of business, research on image congruence in the food business sector, particularly organic food, is relatively scarce. Previous studies could only identify the congruence of certain psychological elements of perceived image. Other functional elements, on the other hand, are mostly overlooked. Considering these issues, this study aimed to examine customer perceptions of themselves as organic food consumers, of organic food products, and of organic food makers and distributors. Using data gathered from nine informants, this study found that organic food consumers held certain images concerning organic food products, their makers and distributors, and themselves. Some images of organic food consumers and organic food products shared the same nature. In addition, some images of organic food consumers, makers and distributors had the same essence. Implications of these findings, together with the limitations of this study and directions for future research, were discussed.

This chapter examines the functioning of the national fresh produce markets (NFPMs) with specific reference to three themes: 1) challenges in the regulatory environment and the importance of social relation, 2) networks in shaping the functioning of the NFPMs alongside the narrow economic relations that exist among the market participants in the NFPMs, and 3) the importance of the informal sector for the NFPMs and the distribution of fresh produce. Specific reference is made to the Johannesburg Fresh case study for the Johannesburg Fresh Produce Market (JFPM). The chapter proposes solutions to the mentioned challenges.

Cooperatives are deemed as vehicles for rural development, especially in developing countries. Smallholder farmers generally experience huge marketing transactional costs, and to reduce this they normally resort to becoming members of cooperatives. In South Africa, most small-scale farmers do not have access to markets. With no markets in place, the transformation would be almost impossible. The study's aim was to assess the determinants of market access in agricultural cooperatives in Limpopo Province. A face-to-face survey was conducted with 146 agricultural cooperatives members. The results showed that market access was associated with age, training, workers in a household, knowledge of the cooperatives principles and act, annual income, and access to funding. The study recommended younger farmers be targeted when designing market access interventions as the impact of job creation is huge amongst this group. Tailor-made training on market access and cooperative principles and acts should be designed. Funding should be made more accessible to small-scale farmers in cooperatives.

This chapter seeks to discuss and assess the concept of commercialization of smallholder farmers, and whether it is a reality or a myth. The focus of the study area was the former Mutale Local Municipality, in Vhembe district of Limpopo Province in South Africa, drawn from 4 areas namely Tshipise, Tshixwadza, Tshishivhe, and Masisi agricultural zones by using a sample of 153 small holders drawn randomly. The problem of the study was that there are more than 4 million households engaged in agriculture in South Africa who in are engaged in subsistence farming. If these farmers can be assisted, they can feed the nation. Data were collected through a structured qualitative and quantitative questionnaire that was administered face-to-face to respondents and the data was analysed by using a computer programme of the SPSS Version 24. The findings suggest that farmers need support in different aspects such as training, extension, and comprehensive support by the government

Climate change is instrumental for agricultural performance, with its production and marketing being the leading cores. Although numerous climate change adaptive strategies have been identified, there is a need to explore how climate change impacts agricultural marketing. The focus of the study was to investigate the agricultural marketing dynamics brought by the impact of climate change among rural-based emerging farmers. The study was conducted in the Vhembe district of the Limpopo province. A

Preface

This is a scholarly book that is well-researched and relevant to scholars, universities, researchers, agricultural policymakers, small-scale farmers, agricultural industries, government agencies and the private sector globally. Most issues concerning global food demand are presented in various chapters depending on the approach and the influence of current discourse, debates, developments, and research from various countries globally. Some of the chapters paid attention to the food demand and supply imperatives as a way of addressing the expected growth in food demand due to the inevitable population increase. Considering that most developing countries rely on smallholder agriculture for food security, the development of the agricultural marketing sector is crucial. The projected demands call for a functional agricultural marketing environment, especially in developing countries. This book addresses the requirements that should be in place for the creation of a conducive agricultural marketing environment. This book provides the reader with pedagogical knowledge of research, insights and strategies on agricultural marketing policy environment, global marketing certification requirements and its implications for developing countries, trends, prospects and barriers in promoting regional trade under AfCFTA, an overview of formal and informal markets in Sub-Saharan Africa, fresh produce market challenges, marketing in the context of climate change, cooperatives role in marketing, organic food market perspective, genetically modified food market and prospects of smallholder commercialization that contributes to the creation of effective agricultural markets for the smallholder farming sector. Chapters in this book articulate the importance and the role of graduates, academics, agricultural economists, and agricultural extension officials in the agricultural marketing sector in alleviating market access problems.

This book aimed to provide relevant theoretical frameworks and the latest empirical research findings in the area of agricultural marketing. The book focused on providing pedagogical content knowledge to address the gaps in Global Agricultural and Food Marketing in a Global Context to advance policy, management and Innovation in the smallholder farming sector. This topic caters to knowledge to combat food security and improve farmers' livelihoods in predominantly developing countries.

The audience of this book includes students, farmers, graduates, national agriculture marketing directorate and research, agricultural economists, academics, researchers, banks, marketing agencies, curriculum developers, agricultural government institutions, private institutions, farmer associations, researchers and the public domain who want to improve their understanding of the strategic role of the agricultural marketing environment, global economy, smallholder farmers, agricultural information systems and individuals as actors in the agricultural marketing environment.

Chapter 1, 'A Systematic Literature Enquiry of the South African Agricultural Marketing Environment Pre-1913 to the Present', was written by Phineus Khazamula Chauke. This chapter presented a systematic literature review of the South African agricultural marketing environment from just before the promulgation of the Natives Land Act of 1913 to the present time. It is common knowledge that the South African agricultural economy has been marred by policy environments that tilted towards advantaging the white commercial farming sector to the detriment of its black subsistence counterpart. In the final analysis, the chapter sought to uncover the extent to which such a policy environment could have impacted the participation of the two farming groups generally, but specifically on black farmers' agricultural production and marketing efforts through a literature enquiry. The participation of emerging black farmers in the South African agriculture sector is still a matter of serious concern, exacerbated by the existence of racially based commercial and subsistence farming. In order to ensure that the research and information retrieval processes were not only accurate but also impartial, the method adopted for this chapter was a systematic literature review method (SLRM). Using this methodology, the chapter is well set to respond to the historical theme of the book through an assessment of the South African agricultural marketing environment commencing prior to 1913, cutting across colonial and self-declared apartheid republican administrations until the achievement of a democratic dispensation in 1994. The findings of the review process are that the sector went through three marketing environmental phases commencing pre-1913 to date. The early 1900s reflected what a competitive non-regulated market environment could achieve even for those with minimal productive capacities while the second phase that commenced with the intensification of discriminatory legislation reflected the negative outcomes thereof. The last phase, covering the democratic dispensation period after 1994, reflected the development of an inclusive policy environment with minimal incorporation of the emerging black farming community. The chapter recommends more concerted efforts at graduating emerging farmers, implementation of existing well-crafted policies and improved monitoring of Communal Property Associations (CPAs). The adoption of emerging technologies and policies meant to upscale their participation could be crucial in accelerating agricultural productivity for emerging farmers and thus increase participation in the marketing environment. The chapter also recommends exploration and investment in new green markets such as the bioenergy market where farmers can participate as feedstock producers, specifically in the small-scale sector as this has a huge potential for ensuring the highly advocated just transition in the renewable energy sector.

Chapter 2, 'Trends, Prospects, and Barriers in Promoting Regional Trade in Pulses/Legumes in the ECOWAS Region Under AfCFTA Regime', was written by Anthony Ojonimi Onoja and Ajuruchukwu Obi. This chapter examines the prospects for promoting trade in pulses and legumes within the ECOWAS region in the context of the AfCFTA protocol, paying particular attention to trends, prospects and barriers and where there are flexibilities that allow for beneficial and productive policy intervention. Trade in agricultural products plays a particularly crucial role in addressing food security, and reduction of poverty. The creation of the continent's ability to feed itself a priority goal in the Malabo Declaration, African leaders shared renewed faith in agriculture and how stepped-up investments in production, marketing and distribution can make a strong contribution towards the attainment of goals of the African Continental Free Trade Area agreement (AfCFTA). The AfCFTA was founded by the African Union in 2018 although effective trading did not commence until 1 January 2021. The expectation is that the Free Trade Area (FTA) will serve the 55 member nations of the AU and affect the lives of the 1.2 billion persons that reside in these countries. The mixed methods approach involving both quantitative and qualitative techniques was applied. The first step was to conduct a descriptive profiling of the trends in the production of pulses in West Africa, with particular emphasis on trade (export and import) using computed tables and graphical analysis. Details of these trends with respect to soybean and groundnut were specifically highlighted. The next step was to examine the barriers to the attainment of regional leader status or top market performers under AfCFTA. A combination of desk research (stylized facts analysis), trend analysis and descriptive techniques were applied based on aggregated time series data collected on the West African region from FAOSTAT. The study found that the African region was consuming more pulses than they were producing as their import bills for this commodity grew geometrically at 8,295.53 percent over a period of six decades. The study recommended that at the country and state levels, as well as investors in the private sector should increase their investment and innovations in this agricultural sub-sector, the potential of this commodity as a driver of growth in agricultural commodities in AfCFTA can be realized given that some major constraints against the production and export performance of these crops are mitigated.

Chapter 3, 'Global Certification Systems of Key Importance in Accessing International Markets for the South Africa Fruit Industry', was written by Ndiadivha P Tempia and Moses H Lubinga. This chapter investigates the outcomes of global certification as a requirement to participate in global international markets by various market actors with a specific focus on the implication of these standards for smallholder farmers. The focus was on assessing the nature of Voluntary Sustainability Standards (VSS) that affect South Africa's producers of horticultural produce (fruits in particular) destined for markets within Europe and Africa. According to the United Nations Forum on Sustainability Standards (UNFSS), VSS are specific requirements that are bound to be met by producers, traders, manufacturers, retailers or service providers in relation to a variety of sustainability measures. Secondary data extracted from the International Trade Centre's (ITC) Standards Map were used. ITC's standards Map provides a trusted and neutral source of information on VSS, audit protocols, codes of conduct, reporting frameworks, as well as company programs on sustainability. The study findings indicated that South Africa's agricultural exports to other African countries are generally subjected to 46 VSS while produce destined for the European Union (EU) must comply with 43 VSSs. By focusing on fruits alone, the EU requires that South Africa's products comply with 40 VSS yet intra-Africa trade has 43 VSS that must be complied with. This suggests that South Africa's agricultural produce, specifically fruits are required to meet more VSS if destined for Africa unlike if products are destined for the EU. When the EU was considered as the supplier of fruits to Africa, 48 VSS were identified and South Africa requires that 45 standards are complied with. For fruits destined for South Africa from other African countries, 51 VSS must be met. The high number of VSS required to be met by African economies to trade with each other is a big concern, most especially because countries are operating under the AfCFTA that aims at smoothening intra-Africa trade. Of the identified VSS affecting South Africa's trade in fruits with Europe and within Africa, eight of the VSS do not address food safety and traceability themes. Most of the enforced VSS addressing different themes may render compliance a challenge, most especially for smallholder farmers that are resource constrained. Therefore it is recommended, given the independent nature of the end goals from the institutions responsible for these VSS, it becomes problematic and impractical to argue that VSS focus on similar thematic areas, e.g., GLOBALG.A.P. Crops and IFOAM should be merged into one standard. Thus, it is prudent that smallholder producers outrightly identify a specific target market and the corresponding standard requirements such that all activities are undertaken in accordance with what the market requires. The government and non-government organisations must support smallholder farmers in adopting and complying with these standards through facilitating capacity-building sessions (training), as well as acquiring certification to meet the standards. As it is beyond doubt that the high cost of certification is a huge stumbling block to smallholder farmers in South Africa. With the possibility of the AfCTA agreement, it is important that African countries work towards reducing these VSS to a bare minimum, the basis upon which trade in fruits is bound to flourish.

Chapter 4, 'Market Attitude Towards Genetically Modified (GM) Food Products: A Developing Economy Perspective', was written by Yajeshwar Seetul and Praveen Balakrishnan Nair. This chapter attempts to investigate the influence of perceived benefits, perceived risks, knowledge, general attitude, and trust on attitude towards GM food products in Malaysia, a fast-developing economy. The authors indicated that there were limited attempts made to determine the antecedents that influence the attitude towards GM food products in the Southeast Asian context. A sample of 301 Malaysian consumers was used in the study using convenience sampling. The authors paid attention to Public/social behaviour as one of the main elements that determine the success or failure of any commercial product launched in the market and food products. They identified several studies conducted to assess attitude, awareness, or acceptance of Genetically Modified (GM) food products, giving a good insight about the perception towards this product. They also indicated that it is important to effectively develop the GM food industry for the government, policymakers, non-governmental organisations and the industry, in general, to seriously consider the market's attitude towards GM food products, while formulating commercialisation strategies and this has implications for the agricultural industry. The authors used Fishbein`s Multi-Attribute Attitude model to investigate the influence of perceived benefits, perceived risks, knowledge, general attitude, and trust on attitude towards GM food products in Malaysia. However, the results showed that the validity of a model varies among regions and populations, as knowledge, which is also part of Fishbein`s-model, was not supported in the Malaysian context. Perceived benefit is the strongest factor influencing attitude towards GM food products, followed by perceived risk. Although knowledge did not influence the attitude of consumers towards GM foods in the study area. The knowledge of genetically modified food is still a major factor to influence the perceptions of the consumers to adopt it as an acceptable product in the market. The authors' findings rec recommended that with accurate knowledge from government organisations, nongovernmental organisations, public media, producers, marketing departments and international organisations, misconceptions can be reduced, which would eventually decrease perceived risk and increase perceived benefits with substantiated information.

Chapter 5, 'Formal and Informal Agricultural Markets in Sub-Saharan Africa', was written by Unity Chipfupa. This chapter evaluated the formal and informal agricultural markets in SSA and their linkages with the smallholder farming sector. Despite some provisions in some formal agricultural markets to accommodate small-scale farmers, their access to such markets is mainly limited by a lack of knowledge, limited access to resources (including technology), poor food safety standards, poor quality of produce and inconsistency in supply. These limitations affect their ability to meet the standards of the clientele serviced by the formal agricultural markets. The study was qualitative in nature employing a review desktop approach on studies with practical experience in supporting agricultural marketing in Africa. Scanning was used to identify relevant issues in the literature, which were analyzed through thematic content analysis. The study found that Informal markets are integral to the survival of the smallholder producers in SSA. Though it is important to promote small-scale farmers' access to formal markets, regulating informal markets out of existence will do more harm than good to the smallholder farming sector, especially given the obvious unaddressed challenges of integrating smallholder producers in the formal markets. The Latin American countries' experiences with the Productive Alliances (PA) approach could offer lessons to SSA governments on effective strategies for supporting smallholder cooperatives whilst promoting ownership and sustainability. Deliberate efforts should be made that promote a symbiotic relationship between the formal and informal agricultural markets. Smallholder farmers' relationship with middlemen is mostly detrimental because farmers lack market information and hence bargaining power in price negotiations. Without any intervention from the gatekeepers, farmers will continue to lose their hard-earned income to the middlemen. The study recommended that policymakers should put measures that seek to protect farmers such as seasonal price setting. The promotion of e-commerce in the sector might also improve the informational capabilities of smallholder farmers. Currently, there are minimal efforts to explore this form of marketing by smallholder farmers in SSA. It is high time that investments are made in online marketing in the smallholder sector. Addressing the challenges of internet connectivity and the cost of data will open a whole new world of marketing to millions of small-holder farmers in SSA. Going forward, SSA governments should take a more active role in supporting e-commerce in agriculture. They should work with UN agencies such as FAO and other international and local organizations that have projects that are trying to stimulate e-commerce in the sector.

Chapter 6, 'Images of Organic Food Products, Consumers, Makers, and Distributors: An Image Congruence Study', was written by Bình Nghiêm-Phú and Jillian Rae Suter. This chapter examines customer perceptions of themselves as organic food consumers, organic food products, and organic food makers and distributors. Research on image congruence in the food business sector, particularly organic food, is relatively scarce and previous studies could only identify the congruence of certain psychological elements of the perceived image. Other functional elements were mostly overlooked. Given its exploratory nature, the study opted to collect and descriptively analyze qualitative data to identify American organic food customers' perceptions of themselves, organic food products, and organic food makers and distributors. In addition to direct interviews, data was collected through YouTube videos and blogs. By analyzing the data gathered from informants, this study found that organic food consumers held specific images concerning organic food products, their makers and distributors, and themselves. Some images of organic food consumers and products shared the same nature. In addition, some images of organic food consumers, makers, and distributors had the same essence. More and more people are consuming organic food. They, of course, hold specific images concerning organic food products, their makers and distributors, and themselves. Some images of organic food consumers and organic food products share the exact nature, such as safety and healthiness. In addition, some images of organic food consumers, makers, and distributors have the same essence, such as social and environmental responsibility. The chapter recommends that with the projection and promotion of proper information, image congruence can be improved to further facilitate future organic food product consumption. In the study, congruence was identified, but the change of congruence with the inputs of additional information could not be explored. The study then recommends future researchers carry out experiments or longitudinal studies to clarify this issue. The abovementioned efforts will help expand the literature on organic food and the involved businesses. The findings have implications for the farmers, agricultural marketing and organic food regulatory authorities in future as this indicates what current and future consumers expect from the organic food market.

Chapter 7, 'Fresh Produce Market Challenges and Opportunities: A Case for the Johannesburg Municipal Fresh Produce Market', was written by Ndiadivha P. Tempia, Elvis Nakana, and Makhanana Malungane. The National Fresh Produce Market (NFPM) performs a crucial role in the distribution of fresh produce, providing a market outlet for farmers, access to produce for retailers and through all of this contributing to economic activity, livelihoods and food security. Keeping these markets functioning effectively as public institutions with the mandate of contributing towards these public goods is vital and worth investing in. These markets, however, face fundamental challenges that are evidenced in their overall loss of market share and the lack of transformation. If the loss of market share continues the competitiveness and viability of these markets could be under threat. This could have extremely negative effects on small-scale farmers who cannot access corporate supply chains on a beneficial basis.

This chapter provides a qualitative description of the Johannesburg Fresh Produce Market (JFPM) as a case study with reference to the functioning and challenges in the regulatory environment. Due to the descriptive nature of the study, no sample size was calculated. The study population covered the full range of market actors including, market agents and traders across all hubs (vegetables, fruit, potatoes and onions), and the Mandela and Unity markets, at the JFPM, market management, government officials, porters, farmers and buyers. Buyers interviewed covered the full range from small-scale informal traders to wholesalers and exporters. Primary data collection involved ethnographic research including interviews, gathering life histories of market actors, and observations. Due to the diverse nature of the economic agents at the JFPM, both semi-structured and structured interviews were conducted. The study indicated that the lack of transformation in the NFPMs is not sustainable in South Africa today and undermines their ability to support needed transformation in the agricultural sector. In view of this, the study recommended a regulatory reform recognising the importance of social relations to the functioning of the markets and factoring this into any interventions. Also valuing the importance of the informal food sector for the markets.

Chapter 8, 'Market Access Enablers for Small-Scale Crop Farmers', was written by Aluwani Tagwi. Cooperatives are deemed as vehicles for rural development, especially in developing countries. Smallholder farmers generally experience huge marketing transactional costs and to reduce this they normally resort to becoming members of cooperatives. Small-scale farmers (SCF) play a significant role in Africa, in Sub-Saharan Africa (SSA). About 80% of the food production comes from SCF which accounts for 90% of food production in the region. However, SSA's storage capacity is constrained, despite recent crop yield advancements, resulting in large food losses. The role of SCF in food production cannot be underestimated Moreover, the role can be amplified if there are opportunities for markets. Small-scale farmers typically face a wide range of complex challenges in developing countries, including high transaction costs for gaining access to markets, and weak bargaining power in the markets. Trends show that the major share of food production will continue to come from smallholder farmers and will remain critical to global population food security immensely. In South Africa, most small-scale farmers do not have access to markets. With no markets in place, the transformation would be almost impossible. The study's aim was to assess the determinants of market access in agricultural cooperatives in Limpopo Province. A face-to-face survey was conducted of 146 agricultural cooperatives members. The results showed that market access was associated with age, training, workers in a household, knowledge of the cooperatives principles and act, annual income and access to funding. The study recommended younger farmers be targeted when designing market access interventions as the impact of job creation is huge amongst this group. tailor-made training on market access and cooperative principles and acts should be designed. Funding should be made more accessible to small-scale farmers in cooperatives. In the future, this study could be expanded to incorporate market players such as retailers and market agencies to gain a broader understanding of the dynamics of market access. The activities between farmer cooperatives and the retail sector should be assessed as to the extent to which these affect farmers' market access and profit.

Chapter 9, 'Commercialization of Smallholder Farmers in South Africa: A Myth or a Reality', was written by Elliot Zwane and Elekanyani Nekhavhambe. This chapter assesses the concept of commercialization of smallholder farmers, i.e., whether it is a reality or a myth. The focus of the study area was Mutale Local Municipality, in the Vhembe district of Limpopo Province. More than 4 million households are engaged in subsistence agriculture in South Africa, if these farmers were to be commercialized, they would contribute beyond food security, grow the economy, and create job opportunities. The study was conducted in the Mutale Local Municipality, Vhembe District of South Africa on a proportionally randomly selected sample of 153 smallholder farmers after clustering them into agricultural zones (Tshipise, Tshixwadza, Tshishivhe, and Masisi) and commodity groupings (vegetables under irrigation, dryland maize and citrus fruit farming). Data were collected through a structured qualitative and quantitative questionnaire that was administered face-to-face to respondents (smallholders and extension officers). Secondary data sources were also used to explore some concepts related to this study._Data were collected through a structured qualitative and quantitative questionnaire that was administered face-to-face to vegetable, maize, and citrus small-scale farmers. respondents and the data. The findings indicated that commercialization is not an easy process, its determinants are based on socio-economic characteristics, and the majority of farmers operate at lower production levels, generate low income from farming, but depend on off-farm income like old age grants. Female-headed households were found to focus more on food security while male-headed households grow cash and high-value crops. Training and financial support were crucial interventions needed by farmers. Access to resources such as mechanization, irrigation water and distance to the markets were some of the challenges. Commercialization policies, effective advisory services and market training were recommended.

Chapter 10, 'Agricultural Marketing Dynamics for Rural Economics in the Face of Climate Change: Insights of Cash-Crop Emerging Farmers', was written by Mpho Tshikororo. This chapter investigated the agricultural marketing dynamics brought by the impact of climate change among rural-based emerging farmers. Climate change is instrumental for agricultural performance, its production and marketing being the leading cores. Although numerous climate change adaptive strategies have been identified, there is a need to explore how climate change impacts agricultural marketing. Agriculture is highly affected by climate change; hence unfavourable climatic changes bring undesired shocks to the sector. It has been noted that climate change will continuously impact agricultural production negatively. It has also been noted that agricultural supply chain systems are disrupted by climate change, with farm production, marketability, processing, retailing, and human consumption being among other vital areas that suffer the most. Agricultural markets have been hampered by the impacts of climate change, with most farmers severely affected at the farm level. Consequently, farmer productivity has been significantly challenged, leading to poor market performance. The study was conducted in the Vhembe district of the Limpopo province. A simple random sampling technique was used to select the 229 emerging farmers, while a structured questionnaire was used to collect data. Discriminant analysis was used to draw a distinct difference between farmers who had observed the impact of climate change on their marketing dynamics from those who didn't. The study results show that unstable selling prices have been a negative impact of climate change on the market. The study recommended the move towards agricultural regulated markets to ensure farmers' protection and sustainability.

Aluwani Maiwashe-Tagwi
University of South Africa, South Africa

Ailweli Solomon Mawela
University of South Africa, South Africa

Phineas Khazamula Chauke
University of South Africa, South Africa

Acknowledgment

It was incredible to see authors coming together to contribute to this book *Global Agricultural and Food Marketing in a Global Context: Advancing Policy, Management, and Innovation*. The need for the book was echoed and supported by many experts globally in the field of Agricultural Economics. The book unpacked different facets of agricultural and food marketing globally, the implementation of agricultural policies, management and innovation concerning genetically modified food production perspective. Different chapters attest to the authors' passion for agricultural marketing and confirm the need for pedagogical content knowledge of agriculture and food marketing and how it affects developing nations. Editors would like to express their gratitude to all the authors for sharing their experiences, and their expertise towards the creation of novel innovative subject content knowledge and the functionality of the global agricultural markets in developed and developing countries to address today's farmers' and consumer needs. We thank all the reviewers who took the time to improve each chapter by sharing their agricultural expertise. Reviewers are thanked for believing in the impact to enhance the course of rural development and inclusive agricultural economy. Their efforts are recognised and highly appreciated.

It is the editors' view that this book's chapters will contribute towards enhancing the knowledge of the smallholder farming sectors through effective implementation of policies, management, and innovative ways of participating in mainstream markets. This scholarly book chapters challenges experts in the field of agricultural marketing and various stakeholders such as the agricultural industry, government, academics, students, and smallholder farmers to continue contributing through empirical work by sharing what needs to be done, what has not worked concerning transformation in the agricultural marketing sector. We would like to express our gratitude to the members of our own families and close friends who endured our absence during the lengthy writing, meetings and editing. Without their support and belief in us, we would not have been able to edit this book. The book project happened at a very difficult time of our lives when we lost loved ones and acquittances to the COVID-19 pandemic. We thank participants, farmers, communities, and stakeholders for their contributions towards this work. We are eternally grateful to God almighty for giving us the strength to complete this scholarly peer-reviewed edited book.

Acknowledgment

Chapter 1
A Systematic Literature Enquiry of the South African Agricultural Marketing Environment Pre-1913 to the Present

Phineas Khazamula Chauke
University of South Africa, South Africa

ABSTRACT

Over a period of time, the South African agricultural sector has been divided into two sub-sectors, i.e., commercial and subsistence farming for whites and black farmers respectively. It also went through three marketing environmental phases, commencing pre 1913 to date. Use is made of the systemic literature review methodology, considered superior to its more unstructured narrative review counterpart. The findings of the review process are that the early 1900s reflected what a competitive non-regulated market can achieve even for those with minimal productive capacities, while the second phase that commenced with intensification of discriminatory legislation reflected the negative outcomes of the system. The last phase, covering the period after 1994, reflected development of an inclusive policy environment with minimal incorporation of the emerging black farming community. The chapter recommends more concerted efforts at graduating emerging farmers, implementation of well-intended and crafted policies and improved monitoring of CPAs.

INTRODUCTION AND BACKGROUND

Commencing in 1994 when the new democratic government seized political and to some extent economic power, an effort was made to reverse both the colonial and apartheid agricultural marketing environment in South Africa (Piesse, et al., 2005; Walker, 2017; Vink, 2012; Spierenburg, 2020). An enabling policy, the Marketing of Agricultural Products Act No 47 of 1996 was passed, not only to level the playing field but also to taint the scale more towards the small-scale and emerging black farming sector. In particular, the Act anchored on one of the four objectives, i.e., "increasing market access for all participants" ((sub-

DOI: 10.4018/978-1-6684-4780-2.ch001

section 2(1) (a)). Needless to say, this sub-section sought to close the host of discriminatory colonial and apartheid legislation. Although the emphasis was placed on a deregulated marketing environment, some support for emerging black farming was unavoidable as witnessed by a number of related strategies such as the land reform programme, especially the promulgation of the Restitution of Land Rights Act No 22 of 1994.

While earlier support to black farmers was provided through the Farmer Support Programme (FSP) (Kirsten, 1994), and Comprehensive Agricultural Support Programme (CASP) (Chabalala, 2008), more recently the Agricultural Industrial Fund was initiated under the state-supported Industrial Development Corporation (IDC) (Reuters Staff- online). As articulated by Kirsten (1994) the FSP was not only introduced in South Africa by the Development Bank of Southern Africa (DBSA) as one of the major agricultural development strategies to support black farmers who have been historically denied access to basic support services", but also in response to the inefficient and costly large-scale project approach that had been adopted until then. Some outcomes of these strategies included the provision of farming information, on-and off-farm infrastructure, production inputs, marketing information, business plan development and enabling finance for acquiring production capital (Msibi et al., 2011). However, as noted by Aliber and Hall (2012) implementing some of the above-mentioned programmes met with little success especially due to high implementation costs and beneficiation to fewer black farmers. It was suggested at that time that resolving the challenge required what was referred to as making concise decisions regarding two paradigms, i.e., either ensuring achievement of "accumulation for the few' or 'accumulation from below' (Aliber and Hall, 2012). The first involved adoption of a strategy that could benefit a few black large-scale commercial farmers, while the latter focussed on benefiting many sustainable black commercial smallholders. However as articulated recently by Mmusi (2020) the challenge seems to revolve around skills transfer to black farmers that could be resolved through the adoption of concerted inclusive efforts at transferring farming skills to black farmers by government, non-government organisations and private sector institutions. Thus, while the above programmes succeeded in providing basic support services, their long-term sustainability hinged on levelling the playing field and uniform application. It is however critical at the commencement of this chapter to contextualize the meaning of the term 'marketing' to avoid ambiguities towards exchanging goods and services, and thus to correctly follow discussions that are a focus of this chapter.

This chapter adopts the definition of food marketing as defined by Kohls and Uhl (1990), i.e.," … performance of all business activities involved in the flow of food products and services from the point of initial agricultural production until they are in the hands of consumers". What is crucial about this definition is that marketing includes production on the farm. The chapter seeks to present an assessment of the South African agricultural marketing environment with a special focus on the period spanning the late 19th century to that which followed the ushering in of a democratic state in 1994. The major aim is of course to present a discourse on the impact of a policy environment that sought to disadvantage the black farming community in favour of its large-scale white commercial farming counterpart.

The latter occurred over a period that spanned almost a century with devastating consequences for the agricultural marketing milieu. Of specific interest though would be assessing the extent to which a new policy directive that was promulgated in 1996 performed in terms of reversing the untenable environment that was created by successive colonial and apartheid policies (Marketing of Agricultural Products Act No 47 of 1996). It must be emphasized here that the Act replaced two major Acts that were passed by the colonial and apartheid administrations, the Marketing of Agricultural Products Act of 1936 and the Marketing of Agricultural Products Act of 1968. Whereas the former focused on support to white farmers through human capital investment, input supply and establishment of control boards the subsequent overproduction that was exacerbated by massive subsidies culminated in overproduction, necessitating the need for agricultural commodity schemes- the major focus of the 1968 Act.

METHOD

In order to ensure that the research and information retrieval processes were not only accurate but also impartial, the method adopted for this chapter was a systematic literature review (SLR) (Fernandez-Aleman et al 2013). Using this methodology, the chapter is well set to respond to the historical theme of the book through an assessment of the South African agricultural marketing environment commencing prior to 1913, cutting across colonial and self-declared apartheid republican administrations until the achievement of a democratic dispensation in 1994. Specifically, Fernandez-Aleman et al (2013) suggested the following procedure in organizing the study, i.e., searching publications from electronic databases; exploration of the topic, abstract and keywords; the completion or partial reading of those articles that were not eliminated from the previous phase and scanning the reference list for articles that will lead to uncovering new studies. As also attested by Okoli (2016), SLR is an explicit methodology especially for explaining the qualitative procedure followed in data collection.

As noted by Xiao and Watson (2019), SLR commences with a search for keywords and phrases related to the topic under investigation. In terms of inclusion criteria, all literature of a historic nature (from both electronic and print media) that focused on participants in farming and land dispossession was targeted. Information on general discriminatory practices outside the agricultural milieu was excluded from the stud. Google Scholar and EBSCOhost were mainly targeted. For this study, such keywords included: African farming in the late eighteenth and early nineteenth centuries, land dispossession, the Native Land Act No 27 of 1913, support to white commercial farmers, the impact of the great dispossession, scheduled and released areas, legislation following the 1948 apartheid general elections, homeland agriculture, land reform, The Marketing of Agricultural Products Acts of 937, 1968 and 1996, the Restitution of Land Rights Act No 22 of 1994, support to African farmers following a new dispensation in South Africa.

Collected information (data) was arranged into the following themes - phases- for further analysis: the agricultural marketing environment pre-1913, the period following the 1948 apartheid-based legislative environment and the agricultural marketing environment that ensued following a democratic dispensation in the 1994 inclusive elections. In line with Iden, and Eikebrokk (2013) methodological imperatives also required the formulation of three critical research questions (RQ), i.e. RQ1: Who were major players in the South African agricultural marketing environment before the implementation of the Natives Land Act No 27 of 1913? RQ2: How did the apartheid policy implemented from 1948 impact black and white agricultural marketing in South Africa? RQ3: What has been the impact of the 1994 democratic political environment on black farming in South Africa? Findings and discussions are presented below.

FINDINGS AND DISCUSSIONS

Phase 1: The Pre- and Post-1913 Marketing Environment

Prevailing Production and Marketing Environment Pre-1913

As reported by Schneider (1984), between the years 1870 and 1930, and despite the discovery of minerals during this period, most of the area that later formed the Union of South Africa was basically an agrarian society. The composition of the farming community during that period was a mixture of black and white farmers that competed with each other for land and local markets in response to an increasing agricultural market base that anchored on an ever-increasing industrial centre (Schneider, 1984). The above is corroborated by the World Bank (1994) which observed that the late 19th century South African agricultural production and marketing environment generally witnessed the co-existence of black and white farming, with the former concentrated mainly inland and the latter more inclined towards producing marketable cash crops along the coastal areas.

Specifically, three types of farming systems at that time were clearly discernible, i.e. large-scale white farming that depended largely on hired labour, indigenous tenant farming mainly for black farmers and widely spread small-scale indigenous farming. The World Bank (1994) also attests that while coastal farming specialized in producing exported horticultural crops and livestock, their interior-based largely black counterparts produced mainly for domestic consumption and only marketed surpluses. Inter-trade between the interior and coastal operations also existed, especially as regards ammunition, transport equipment and textile. However, in some areas of the then fragmented country such as the Cape, policies to discriminate against farmers along racial lines had commenced. Paramount to such policies was the Glen Grey Act of 1894 promulgated under the leadership of Cecil John Rhodes. The focus of the Act was Glen Grey District in the Pondoland (now Eastern Cape) to quell the superiority of blacks over their white farmers' counterparts. As holders of the franchise, the latter lobbied government not only to curtail land availed to black farmers but also to impose other restrictive measures such as denial of mortgages and forced communal settlement on small pieces of land.

Prior to the above and as a consequence of the discovery of minerals such as gold and diamonds in the interior around the late 19th century, especially in the 1870s and 1880s, significant impacts on agricultural production and marketing in South Africa ensued (Gordon, 1980, World Bank 1994, Higgins 2001 and Van Zyl et al, 1996). In particular and due mainly to their low technology production inputs and collaborative collective ploughing and harvesting efforts, black farmers responded more positively to the demand for food that emanated from the diamond and gold sectors (use of animal-drawn carts for transportation, large-scale cooperative participatory production processes etc.). It was a logical follow-up therefore that negotiated Anglo-Boer war efforts also centred around the promulgation of policies that would discourage African production and marketing capabilities through instruments such as taxation on road infrastructure, land use and rearing of livestock (Greyling et al, 2018).

To appease white farmers the colonial government instituted discriminatory policies that availed inadequate land allocations to black farmers (artificial land shortage strategies) and thus encouraging them to move out of active farming to pursue job opportunities on large-scale commercial white farms and other industries in towns and cities (Van Zyl et al 1996). As noted by SA History (online) preceding the formation of the Union government in 1910 a commission, led by Godfrey Lagden, did not only recommend the establishment of reserves for black people under the custodian of tribal chiefs, but also the overall state administration of such reserves. The commission's recommendations were immediately followed by the passing of the Land Settlement Act of 1912 which set aside close to 167 000 hectares of land for sole settlement of whites as a precursor to the Natives Land Act No 27 of 1913 that set aside 93% of Union land surface for white occupation.

The Promulgation of the Natives Land Act No 27 of 1913 and Its Immediate Impact on Farming

According to Vink (2019) "... the marketing of most agricultural products in South Africa was subject to intervention by statutory bodies over a period of some 60 years. At the time of the promulgation of the first legislation, there was a vigorous debate on the predicted economic effects; in 1983 one of the architects of these instruments published a justification on these pages, followed in 2000 by a 'post-mortem,' which confirmed most of the negative predictions that had been made. The purpose of this article is to revisit this debate and to provide a first assessment of the long-term impacts of the legislation". The present article is therefore appropriate to look at developments since then.

As observed by Plaatjie (1916), on the 19th of June 1913 South African black people assumed the status of foreigners in the land of their birth following the passing of the Natives Land Act No. 27 of that year. However, at that time the dispossession was still very much theoretical as the geographical area for black occupation was not only still blurred but many black people were still scattered all over the country. To achieve a clear reflection of racial geographical demarcation, the Beaumont Commission was set up three months later to determine boundaries for black and white occupation. The outcome of the commission was only released three years later in 1916. While the Act had devastating consequences for black inhabitants, some white scholars' view is however that initially the Act had limited impact, save for undermining the status of black tenants on white-owned farms and having noted its positive outcome as cementing traditional authority in rural areas (Beinart and Delius, 2014).

The above soft-line view of assessing the impact of the Natives Land Act of 1913 is however countered by more radical views (SA History online) that saw the immediate impact of the Act as having been the cornerstone of other legislation that later dispossessed and subjugated blacks under both colonial and Apartheid rules, congesting them initially into only 7% of the country's land surface (increased to 13% in 1936), preventing blacks from active participation in sharecropping activities to make way for their white counterparts (alleviating the plight of poor whites), preventing skilled and semi-skilled urban blacks from the effective competition with white labour and closing of livelihood avenues for the former category of workers.

While the Act created a dire situation for black farmers in particular, their white equivalents benefitted immensely from this legislation. The most immediate predecessor Act that sought to address the plight of poor white farmers was the Land Bank Act of 1912, a legislation that was aimed at weaning the latter from dependence not only on selling their labour but also to compete with their counterparts for sharecropping opportunities. In effect, the Act sought to enable white farmers to assume the status of successful commercial farmers that could access highly productive land at subsidised low costs. To further strengthen their agricultural productive abilities, the Cooperative Society Act was passed in 1922 to avail inputs and other productive capital. In response to poor produce markets that emerged from the Great Depression of 1929 (resulting in very low commodity prices) and following the exertion of pressure on the government by the farming community that ushered in the pro-Afrikaner National Party government around that time, the Marketing of Agricultural Products Act of 1937 was passed, especially to ensure a regulated marketing environment for white farmers. Implementing the latter Act, produce marketing schemes - through which specific produce could only be marketed – were initiated. It must be noted that the Marketing Acts of 1937 and 1968 were meant to promote the divide between white and black farming through massive support for the former and complete ignorance of the latter (Greyling et al., 2015). Specifically, these Acts sought not only to support white farmers but also to ensure that the two were separated socially, geographically, and economically. A study that also looked at the Free State Province, conducted around 2015 to assess the impact of a deregulated marketing environment on black farming, noted that the latter was still highly deprived in terms of land ownership (only 12% of land under black occupation), with 76% of the land restitution budget having been utilised for financial compensation (rather than the acquisition of landed properties). Generally, the study found that the land reform programme failed mainly due to the adoption of the group and cooperative approaches in land ownership and operation.

Under the Act, agricultural commodity control boards were established. Most agricultural produce could only be marketed through these boards. As ascertained by Chauke (2021), the boards' appointed agents, especially agricultural cooperatives, to be structures through which farmers (e.g., maize producers) could deliver their produce in exchange for a pre-determined floor (lowest) or ceiling (highest) price. Consequently, the main challenge facing white farmers was to ensure increased production of agricultural produce without having to be concerned about market access issues. The general failure of a regulated agricultural environment could best be explained by Rees (1979) who observed:

"...the relative powerlessness of marketing boards to achieve significant variations in market prices unless supported by political interest groups" (rees, 1979).

Phase 2: Agricultural Marketing Policy Environment Following the National Party Government From 1948

White Commercial Agricultural Marketing Policy After 1948

Perhaps as a prelude to this subsection, it is critical to reflect that one main objective of the Apartheid policy that was earnestly implemented in 1948 was to provide abundant African labour to the white commercial farming sector (Posel, 1987). Thus, with a lack of own land ownership rights, at best, black farm labourers could only practise backyard subsistence farming mainly for their own consumption and never for the market that was ring-fenced for their white counterparts.

Consequently, during the 1948 Whites-only general elections of the Apartheid National Party (NP) government emerged victorious, more so in that it propagated for a more racially inclined and segregated policy environment that would immensely benefit white people, especially the white farming community. Schirmer (2001) observed that whereas the intention of the NP was to immediately embark on the new policy shift, it took a number of years for it to realize that objective. It was only in the 1960s that white farmers experienced massive government support, especially in the form of subsidies, and mentoring mechanisms to ensure positive returns from farm capital investment, in the process wasting substantial valuable resources that could have been used to develop other sectors of the economy (Schimer, 2001).

Despite the provision of productive assets such as inputs by the state to the white commercial sector, it still depended to a large extent on black labour tenancy that was still characteristic of the rural landscape up to the mid-1960s when forced removals commenced in earnest. Duncan (1991) noted that at that time, as the apartheid state was still struggling to support the white commercial farming sector, black labour provision was still of paramount importance in ensuring sectorial success. More concerted state support for white commercial farming occurred in the 1970s. The outcome thereof was an increase in output, employment, wages, and exports with more successful and thus profitable farmers consolidating their landholdings (Vink, 1999). Specifically, the state did not only provide professional support to white farmers but also ensured a farmer cooperative-controlled agribusiness sector that, besides providing inputs to farmers, also processed and distributed products on their behalf.

As noted by Vink (1991) "...superficially, the sector seemed to be successful" but proved otherwise by events that took place in the subsequent decade (the 1980s) which ".... saw the beginning of a process of state withdrawal from the sector that continues to this day".

From the foregoing, it can be insinuated that massive support to white commercial farmers also yielded negative consequences for the country's agricultural sector specifically, but also for its entire economy generally. Seventeen years before the complete demise of a regulated marketing environment in 1996, Rees (1979) concluded that, as far as the subsidization of agricultural produce was concerned, such could only be justified as political responses to farmer's demands rather than on economic principles, and that such needed to be discontinued. The centralization and control (price fixing) were only a haphazard undertaking and in no way could such control measures guess the actual market price. Quite importantly as further noted by Rees (1979) and based on the theory of economic regulation rather than maximization of economic welfare, marketing boards were established as a mechanism to redistribute such wealth. The question therefore could be what the achievement of a deregulated policy environment in South Africa has been and that will be responded to in the following sections.

Homeland Agricultural Development Corporations

Perhaps the performance of black farming and produce marketing during the segregation era could best be expressed through observations from some former South African 'homelands', a term derived to designate the ten (10) ethnically segregated black reserves of Bophuthatswana, KwaZulu, KwaNdebele. Kangwane, Lebowa, Gazankulu, Transkei, Ciskei, Venda and Qwaqwa. Many of the homeland areas were characterized by locational and marginal agricultural productive disadvantages. Bophuthatswana for example experienced semi-arid climatic conditions for crop production (Molatlhwa, 1976). A study conducted in Gazankulu, one of the former homelands situated mostly in present-day Limpopo Province, but also incorporating substantial portions of the present Mpumalanga Province, revealed several challenges that impacted agricultural production and thus produce marketing. It should be recalled that the initial and overriding goal of the Natives Land Act of 1913 was to create reserves for the resettlement of black people.

Schneider (1984) noted that the implementation of the homeland policy had negative consequences for homeland agriculture. Specifically, reserve land that was initially set aside for general settlement of black people under subsistence farming was altered to assume the function of not only settling Black ethnic units but also as a means of generating the necessary income for its inhabitants through the adoption of 'commercial agricultural' practises (Schneider, 1984). The latter was to occur in the midst of a low skills base and land scarcity. Achievement of such an ambitious farming policy required the demarcation of larger farming units to a few elites, usually politicians and influential government employees, and some private sector stakeholders, especially on land transferred from white farmers under the Natives Land Trust Act of 1937. To ensure the 'successes of this new model, managerial and other input costs were largely drawn from state resources in the form of seconded officials that drew their salaries and production inputs (fertilizers, pesticides, herbicides etc) from the state. Generally, several studies that have been conducted on former homeland agriculture revealed several challenges. These included a lack of investment capital (Porter et al, 1997), low levels of production due to deterioration in natural resources and overstocking in homelands such as the Transkei (Bembridge, 1987), pressure on agricultural resources as a consequence of high population densities (Simkins, 1981), and lack of facilities such as credit, transport and land rights (Lahiff, 1997).

Initiation and Achievement of Development Corporations

Under the auspices of the Bantu Investment Corporation Act of 1959, the National Party government initiated and capitalised individual entrepreneurs and development corporations in each of the 10 former homeland areas (Davenport 1987). That policy imperative was not only promulgated to legitimise the apartheid policy but also to curb the migration of surplus black labour from reserves (homelands) (Tomlinson et al 1984). The National Party government, therefore, established Development Corporations to assume a pivotal role in development initiatives such as industrial and agricultural development. Before 1981 financial support for these amounted to more than R500m for the four independent homelands of Transkei, Ciskei, Bophuthatswana, and Venda, and only R173m for the remaining six (6) self-governing homelands that were reluctant to assume 'independence.

The demise of these projects is however common knowledge as their initiation was meant to achieve the goal of an untenable ethnic separate development policy. In terms of their contribution towards job creation and participation in the marketing space, some homeland development corporations contributed significantly, especially in the export market for high-value fruits. It is for that reason that some authors such as Mann (2001) held their demise as having been unfortunate. However, a comprehensive study of the contribution of development corporations in the KwaZulu homeland attested to their dismal failure in absorbing the rural unemployed, in the process contributing towards massive outmigration to nearby 'South African' towns and cities (Lenta and Maasdorp, 1984).

Phase 3: The Farming Policy Environment That Followed the 1994 Democratic Dispensation

The Marketing of Agricultural Products Act (MAP) No. 47 of 1996

According to Section 2(2) of the MAP Act No 47 of 1996, the legislation was promulgated to achieve four objectives, i.e., increasing market access to all market participants; promotion of the efficiency of the marketing of agricultural products; optimisation of export earnings from agricultural products and enhancement of the viability of the agricultural sector. The Act also established a ten-member National Agricultural Marketing Council (NAMC) (Section 3 of MAP Act) to oversee the activities of the country's agricultural marketing environment. At the time of the Act's promulgation in 1996, South Africa faced a dual farming sector, i.e., large-scale white commercial farming and small-scale largely indigenous black farming sectors. The former was characterised by a sector that had emerged from massive public sector support and thus highly efficient infrastructure and productivity levels. Its counterpart did not only consist of indigenous small-scale farmers that had lost productive skills over time due mainly to small farming units (usually less than a hectare) and lack of public and private sector support but also located within congested marginal productive homeland areas. The latter was therefore largely excluded from effective farming participation and thus targeted beneficiaries of the Act in terms of increasing market access (the first objective of the Act).

To ensure the continued productivity of largely white commercial farmers, commodity Trusts were established to replace the Commodity Control Boards, especially to manage the newly deregulated marketing environment. One critical component of the Trusts was the collection of statutory levies not only for providing some support to commercial farmers but also to apportion a certain amount to fund the emerging black farming sector. The above is asserted by NAMC (online), which observed that Trusts inherited funds from most of the former Control Boards and that their activities are guided by Deeds of Trusts adopted for each. Representation is by respective Industry Trust members and the Ministry of agriculture on a 50/50 basis. NAMC (online) also confirms that statutory levies were not only meant to benefit the various Trusts but were also conditional to the allocation of a certain portion towards transformation (i.e., beneficiation of the emerging black farming sector operating within the realm of each Trust).

As of 2009 the target set for transformation emanating from statutory levies was 10%, which had increased to 20% in 2015 (NAMC, 2017/2018). Some functions targeted for achieving transformation were enterprise development and market information provision. Over time achievement in terms of ensuring market access to all stakeholders, especially the emerging black farming sector was reported in general terms. The NAMC Annual Report for 2009 recorded that in order to achieve the objectives of assisting low-income agribusinesses and thus to improve their market participation, NAMC was partly required to investigate statutory measures and to provide continuous market and economic research so as to improve the efficiency of the market, and in the process optimise export earnings (NAMC, 2009). The majority of Industry Trusts, therefore, operate under various statutory measures. Participating Trusts are required under the Act not only to be subjected to compulsory registration but also to keep records and returns, payments of levies and provision of records of produce exports to the NAMC on an annual basis (NAMC-online).

More decisive guidelines for achieving sectorial transformation were adopted by the NAMC in 2018 (NAMC Annual report, 2018/2019). The guidelines advanced four (4) pillars and clear activities for achieving transformation (some reflected in brackets below), i.e. (a) enterprise development (aimed at funding black-owned agribusinesses for procuring production infrastructure, materials, inputs, vaccinations, market access etc); (b) Skills Development pillar (funding training and mentoring of farmers by SETA accredited providers etc.), (c) management control (funding positions/appointments of transformation managers within black-owned businesses etc); and (d) the Ownership Pillar (purchasing of equity in existing enterprises for black designated groups etc.). Expectations were that as from the 2019/20 annual report, the NAMC would account for achievements, especially addressing activities achieved under each pillar.

Indeed the NAMC (2020) reported significant improvements in transformation, especially in enterprise development with an allocation of almost R68 mil constituting close to 54% of the transformation levy as well as in skills development (almost R23.5 mil constituting almost 20.2%). The latter two pillars surpassed anticipated guidelines of 50% and 20% respectively. Although below the expected guideline (i.e., about 21.4% instead of 30%) a cluster of employment equity, management control, socio-economic development and ownership also made significant strides. Also, a clearer realisation of the pillars would have emerged had the original four identified pillars been strictly adhered to. The latter could have reflected below-expected performance in the clustered management control and ownership pillars. The

NAMC is however commended for this new focus on transformation pillars as this would have the impact of a concerted effort at reporting progress achieved in an inclusive agricultural sector.

Some Contemporary Study Findings on Smallholder and Emerging Farming Within the South African Agricultural Marketing Environment

A substantial literature on factors that impacted small-scale farmers' effective participation in the production and marketing environment in South Africa post-1994 abounds. A study conducted by Thamaga-Chitja and Morojele (2014) on the impact of institutional dynamics on the socio-economic situations of smallholder farmers uncovered that the policy environment and agroclimatic zones are not conducive for effective market participation. The latter observation was exacerbated by the dual nature of farming in the country, characterised by the marginalisation of the smallholder farming sector. Thamaga-Chitja and Morojele (2014) suggested that in order to reduce poverty and inequality the country needed to adopt an asset–building approach that would be linked to social protection and institutional readiness.

Other studies have revealed a very close link between land dispossession and poverty in South Africa (Modise and Mtshiselwa, 2011; Maylan, 1986 and Stats SA, 2012). Assessing statistical data prior to and after the promulgation of the Natives Land Act of 1913, Modise and Mtshiselwa (2011) concluded that poverty levels were much lower in the former period than the latter, thus ascribing the ensuing poverty to land dispossession and alienation of black inhabitants. In contrast to that era, recent years have witnessed massive investment in support of the transformation agenda for emerging farmers in South Arica (Sebola, 2018). However, most of the qualifying beneficiaries possess little experience and interest in farming as attested by the many failed agricultural projects. Sebola (2018) further observed that black South African farmers remained agriculturally underdeveloped despite the huge investments that threaten future food security.

In a study by Obi and Maya (2021) on the adoption of Climate Smart Agriculture (CSA), a new concept that has also been suggested to enhance effective participation of smallholder farming, asset fixity was found to be a hindrance to farmer's adoption of new farming practices. The authors suggested awareness campaigns by both public and private sector agencies as a strategy to increase access to facilities such as credit and information, factors that could enhance adoption rates. The latter could also contribute significantly to increased food security and living standards for rural dwellers (Obi and Maya, 2021). A recent study related to the impact of climate change on smallholder farming, especially as regards mitigation strategies against drought, noted that while few farmers adopted strategies such as the utilisation of boreholes, purchasing of fodder and transporting water from rivers, the majority were unaware of such mitigation strategies (Pili and Ncube, 2022).

A study by Mpetile and Chinyamurindi (2021) focussed on the impact of farming as a career entrance into agriculture (entrepreneurial activity), the transition from subsistence to emerging and commercial farming, integration of emerging farming into commercial farming through the land reform programme and farmer support strategies. The study's major findings were that enterprise development in farming was impacted by several dynamics such as financial availability, economic sustenance, and community and socio-economic factors. As noted by Gwiriri et al, (2021) many black farmers have reflected a lack of real 'transition' from subsistence farming practises, relying instead on increased livestock holdings and government social grants for survival. A critical policy imperative that was initially sought as a strategy to immediately transform smallholder farming in South Africa – and thus convert them into entrepreneurial participants - was an equitable redistribution of land through the land reform programme.

A study conducted by Gwiri et al (2019) to investigate the transformation of land reform livestock beneficiaries revealed the existence of three broad categories, i.e.: a large group occupying private farms but remaining trapped in subsistence farming; small/medium-scale commercial producers that approximate 'emergent' farmers; and an elite group of fully commercialized farmers with access to capital.

A comprehensive study by Zantsi (2021) identified several challenges, inclusive of inappropriate identification of potential beneficiaries and misconceptions regarding transitions from subsistence/ emerging to commercial farming. Other challenges included a lack of detailed information regarding smallholder farmers, failure to place beneficiaries into clearly demarcated segments and categories, assumption of smallholder homogeneity, untenable larger landholdings and risks and opportunities associated with being a land reform beneficiary. One critical failure of the redistribution programme was the failure to subdivide redistributed farms into manageable units which could have, amongst others, impacted positively to farm income. A further study by Zantsi et al (2021) suggested distinct differentiation in farming types and tailored support to land reform beneficiaries.

Sihlobo and Kirsten (2021) have presented a chronology of legislation that was aimed at levelling the playing field for effective participation of black farmers in the South African agricultural environment, commencing with the White Paper on Agricultural Policy (1995), Broadening of Access to Agriculture Thrust (1996), Land Reform for Agricultural Development (2000), Strategic Plan for South African Agriculture, Pro-Active Land Acquisition Strategy (2006), Agriparks Initiative (2015) and Operation Phakisa (2017). Whereas these policies were properly designed, drafted, and launched, implementation thereof has always been erratic. Ironically the new policy direction benefitted white commercial farming quite substantially. As reported by Sihlobo and Kirsten (2021), the deregulated marketing environment resulted in the increased value of agricultural produce for the mostly white farming communities. The latter was due largely to increased productivity and thus growth in traditional export and new markets. The authors demonstrated that thirty years after crafting but the minimal implementation of inclusive policies, the dualism that existed in 1994 – continued unabated. Consequently, black farmers are still excluded from beneficial policies identified by Sihlobo and Kirsten (2021) above.

An assessment of the South African marketing environment will be incomplete if it fails to provide a brief review of the impact of the Covid-19 pandemic, especially on black farmers who are negatively impacted by racial and wealth inequality (Wegerif et al, 2021). As noted by Wegerif et al, (2021), the pandemic had not only resulted in global increases in food prices but also in food insecurity. Specifically, an investigation of some fresh produce black farmers in South Africa has shown that in the midst of a growing agricultural sector, black farmers did not only fail to receive adequate support to mitigate the effect of Covid-19 but also experienced reductions in food production and income generation, coupled with massive farm job losses. A study conducted by Béné (2020) has attested to a number of negative impacts emanating from the Covid-19 pandemic, inclusive of food insecurity due to loss of income and reduced capacity of farmers to participate effectively in food production. Other critical impacts of Covid 19 were identified as the availability of labour and reduced resilience of smallholder farmers to participate in crop and livestock farming due mainly to their dependence on human labour (Béné. 2020).

DISCUSSION

The aim of this chapter is to present a systematic literature review of the South African agricultural marketing environment from just before the promulgation of the Natives Land Act of 1913 to the present time. It is common knowledge that the South African agricultural economy has been marred by policy environments that tilted towards advantaging the white commercial farming sector to the detriment of its black subsistence counterpart. In the final analysis, the chapter sought to uncover the extent to which such a policy environment could have impacted the participation of the two farming groups generally, but specifically on black farmers' agricultural production and marketing efforts through a literature enquiry covering three distinct phases, i.e. the first phase covering the era before and after 1913, the second phase that followed the political and economic takeover by the more racially and ethnically inclined apartheid National Party government from 1948 and the third phase that covered the period from 1994 when a new democratic dispensation was achieved in South Africa.

Although the agricultural policy environment was more biased towards white farmers mostly towards the end of the nineteenth century and early 20th century, the composition of the farming community during that period was a mixture of black and white farmers that competed with each other for land and local markets. While black farmers mostly operated inland, the more export-oriented commercial white farmers practised most of their farming activities along the coast. However, the discovery of minerals, especially gold and diamond internally, altered the above farming arrangement with each of the above responding to increased demand for agricultural produce by the emerging mineral and associated industrial sectors.

The superiority of the black farming community, especially through its ability to apply low input cost technologies, coupled with collective resource combination, compelled white farmers to influence the colonial government to promulgate discriminatory policies towards the effective participation of the former. Discriminatory policies were particularly used as a condition for the end of the Anglo-Boer War that was raging towards the end of the nineteenth century and the formation of the Union Government in 1910. Paramount to such a legislative framework was the Natives Land Act of 1913 which effectively dispossessed black farmers of their rights to land ownership and increasingly required them to be landless labour tenants on white farms. Black people could only practise farming in small demarcated and highly congested reserve areas that covered only 13% of the country's land surface in 1936 (increased from about 7% in 1913). Faced with the above situation, and in line with the policy imperative, inhabitants of reserve areas had to offer their labour either in commercial white farms or in other industries especially in towns and cities, in effect removing them from participation in the agricultural produce and marketing environment. As attested in this chapter land dispossession resulted in increased poverty levels for black people.

In order to ensure increased production and marketing of agricultural products by white commercial farmers, the state instituted three policies i.e., the Land Bank Act of 1912, the Cooperatives Societies Act of 1922 and the Marketing of Agricultural Products Act of 1937. Through the Land Bank Act of 1912, the state was able to transfer large quantities of land to poor white farm labourers and sharecroppers while the promulgation of the Cooperatives Societies Act of 1922 availed the necessary inputs to white farmers. The setting in of the Great Depression in 1929 with its devastating impact on produce prices prompted the promulgation of the Marketing of Agricultural Products Act in 1937 effectively removing the farmer from concerns of produce marketing and thus concerted efforts at the production of agricultural commodities. Various schemes and control boards were set up to manage the marketing aspect with farmers receiving a price that could not only sustain them but also encourage increased production. Thus, the first phase was characterised by the promulgation of policies that not only ensured the removal of black farmers from effective participation in agricultural and marketing activities but also increased the availability of agricultural land to the white farming community and massive support to their farming efforts.

The racially and ethnically aligned election manifesto of the National Party government in 1948 set the tone for the agricultural production and marketing environment that characterised phase 2, i.e., continued isolation and subjugation of black farming and massive support to white farmers. Whereas the period before 1970 was characterised by policy development towards increased support to the white farming community, including those on forced removals of black people to ethnic homelands, the period also witnessed increased professional and input support to the latter, resulting in superficial increases in agricultural output, farm employment, wages and consolidation of farming units. The untenable support could however not be sustained as the 1980s witnessed continual decreases in state budgetary allocations to white farming and complete withdrawal towards the end of the apartheid rule.

The third phase that commenced with ushering in a new democratic dispensation in 1994 brought in heightened hope for an inclusive agricultural marketing environment for South Africa. New policies that sought to redress the skewed participation of black farmers included the Restitution of Land Rights Act No 22 of 1994 and the Marketing of Agricultural Products Act No 47 of 1996. The latter Act was not only aimed at the restoration of land to those that were unfairly dispossessed of this crucial resource but also to restore their dignity and equal access to land and its resources. The latter Act was specifically designed to remove the regulated marketing environment and ensure equal participation of all farming groups in the country, especially black smallholder and emerging farmers. To ensure continued and competitive productive and marketing efforts by the white-dominated agricultural sector, Control Boards were replaced by the initiation of Agricultural Commodity Trusts that not only managed the huge assets previously owned by the latter but also collected statutory levies from farmers for enhancing their human capital, production efforts and produce marketing activities. A portion of the levy income was targeted at transforming the emerging black farming sector, although with minimal success.

CONCLUSION AND RECOMMENDATIONS

It was hoped that after many years of implementing a policy that was characterised initially by inclusivity, increased participation of black farmers will necessarily emerge. Almost thirty years (as of 2023) after a democratic dispensation for South Africa, juxtaposed with an enabling and inclusive policy environment, the situation still remains appalling for black farming. Not only is land ownership and occupation still highly skewed against black farmers, but the participation of the latter in the agricultural marketing environment, especially in global exports, is either non-existent or far lower than anticipated. While South Africa is known as the world's best in policy formulation, implementation thereof seems to be the major challenge. The Marketing of Agricultural Products Act No 47 of 1996 is quite explicit on achieving an equitable and inclusive agricultural marketing environment in the country, especially as espoused in Section 2 thereof. Specifically, Section 2a of the Act calls for levelling the playing field and thus ensuring the effective participation of all farming stakeholders.

This chapter aligns itself with suggestions put forward by Sihlobo and Kirsten (2021), especially the need for government to enhance the participation of all farming stakeholders through support to a unified agricultural organisation and thus the dismantling of the numerous fragmented and racially oriented formations. As stated above, the National Agricultural Marketing Council is performing a sterling job of ensuring that the 20% statutory levy meant for the development of the emerging black farming sector is clearly outlined and implemented following a standardised reporting framework with clear achievable objectives. However successful black farmers should increasingly be incorporated into the various commercial commodity marketing opportunities.

In short, each commodity organisation should be able to clearly articulate the number of emerging black farmers that graduated into commercial farming and thus contributing to the growth of the levy collection base. Adoption of emerging technologies and policies meant to upscale their participation could also be crucial in accelerating agricultural productivity for emerging farmers and thus increased participation in the marketing environment. Some of these technologies include the inclusion of some insects into the ration of livestock (Chia et al., 2019) due to their high quality and lower cost; increased participation in Black Economic Empowerment ownership opportunities (Bowman, 2019) and more inclusion of small-scale farmers in sectors such as sugarcane farming (Dubb, 2020) due largely to this sector's ability to create jobs, Success in achieving the above will, however, require that barriers that particularly impact on smallholder and emerging farmers be mitigated. Such barriers include climate change, access to land, general management, adherence to food safety standards ((Miyen et al, 2019), and adoption of digital technology (Smidt, 2018). Also, in the context of climate change global footprint reduction mandates and the current energy insecurity, there is a need to explore and invest in new green markets such as the bioenergy market where farmers can participate as feedstock producers, specifically in the small-scale sector as this has a huge potential for ensuring the highly advocated just transition in the renewable energy sector (Tagwi, 2022; Tagwi and Chipfupa, 2022).

The many well-crafted but unimplemented policies point to unnecessary use of the sector for scoring political goals. As all the policies were mainly formulated by the ruling party it can perhaps be safely assumed that each new administration attempts to come up with new policies that would enhance its own political ambitions rather than a concerted effort at improving the implementation of existing policies for the beneficiation of the country's citizens. Perhaps the best indicator of poor implementation of well-intended policies could be in the Land Restitution Programme, one critical component of the South African Land Reform Programme. Whereas substantial land has been restored to African beneficiaries at exorbitant state cost, monitoring mechanisms to ensure the productivity of such land is at best non-existent.

Communal Property Associations - governance structures meant to oversee the overall development of such land - are simply left on their own to mismanage previously highly productive land. Moreover, beneficiation is centred on a few individuals, usually, the management committee that continues to flout CPA constitutional guidelines, especially the arrangement of annual general meetings, reporting to ordinary members and compliance to statutory regulations, especially submission of annual reports and presentation of financial statements to members. More importantly, once elected, such CPA directors remain in leadership positions forever, contrary to constitutional requirements. Left unchecked and without a concerted assistance programme, these organisations will not only remain cash cows to a few influential beneficiaries but as structures for rent-seeking stretching to the highest level of state administration. Concerted efforts must be re-directed towards CPA management models to ensure that CPAs makes meaningful contributions to farmers livelihoods especially in rural areas.

REFERENCES

Aliber, M., & Hall, R. (2012). Support for smallholder farmers in South Africa: Challenges of scale and strategy. *Development Southern Africa*, *29*(4), 548–562. doi:10.1080/0376835X.2012.715441

Beinart, W., & Delius, P. (2014). The historical context and legacy of the Natives Land Act of 1913. *Journal of Southern African Studies*, *40*(4), 667–688. doi:10.1080/03057070.2014.930623

Bembridge, T. J. (1987). Aspects of cattle production in Transkei. *South African Journal of Animal Science*, *17*(2), 74–78.

Béné, C. (2020). Resilience of local food systems and links to food security–A review of some important concepts in the context of COVID-19 and other shocks. *Food Security*, *12*(4), 805–822. doi:10.100712571-020-01076-1 PMID:32837646

Bowman, A. (2019). Black economic empowerment policy and state–business relations in South Africa: The case of mining. *Review of African Political Economy*, *46*(160), 223–245. doi:10.1080/03056244.2019.1605587

Chabalala, B. R. (2008). *An investigation of the success of comprehensive agricultural support programme projects on the farms allocated under land redistribution for agricultural development: a case study approach in the Waterberg District, Limpopo Province* [Doctoral dissertation, University of Limpopo].

Chia, S. Y., Tanga, C. M., van Loon, J. J., & Dicke, M. (2019). Insects for sustainable animal feed: Inclusive business models involving smallholder farmers. *Current Opinion in Environmental Sustainability*, *41*, 23–30. doi:10.1016/j.cosust.2019.09.003

Dubb, A. (2020). Commodity Study: Small-Scale Sugar Production.

Duncan, D. (1991). The State Divided: Farm Labour Policy in South Africa, 1924–1948. *South African Historical Journal*, *24*(1), 67–89. doi:10.1080/02582479108671686

Gordon, S. (1988). *Land Reform in Post-apartheid South Africa: Some Lessons from Zimbabwe* [Doctoral dissertation, University of Cape Town].

Greyling, J. C., Vink, N., & Mabaya, E. (2015). South Africa's agricultural sector twenty years after democracy (1994 to 2013). *Professional Agricultural Workers Journal*, *3*(1), 10.

Greyling, J. C., Vink, N., & van der Merwe, E. (2018). Maize and Gold: South African Agriculture's Transition from Suppression to Support, 1886–1948. In Agricultural development in the world periphery (pp. 179-204). Palgrave Macmillan, Cham.

Gwiriri, L. C., Bennett, J., Mapiye, C., & Burbi, S. (2019). Unpacking the 'emergent farmer' concept in agrarian reform: Evidence from livestock farmers in South Africa. *Development and Change*, *50*(6), 1664–1686. doi:10.1111/dech.12516

Gwiriri, L. C., Bennett, J., Mapiye, C., & Burbi, S. (2021). Emerging from Below? Understanding the Livelihood Trajectories of Smallholder Livestock Farmers in Eastern Cape Province, South Africa. *Land (Basel)*, *10*(2), 226. doi:10.3390/land10020226

Higginson, J. (2001). Hell in small places: Agrarian elites and collective violence in the Western Transvaal, 1900-1907. *Journal of Social History*, *35*(1), 95–124. doi:10.1353/jsh.2001.0084

Iden, J., & Eikebrokk, T. R. (2013). Implementing IT Service Management: A systematic literature review. *International Journal of Information Management*, *33*(3), 512–523. doi:10.1016/j.ijinfomgt.2013.01.004

Kirsten, J. F. (1994). *Agricultural support programmes in the developing areas of South Africa* [Doctoral dissertation, University of Pretoria].

Kohls, R. L., & Uhl, J. N. (1990). *Marketing of agricultural products*. Prentice Hall.

Lahiff, E. P. (1997). *Agriculture and rural livelihoods in a South African 'homeland': A case study from Venda*. University of London, School of Oriental and African Studies.

Lenta, G., & Maasdorp, G. (1984). Food production in the homelands: Constraints and remedial policies. .

Mafole, M. I. (2018). Memorialising the Pan-Africanist Congress of Azania1. In Social Memory as a Force for Social and Economic Transformation (pp. 257-266). Routledge.

Mann, R. (2001). Homeland agriculture in the era of liberalisation: High-value fruit exports from parastatal projects. *The South African Geographical Journal, 83*(3), 214–220. doi:10.1080/03736245.2001.9713739

Maylan, P. (1986). *A history of the African people of South Africa, from the early iron age to the 1970s*. David Phillips.

Mmushi, M. P. (2020). *Towards an appropriate model for skills transfer to land reform beneficiaries in South Africa* [Doctoral dissertation, University of Pretoria].

Modise, L., & Mtshiselwa, N. (2013). The Natives Land Act of 1913 engineered the poverty of black South Africans: A historico-ecclesiastical perspective. *Studia Historiae Ecclesiasticae, 39*(2), 359–378.

Molatlhwa, T. M. (1976). Agriculture in Bophuthatswana. In *Strategy for Development* (pp. 73–89). Palgrave Macmillan. doi:10.1007/978-1-349-02896-2_6

Mpetile, Z., & Chinyamurindi, W. (2021). Motivational factors into agriculture as a career path: narratives of Black emerging farmers as entrepreneurs in South Africa. *Journal of Enterprising Communities: People and Places in the Global Economy.*

Msibi, M.L. and Dlamini, P.Z., (2011). Water allocation reform in South Africa: History, processes and prospects for future implementation. *Water Research Commission Report, 11.*

Myeni, L., Moeletsi, M., Thavhana, M., Randela, M., & Mokoena, L. (2019). Barriers affecting sustainable agricultural productivity of smallholder farmers in the Eastern Free State of South Africa. *Sustainability, 11*(11), 3003. doi:10.3390u11113003

National Agricultural Marketing Council. (2018). *Annual Report, 2019*. Agricultural Marketing Council. amc.co.za/wp-content/uploads/2020/01/2018-19-NAMC-ANNUAL-REPORT-13-09-2019-Low-Reso.pdf

National Agricultural Marketing Council. (2020). *Annual Report, 2019/20*. Agricultural Marketing Council. https://www.namc.co.za/wp-content/uploads/2021/04/NAMC-ANNUAL-REPORT-2019-20.pd

National Agricultural Marketing Council. (n.d.). *Statutory Measures*. NAMC. https://www.namc.co. za/wp-content/uploads/2017/09/Statutory-Measure-Profile.pdf

National Agricultural Marketing Council. (2008/9). *Annual Report, 2008/09*. Agricultural Marketing Council. https://www.namc.co.za/wp-content/uploads/2017/10/ANNUALREP2009FINAL PRINTED 300909.doc.pdf

Obi, A., & Maya, O. (2021). Innovative Climate-Smart Agriculture (CSA) Practices in the Smallholder Farming System of South Africa. *Sustainability*, *13*(12), 6848. doi:10.3390u13126848

Pili, O., & Ncube, B. (2022). Smallholder farmer coping and adaptation strategies for agricultural water use during drought periods in the Overberg and West Coast Districts, Western Cape, South Africa. *Water SA, 48*(1 January).

Plaatjie, S. (1916). Native Life in South Africa before and since the European War and the Boer Rebellion. *SA History*. http://www.sahistory.orf/archives/native-life-south africa-and-european-war-and-boer-rebellion-sol-t-plaatjie

Porter, G., & Phillips-Howard, K. (1997). Agricultural issues in the former homelands of South Africa: The Transkei. *Review of African Political Economy*, *24*(72), 185–202. doi:10.1080/03056249708704252

Posel, D. (1987). The meaning of apartheid before 1948: Conflicting interests and forces within the Afrikaner nationalist alliance. *Journal of Southern African Studies*, *14*(1), 123–139. doi:10.1080/03057078708708162

Rees, D. J. (1979). Regulation of South African agriculture: performance of agricultural marketing boards..

Schirmer, S. (2001). *Market regulation and agricultural development*. Econometric Research Southern Africa, University of the Witwatersrand.

Sebola, M. P. (2018). Financing emerging black farmers for agricultural development in South Africa: A wasteful and unworkable model for creating black farmers. *TD: The Journal for Transdisciplinary Research in Southern Africa*, *14*(1), 1–7. doi:10.4102/td.v14i1.555

Sihlobo, W., & Kirsten, J. (2021). Agriculture in South Africa. The Oxford Handbook of the South African Economy, 195.

Simkins, C. (1981). Agricultural production in the African reserves of South Africa, 1918–1969. *Journal of Southern African Studies*, *7*(2), 256–283. doi:10.1080/03057078108708028

Smidt, H. J., & Jokonya, O. (2022). Factors affecting digital technology adoption by small-scale farmers in agriculture value chains (AVCs) in South Africa. *Information Technology for Development*, *28*(3), 558–584. doi:10.1080/02681102.2021.1975256

Statistics South Africa. (2012). *Social profile of vulnerable groups in South Africa: 2002-2011*. SSA.

Tagwi, A. (2022). The Impacts of Climate Change, Carbon Dioxide Emissions (CO2) and Renewable Energy Consumption on Agricultural Economic Growth in South Africa: ARDL Approach. *Sustainability*, *14*(24), 16468. doi:10.3390u142416468

Tagwi, A., & Chipfupa, U. (2022). Participation of Smallholder Farmers in Modern Bioenergy Value Chains in Africa: Opportunities and Constraints. *BioEnergy Research*, 1–15. doi:10.100712155-022-10451-z

Thamaga-Chitja, J. M., & Morojele, P. (2014). The context of smallholder farming in South Africa: Towards a livelihood asset building framework. *Journal of Human Ecology (Delhi, India)*, *45*(2), 147–155. doi:10.1080/09709274.2014.11906688

Van Zyl, J., Kirsten, J., & Binswanger, M. P. (1996). Agricultural land reform in South Africa: Policies, markets, and mechanisms. *Agris.* https://agris.fao.org/agris-search/search.do?recordID=XF2015022733

Vink, N. (1999). South African agriculture in the 1970s. *South African Journal of Economic History, 14*(1_2), 90-113.

Vink, N. (2012). The long-term economic consequences of agricultural marketing legislation in South Africa. *The South African Journal of Economics*, *80*(4), 553–566. doi:10.1111/j.1813-6982.2011.01302.x

Wegerif, M. (2021). The impact of Covid-19 on black farmers in South Africa. *Agrekon*, 1–15.

World Bank. (1994). South African agriculture: structure, performance, and options for the future. *Working Paper 12950.*

Zantsi, S. (2021). Unlocking the potential of the emerging smallholder farming sector in South African agriculture: An agent-based approach.

Zantsi, S., Pienaar, L. P., & Greyling, J. C. (2021). A typology of emerging farmers in three rural provinces of South Africa: What are the implications for the land redistribution policy? *International Journal of Social Economics*, *48*(5), 724–747. doi:10.1108/IJSE-10-2020-0728

Chapter 2
Trends, Prospects, and Barriers in Promoting Regional Trade in Pulses/Legumes in the ECOWAS Region Under AfCFTA Regime

Anthony Ojonimi Onoja
College of Agriculture and Environmental Sciences, University of South Africa, South Africa

Ajuruchukwu Obi
College of Agriculture and Environmental Sciences, University of South Africa, South Africa

ABSTRACT

Trade in agricultural products plays a particularly crucial role in addressing food security and reduction of poverty. By making the ability of the continent to feed itself a priority goal in the Malabo Declaration, African leaders shared renewed faith in agriculture, and showed how stepped-up investments in production and distribution can make a strong contribution towards the attainments of goals of the African Continental Free Trade Area agreement (AfCFTA). Food security goes beyond quantity of food available to inclusion of nutritious diets in households' menu. These commodities are vital to the attainment of food and nutritional security, soil conservation, and, if commercially produced in efficient systems, can boost ECOWAS participation in African trade. This chapter therefore examines the prospects for promoting trade in pulses and legumes within the ECOWAS region in the context of the AfCFTA protocol, paying particular attention to trends, prospects and barriers and where there are flexibilities that allow for beneficial and productive policy intervention.

INTRODUCTION AND BACKGROUND

The African Continental Free Trade Area (AfCFTA) was founded by the African Union in 2018 although effective trading did not commence until 1 January 2021. The expectation is that the Free Trade Area (FTA) will serve the 55 member nations of the AU and affect the lives of the 1.2 billion persons

DOI: 10.4018/978-1-6684-4780-2.ch002

Table 1. General objectives of AfCFTA S/No. objective alignment with SDG

1 Create a single market for goods, services, facilitated by movement of persons in order to deepen the economic integration of the African continent and in accordance with the Pan African Vision of "An integrated, prosperous and peaceful. Africa" enshrined in Agenda 2063	Goals 1, 2, 3, 9, 17
2 Create a liberalized market for goods and services through successive rounds of negotiations	Goal 17
3 Contribute to the movement of capital and natural persons and facilitate investments building on the initiatives and developments in the State Parties and RECs	Goals 16 & 17
4 Lay the foundation for the establishment of a Continental Customs Union at a later stage	Goal 16
5 Promote and attain sustainable and inclusive socio-economic development, gender equality and structural transformation of the State Parties	Goals 1, 2, 3, 5, 10
6 Enhance the competitiveness of the economies of State Parties within the continent and the global market	Goal 16
7 Promote industrial development through diversification and regional value chain development, agricultural development and food security	Goals 1, 2, 3, 9
8 Resolve the challenges of multiple and Goal 17 overlapping memberships and expedite the regional and continental integration processes.	

that reside in these countries. The agreement aims to boost trade among African countries in a manner that not only strengthens them individually but serves as a rallying platform for the entire continent to participate in the global market from a position of strength (African Union, 2018; African Union, 2018; The Nigerian Economic Summit Group, 2019). It is expected that trade will produce beneficial spin-offs for the participating countries and enhance their trading positions and voice on the global stage. AfCTA has been conceived as an important component as well as an instrument for attaining the goals of the African Union's Agenda 2063 which encapsulates Africa's vision of how it wishes to develop going forward. Table 1 summarizes the general objectives of AfCFTA as articulated at its inception and reveals the underlying linkages and connections to several other core global and regional initiatives.

Within a broader developmental context, the goals of AfCFTA align closely with those of the Sustainable Development Goals (SDGs) in respect to food and nutrition security as well as generating enhanced well-being for the people through regional integration. Further, the general objectives of the AfCFTA also address core issues underlying group dynamics to enhance the contributions of existing and new institutions to the Agenda 2063 goals. The Comprehensive Africa Agricultural Development Programme (CAADP) of the African Union similarly aims to promote more inclusive and sustainable development that focuses on expanded agricultural investment (African Union, 2021)

In order to develop a useful framework for understanding the trends, prospects and barriers to a strong trade regime for pulses and legumes in the ECOWAS region, it is necessary to present some key statistics. Available data generated by FAO suggest that West African production and trade in pulses has increased over the last decade (FAOSTAT, 2021).

It is estimated that annual increases in the production and trade of pulses could be in the order of 10 percent, with Benin and Nigeria being the biggest players. Awareness of the nutritional and ecological benefits of the pulses appeared to have been the key driver of these increases in the production of pulses, representing another trade and income opportunity, even as Intra-African agricultural trade had been very low. Nevertheless, it would seem that there is a sizeable unmet market demand for pulses, especially since the late 2000s. However, exports of pulses from the ECOWAS region are not meeting up with local and external demands for the commodity. For instance, in 2017 net exports of pulses recorded were in deficits of 565,737,000 tonnes, which is a far cry from available 1961 data that show that net exports stood at 159,972,000 tonnes globally while African and West African net exports were 2,294,000 tonnes and 73,158,000 tonnes, respectively. The share of West African imports in African total imports as of 1961 was 7.89 percent while the export share of West Africa in African exports of pulses was 3.87 percent at the same period. These indicators declined drastically over the decades as the share of imports of pulses fell to 2.63 percent and the share of West African exports in pulses in Africa fell to 1.6 percent in 2017. These scenarios represent an opportunity to key into AfCFTA regime by pulse producers and marketers in West Africa.

The monetary values of the net exports of pulses at the global and West African levels present an equally interesting picture of the changing trends and prospects. For instance, whereas in 1961, the values showed substantial deficits in the order of US$8,445,000 and US$634,000 globally and for West Africa, respectively, and net exports from the entire African continent showed a surplus of US$14,269,000, by 2017 they were all trending upwards. The net export value of pulses for the West African trade region rose to US$21,102,000, while the global and intra-African trade export balances for the commodity stood at deficits of US$13,459,995,000 and US$1,271,046,000, respectively (FAOSTAT, 2021). The West African region's import of pulses and legumes rose from 7,656,000 tonnes in 1961 to 43,408,000 tonnes in 2017. These represent 7.89 percent and 2.63 percent of the West African share of the African imports of pulses for 1961 and 2017, respectively. West African region's export of pulses and legumes rose from 9,950,000 tonnes in 1961 to 21,368,000 tonnes in 2017. These represent 3.87 percent and 1.60 percent, respectively, of the West African share of African exports of pulses and legumes. The total value of exports as of 2017 was $1,057,000 (FAOSTAT, 2021).

The focus on the West African sub-region is an important one. For many years, the West African sub-region that hosts the Economic Community of West African States (ECOWAS) has engaged the attention of the international community for various uncomfortable reasons bordering on insecurity, poverty, looming hunger and starvation (Food Crisis Prevention Network, 2022). While existing literature highlights the high incidence of crimes and misdemeanours, and illegal cross-border trading activities, among others, none of these issues come as close as the spate of insurgencies, terrorist activities and banditry in disrupting production systems and supply chains. The latter has caused more hardships to communities and residents in the most affected areas of the region. In early April of 2022, a consortium of 11 aid agencies operating in the Sahel and Lake Chad belt projected that by June 2022, as many as 38 million persons in the region would be hungry (Food Crisis Prevention Network, 2022). What has happened is that the incessant raids and attacks on local farming populations have caused the land to go uncultivated over large areas, leading to severe food shortages (Oxfam Nigeria, 2022). Again, due to the preponderance of small and tiny national markets and weak resource bases, the financial muscles within each country to deal with these security challenges and their human welfare consequences are often non-existent. Again, weak economies in the region mean that the necessary infrastructure for the effective distribution of goods and services are either under-supplied or completely missing (Obokoh and Goldman, 2016).

Collapse of Infrastructure: Trade

In the foregoing atmosphere of the virtual collapse of the social, economic and physical infrastructure on which the local production of basic foods can be built, it is necessary to examine how trade steps into the breach to ensure that livelihoods disruption is eliminated or minimized. To what extent does trade in the basic foods ensure that the system continues to move legumes and pulses across the region from surplus areas to deficit areas? What lessons can be drawn for the successful operation of the FTA by the manner in which legumes and pulses have trended within the region as revealed by additional data?

The emerging ample opportunities for market access and wealth creation under AfCFTA regime for pulses are not well studied. While some trade flows for African pulses are discernible, it is still not clear what the long-term prospects for the commodities are as revealed by trend analysis. It is also not clear what the prospects and outlook are and what barriers need to be scaled in order to optimize the contribution of the commodity to the attainment of wider AfCFTA goals. Again, and perhaps most importantly, it is not clear which West African countries lead in pulse production and trade and where policy focus needs to be directed to ensure that the commodity rises to its full potential within a regional context.

The importance of pulses and legumes in the farming system has been well-covered in the literature. Pulse as defined by Rahmanian and Batello (2018) and FAO (1994) refers to leguminous crops which produce dry grains used for food and feed. The specific crop species that make up the pulses are characterized and classified by Snapp et al (2018) and FAO (1994). Legumes are members of the *Leguminosae* family, regarded as the third-largest flowering plant family in the world. They are made up of species with unique characteristics that make very vital contributions to sustainable livelihoods, nutrition and sustainable production systems. According to Teferra (2021), pulses are unique in their ability to simultaneously achieve three key developmental goals: reducing poverty, improving human health, and enhancing ecosystem sustainability. With respect to poverty reduction, ICRISAT (2021) studies have established the crucial role played in poverty alleviation by short-duration varieties such as chickpeas and lentils. The advantage of these varieties, particularly for poor farming households, is that several crops are feasible within the year, making it possible for more revenue to be realized by these households. According to Tothill (1986), legumes raise the productivity of farming through their natural ability to fix soil nitrogen, which is at once a poverty reduction feature as well as a matter of environmental sustainability. As a result of their ability to fix atmospheric nitrogen, the legumes are able to build up soil fertility and also provide forage and mulching material.

LITERATURE REVIEW

Pulses have been prioritized in a number of farming systems for their value in family nutrition, income, sustainable agriculture and agro-ecology which make them key components of an agricultural transformation strategy (Snapp, Rahmanian and Batello, 2018). In the age of climate change and the increased need for conservation measures to enhance system resilience, the value of these crops cannot be overstated. The ability of these crops to thrive in difficult environments, including those facing severe water shortage and poor soils, make the pulses an indispensable companion for fighting poverty in many parts of sub-Saharan Africa (Snapp, et al, 2018; Vidigal, et al. 2019). With respect to the improvement of human health, pulses and legumes have assumed a particularly vital role. On one level, they contain proportionately cheaper proteins and essential minerals such as iron and zinc that are implicated in diverse health conditions that generally have nutritional origins (Merga and Haji, 2019). Lately, human health concerns are gaining more recognition as more evidence is available that human diets have practical implications for the incidence and treatment of a number of existing and emerging health conditions. It is widely held that healthy eating is a key tool for the prevention, delay, and cost-effective management of heart disease, type 2 diabetes, and a wide range of chronic diseases. Increasingly, healthy eating has come to be synonymous with more vegetarian diets as more people migrate from meat consumption (Centre for Disease Control, 2021). Nutritionists and medical practitioners are thus commonly prescribing a balanced diet that invariably consists of fruits, vegetables, and whole grains, in combination with small quantities of lean meats, and low-fat dairy products. A report by Rawal, and Navarro (eds. 2019), indicates that between 2011 and 2014, the annual global production of pulses was about 77 million tonnes (31.17%). Out of this, the production of dry bean constitutes about 24 million tonnes i.e. (31.17%), chickpea production, about 13 million tonnes (16.88%), dry pea production, about 11 million tonnes (14.29%) and cowpea production, approximately 7 million tonnes (9.09%). Lentil annual production estimated for the same period stood at approximately 5 million tonnes (6.49%), while that of pigeon pea (5.19%) and faba bean (5.19%) stood at around 4 million tonnes each. The combined annual production of all other pulses – including bambara bean, lupins, vetches and pulses which were not separately classified – hovered at around 7 million tonnes (9.09%). Between 1971 to 2013 the total global exports of the following legumes are as follows: Dry beans rose from 636,000 tonnes to 4,235,000 tonnes; Chickpeas 122,000 to 1,564,000 tonnes. Dry peas 600,000 to 4, 516,000 tonnes, Faba beans 158,000 to 780,000 tonnes; Lentils 153,000 to 2,230,000 tonnes while Other pulses 8,000 to 1,000 tonnes. Joshi and Rao (2016), while discussing the global and regional pulse economies trends and outlook notes that the pulse sector was undergoing dynamic changes at the global, regional and countries levels in a bid to satisfy the challenge of rising demand and supply gap while facing sluggish growth impulses production. It was in the realization of pulses' relevance in human diets that, 2016 was declared the International Year of Pulses (IYP).

This report which captured the dynamics of the pulse sector during the last three decades to 2016 examined the pulse supply, demand, uses, trade, prices, and outlook which researchers and policymakers would find useful in making more informed decisions related to the sector. Globally, it notes, the average share of pulses stood at only 5 percent of the total protein consumption while its contribution to various developing countries ranged between 10 and 40 percent. In order to meet the rising demand and increase their per capita availability, many countries made many efforts to improve the scale of production and explored trade opportunities to augment their domestic supplies. Between 1980 and 2013, global pulses production rose at an annual rate of 1.3 percent with two remarkable two phases of pulses production. First, the report notes that there was almost a period of stagnation in the production of pulses during the1990s after which production increased sharply till 2005. Most of this increased production emerged from developing countries where both area and yield growth (from a low base) contributed to the production. Regarding the trend for developed countries, areas well known for growth in pulse production shifted from Europe to North America and Oceania but in the developing countries, there emerged two new hotspots of production in Eastern Africa and Southeastern Asia (Myanmar).

Fabini (2022) indicates that the current expansion of trade strategies across SSA countries provides growing market opportunities for the agri-food sector. However, it observes that price signals and changes across domestic and regional continental markets remain key factors necessary to boost the success and performance of these markets, including pulses markets. The study used prominent food categories of the agricultural sector in the region, to gauge the market dynamics between pulses, exports, and other staple convenience food. It found heterogeneous market effects across commodities horizontally and vertically for value-added product forms in the food chain. "In domestic intra-regional markets, price transmission is about as high for regional continental markets, while the extent for inter-regional domestic markets varies by the commodity niche on the value chain." It, therefore, concludes that tailored market development policies remain a better strategy to improve the overall functioning of agri-food markets and improve the convergence of markets and trade integration as well as expansion.

The ECOWAS has a big role to play in the attainment of the goals of AfCFTA. ECOWAS has been playing a great role in agriculture development in the region as can be inferred from its report (ECOWAS, 2016) and findings from this study can be useful to the realization of its mandates. The ECOWAS report notes that the body is fully involved in the agricultural and food self-sufficiency drives of citizens of the Region. It implements its activities in the agricultural sub-sector using a four-prong approach embedded in a programme called the Regional Agricultural Investment Programme (RAIP). In summary, the programme aims at:

- Promoting strategic projects to boost food security and food sovereignty,
- Promoting a global environment conducive to regional agricultural development,
- Reducing food vulnerability and the promotion of sustainable access to food, and
- Governance, coordination and monitoring-evaluation of ECOWAS implementation.

Of the fifteen (15) ECOWAS member States, Burkina Faso, Cape Verde, Mali, Niger, and Senegal devoted 10 percent of their national budgets to investments in agriculture by lending support to farmers in the form of inputs such as seeds, fertilizers, foods, cattle, as well as farming implements including ploughs, harrows, and other tools to farmers. The creation of a global environment conducive to the development of a sector and a regional agricultural market is at the core of the RAIP centre of activities. The implementation of the activities of this important intervention component of the organization enabled the implementation of the ECOWAS Regional Agricultural Information System (ECOAGRIS) which was concerned, in its first phase, with seven (7) member states (Benin, Burkina Faso, Côte d'Ivoire, Ghana, Mali, Niger, and Senegal). It enabled the interconnection of the national agricultural information systems. The eight countries (Cape Verde, the Gambia, Guinea Bissau, Guinea, Nigeria, Liberia, Sierra Leone, and Togo) were not integrated into the ECOAGRIS system to ensure the timely acquisition of statistics on Agriculture and Trade of agricultural products within ECOWAS. Another relevant achievement of ECOWAS in promoting agricultural trade was through the signing of partnership agreements with nine socio-professional organizations and Civil Society (POSCAO) with regional bases and national functional platforms for boosting the Agricultural sector booster Plant and animal productions in the region. These organizations are responsible for aiding ECOWAS in establishing observatories on peasant agriculture, cross-border grazing, cattle rearing within ECOWAS, and the collection of statistics within the territorial space of ECOWAS as well as the Economic Community of Central African States (ECCAS) countries, especially Cameroun and Chad, which practice a lot of trade in cattle with the Federal Republic of Nigeria.

METHODOLOGY

The mixed methods approach involving both quantitative and qualitative techniques was applied. The first step was to conduct a descriptive profiling of the trends in the production of pulses in West Africa, with particular emphasis on trade (export and import) using computed tables and graphical analysis.

Details of these trends with respect to soybean and groundnut were specifically highlighted. The next step was to examine the barriers to the attainment of regional leader status or top market performers under AfCFTA. A combination of desk research (stylized facts analysis), trend analysis and descriptive techniques were applied based on aggregated time series data collected on the West African region from FAOSTAT (FAO, 2021).

RESULTS AND FINDINGS

Trends and Prospects in West African Pulses Production

The production, area cultivated, yield and yield gaps trends for West Africa compared to the African and World trends are charted in Figures 1-3. Pulses production in West Africa has been characterized by low and sometimes declining productivity compared to the African and global average growth trends (Figure 1). Figure 2 indicates that total global production (in tonnes) of pulses from 1961 to 2020 trended upwards. The African production outlook over the same period similarly trended upwards. However, despite the upwards trends in total production in the two geographical areas, marked differences in the yield of pulses were observed. This trend is acknowledged by FAO (2017) too which observes a marked gap between farm yields and potential yields in African food crop production including pulses. FAO (2017) attributes the gap to challenges such as inadequate adoption of more efficient and productive technologies, absence of market integration and gender inequalities in smallholder farming.

Measured in hectogram per hectare (hg/ha) (a measure of productivity), the productivity results reveal consistently very low to flat trends. On the basis of a differential analysis between the yield of pulses in West Africa and the rest of the world, it was shown that the yield gap for the World and Africa exhibited an upward trend up to 1970 after which it decreased, mildly in 1980 and more precipitously in 1990 before beginning to rise till around 2001 (see Figure 1). Subsequently, it dipped slightly and recovered before declining steadily till 2020. The highest global productivity growth was observed in 2001 when it stood at 8,702 hg/ha at a time when the West African yield had an average value of 4676 hg/ha, well below the African yield of 6373 hg/ha. By 2020 the yield of pulses in the world had decreased to 7,503 hg/ha, the African yield to 6,407 hg/ha and the West African yield trailing at 4475 hg/ha.

The picture of the land area cultivated is presented in Figure 2. Data were collected from FAOSTAT for the period 1961 to 2020 and projected to 2030 in the graph (Figure 2). According to the results, the land area harvested for pulses rose from 370,774 hectares (ha) to 802,004 ha, representing a growth rate of 116 percent over the period.

In order to effectively assess the impact of the land area cultivated on the production and yield of pulses, data on pulses production and yield were plotted alongside those of the land area (Figure 2). As the results show, the productivity growth for West African pulses rose by a dismal 0.01 percent during this period (1961-2020), comparing unfavourably with the global value which rose to 50 percent and the African value at 27 percent. The trend of low agricultural production in Sub-Saharan Africa has, in recent times, been attributed to factors such as climate change, poor soil quality, slavery and disease (Bjornlund, Bjornlund and Van Rooyen, 2020)

Figure 2 also shows that the actual total production of pulses in the West African region grew during the period from 164,715 tonnes in 1961 to 359,734 tonnes in 2020, representing a growth rate of 118 percent. The implication of the nature of this trend of production growth in relation to the area cultivated

Figure 1. Trend in pulses production yield and gaps in west africa compared to africa and the world

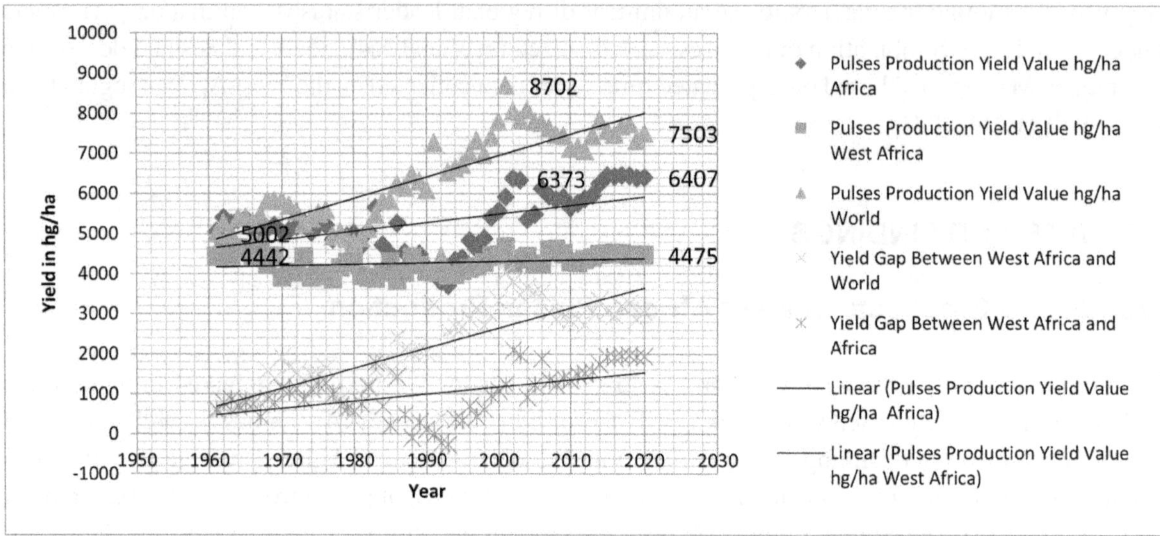

is that West African pulses growth is buoyed mainly by land area expansion but the real productivity growth remained flat and extremely low (0.01%), almost zero growth. There is, therefore, a yawning gap and potential for productivity growth of pulses that appears not to have been tapped in West Africa (see Figure 3).

The observed trends as reported in the foregoing mirrors the situation with respect to yield gaps in agriculture over much of Africa (Nini-Pratt et al., 2010; and African Development Bank (AfDB) Group, 2014). Nini-Pratt et al. (2010) further attribute the poor trend in African agricultural productivity growth to abiotic and biotic factors. They identified radiation, water, temperature and nutrients as the major abiotic constraints to the productivity growth of agriculture in Africa (Nini-Pratt et al., 2010).

Figure 2. Trend in pulses output (tonnes), yield (hg/ha) and land area harvested (ha) growth

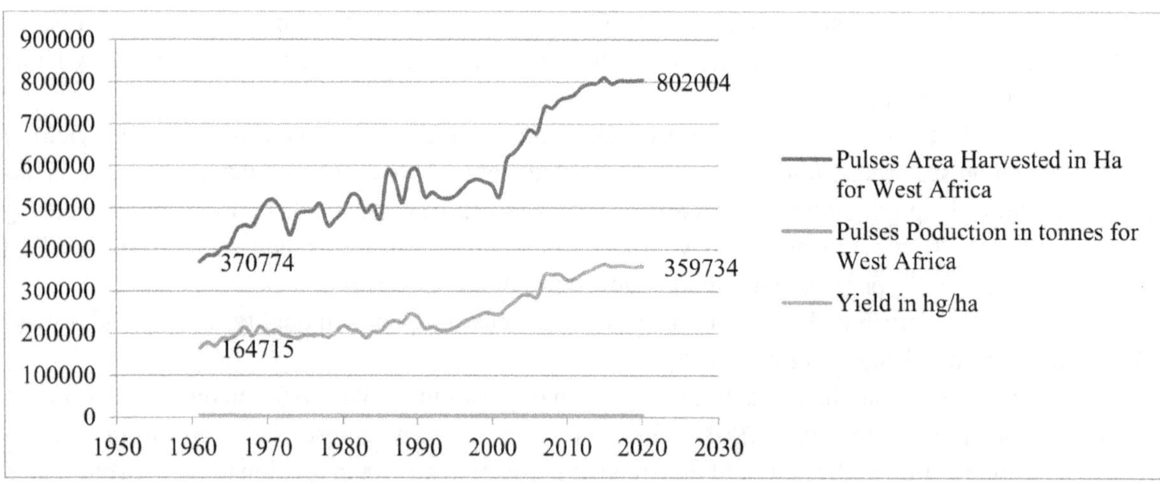

Figure 3. Trend in yield gap growth between west african pulse production and world

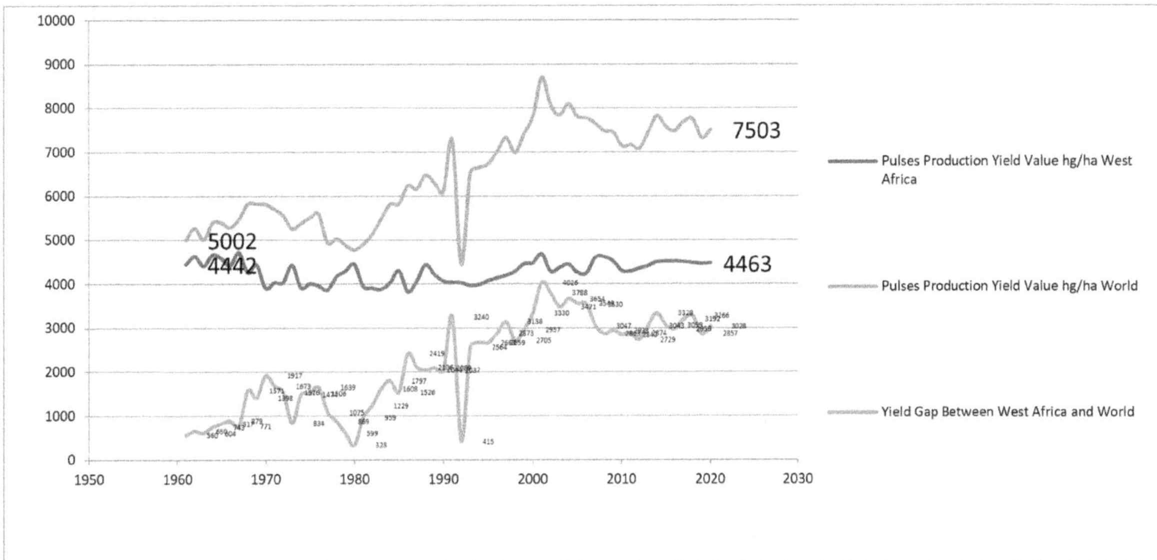

The biotic factors that have constrained agricultural growth on the continent include infestation of weeds, pests, insects and pathogens (Nini-Pratt et al., 2010). Nini-Pratt et al., (2010) emphasize that due to the rain-fed nature of agricultural systems in Africa, dominated by smallholder farmers, yield losses caused by water stress or scarcity remained largely the most constraining factor to African farmers in the Sahel and northern semiarid regions. It was observed that while other West and Central African countries seem to have adequate rainfall, the distribution and timing of rainfall in the ECOWAS region remain unpredictable thus increasing the risk of yield loss in the region. By 2020, a yield gap of 2,028 hg/ha prevailed between West African and global trends. On the other hand, the yield gap between West African and continental African farms averaged 1,932 hg/ha. More efforts are required in the agronomic practices and policy spheres to meet these targets. This trend agrees with World Bank (2018) which observes a large variation in labour productivity between SSA and the advanced economies. The trend was explained by inefficiencies in adopted technologies. The observed inefficiencies in resource allocation in most farms as well as manufacturing firms in SSA were associated with the misallocation of human capital. These, they note happen when "policies and institutions distort the allocation of talent by delivering inefficient occupational choices (which either leads to more informality of a slower structural transformation process) and affecting producers' decisions to invest in new technologies or methods of production as well as their decisions to enter or exit the industry". The findings above corroborate the findings of Rawal and Navarro, (2019, eds.) whose analysis notes that from the 1970s, two phases of high growth in the production of pulses were recorded. The report noted that the first phase, which occurred in the 1980s, witnessed an annual production increase of pulses from about 36 million tonnes at the beginning of the decade to about 48 million tonnes by the end of the 1980s. In that period, they note that the growth of pulses production was largely buoyed by dry pea production. Increased dry pea production over the 1980s was a result of a simultaneous increase in yields (at a rate of 5.3 percent per annum) and expansion of area (at a rate of 4.2 percent per annum). In the report's triennium which ended in 1991, at about 2 tonnes per hectare, the average global yield of dry pea was higher than the average yield of all other pulses. The period of high growth of production in the 1980s was followed by a period of stagnation in the decade of the 1990s.

On the whole, the gap in production coupled with reasonably available land as indicated by the rising trend in land area expansion for pulse production represent an opportunity for pulse producers and marketers in West Africa to key into the AfCFTA regime. This will mean adopting more innovative, climate-smart and highly efficient agricultural practices to be able to optimize the resources available to the region in dominating African pulses trade for inclusive and sustainable green growth of their economies, especially under the AfCFTA regime.

Trends and Potentials in West African Pulse Trade From 1961 to 2020

The trend in trade, including imports and exports is discussed in this section. In addition, a comparative analysis is presented to show how the relevant aggregates compare across regions and over time. The information and data related to these questions are presented in Table 1 and Figures 4 to 14. Examining the import trends and gauging the import dependency trends, quantitative data on import volumes are plotted against years from 1961 to 2020 and presented in Figures 4 and 5. This information is then related to Tables 1 and 2 to shed more light on the import trends for West Africa in relation to the rest of Africa and the World. In Figure 4, it is observed that the African import trend of pulses in West Africa remained on the upward ascent from 1961 to 2020 while it was relatively very stable within West Africa itself. This could mean that there is relative self-sufficiency in West Africa for pulses compared to the rest of the African continent where the trend suggests a rising appetite for the import of pulses. Here too could also lie the potential of West Africa as a potential giant in the African pulse market. The growth rate for the volume of imports of pulses for West Africa over the period stood at 150.7 percent contrasting sharply with the West African growth estimated at 73.6 percent from 1961 to 2020. The total actual imports for West Africa rose from 7,656 tonnes in 1961 to 63,983 tonnes in 2020 (Figure 4).

The share of West African imports of pulses over the entire period also decreased remarkably. From Table 2 and Figure 5, it can be observed that for the period, 1960 to 1970 the share was 7.67 percent. It then decreased rapidly during the 1991-2000 decade to 3.74 percent and further to 3.42 percent from 2011 to 2020. This trend may be attributed to the structural transformation of the agricultural and pulse industries over the period as it is noticed that by the 1990s, most West African nations had embraced the Structural Adjustment Programme (SAP) and had liberalized their markets around commodities in which they had comparative advantage, including the pulses. In fact between 1981 and 1990 when SAP got introduced in the region, the share of imports of West African pulses in African total imports had already decreased to 3.47 percent (See Figures 5 and 6 and Table 1). By 2020, the market share had fallen further to 3.42 percent.

Regarding the monetary values of the imports of pulses into West Africa, the trend is depicted in Figure 7 and its growth rate is shown in Figure 8. During the entire period under review, the growth in African imports of pulses rose geometrically while that of West Africa remained stable and close to its 1961 value (US$1,391,000) with an export value of pulses at US$22,089,000 as at 2020. The implication is that African import expenditures on pulses are growing rapidly or rather geometrically at 8,295.53 percent over a period of six decades. There is therefore immense potential for investments aimed at closing this demand gap for pulses. West Africa can key into this potential to improve its foreign exchange receipts, and improve the poverty profile of their countries while addressing food security issues in the region as its own import growth for the pulses is more stable (1,633.23%) compared to the African growth rate.

Figure 4. Trend in pulses import for West African and Africa

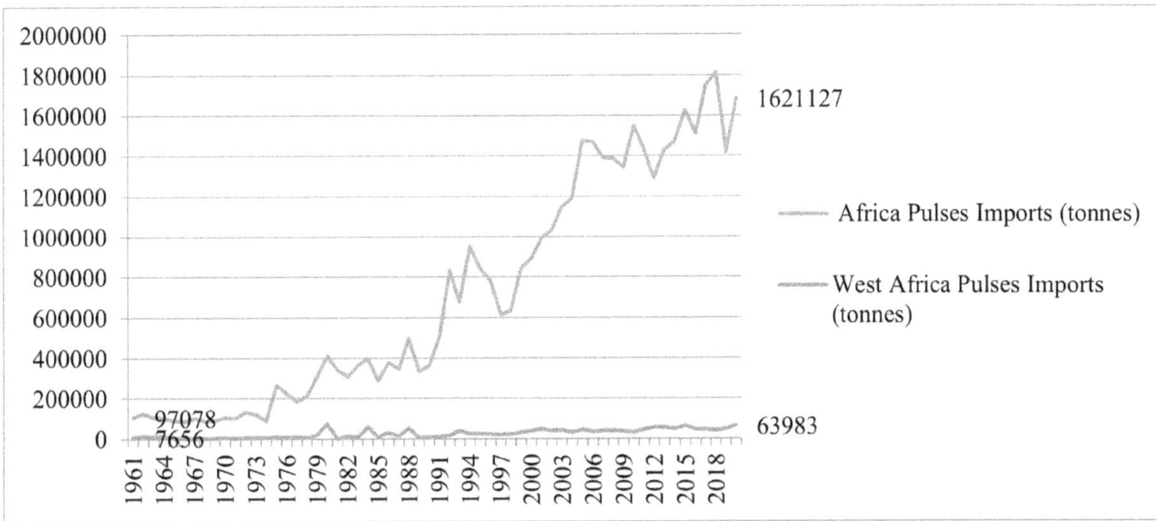

To further corroborate this outlook, Figure 8 presents data for the six decades under review that show a dramatic drop in the share of expenditures on pulses imports of West Africa in the continental total expenditure on pulses imports. This is an indication that West Africa is gradually improving its portfolio in terms of self-reliance in pulses production or trade. The proportion of West African expenditure on imports of pulses dropped from 9.43 percent between 1961 and 1970 to an impressive level of 2.02 percent between 2011 and 2020.

In analyzing the trends of pulses exports in West Africa and in Africa, it can be observed from the chart in Figure 9 that the volume of exports (in tonnes) for Africa fluctuated briefly between 1960 and 2004

Figure 5. Share of West African imports volumes of pulses in african pulse total imports (%)

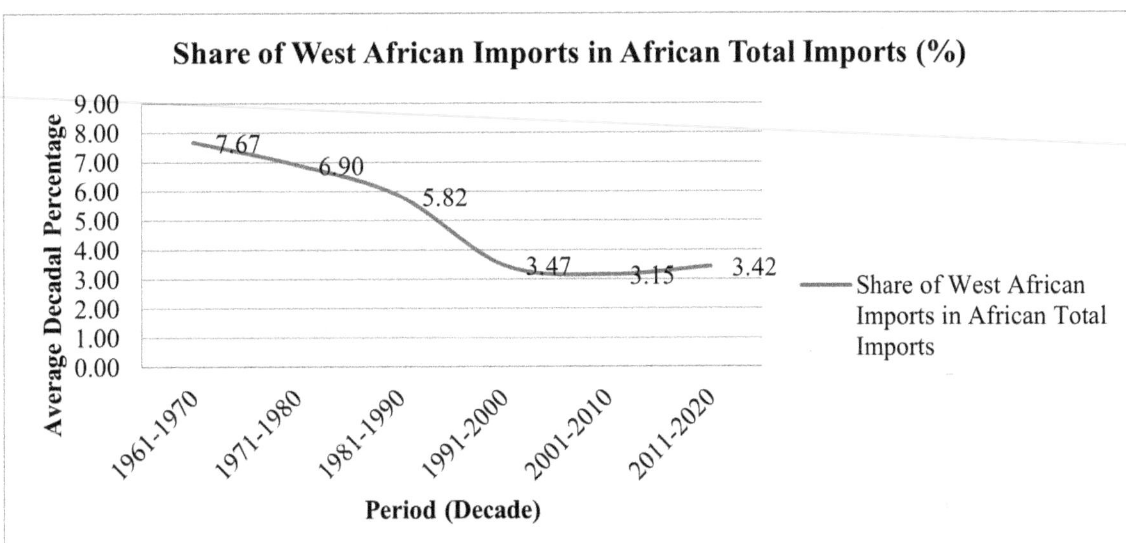

Figure 6. Share of West African imports volumes of Pulses in African pulse total imports

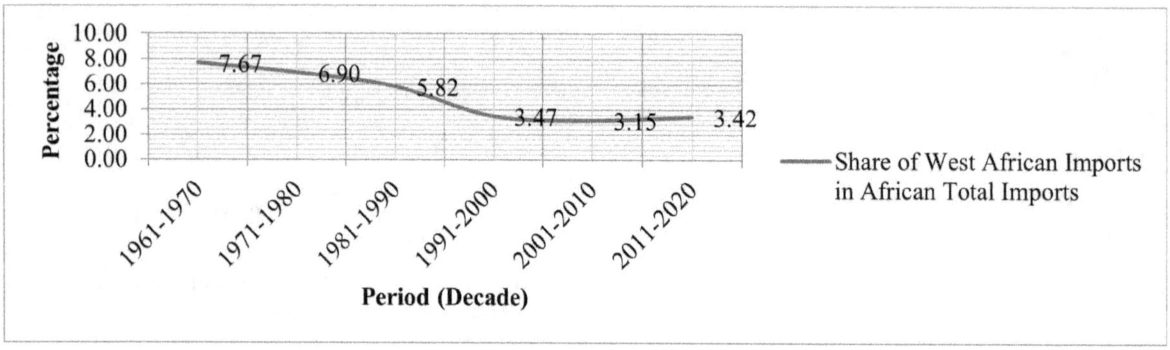

Table 2. Share of West African imports volumes of pulses in african pulse total imports

Decades Code	Share of West African Imports of Pulses in African Pulse Total Imports (%)
1961-1970 1	7.67
1971-1980 2	6.90
1981-1990 3	5.82
1991-2000 4	3.47
2001-2010 5	3.15
2011-2020 6	3.42

but was relatively stable till 2005 when a tremendous spike of exports began to be observed up to 2017 when it peaked and began a steady decline till 2020. In contrast to this trend, the West African volume of exports of pulses, unfortunately, had remained stable and defied growth with a negative growth rate of 1.85 percent which in real terms started with an export volume of 25,119 tonnes in 1961 and ended

Figure 7. West African and African values of pulses imports 1961-2020 in US dollars

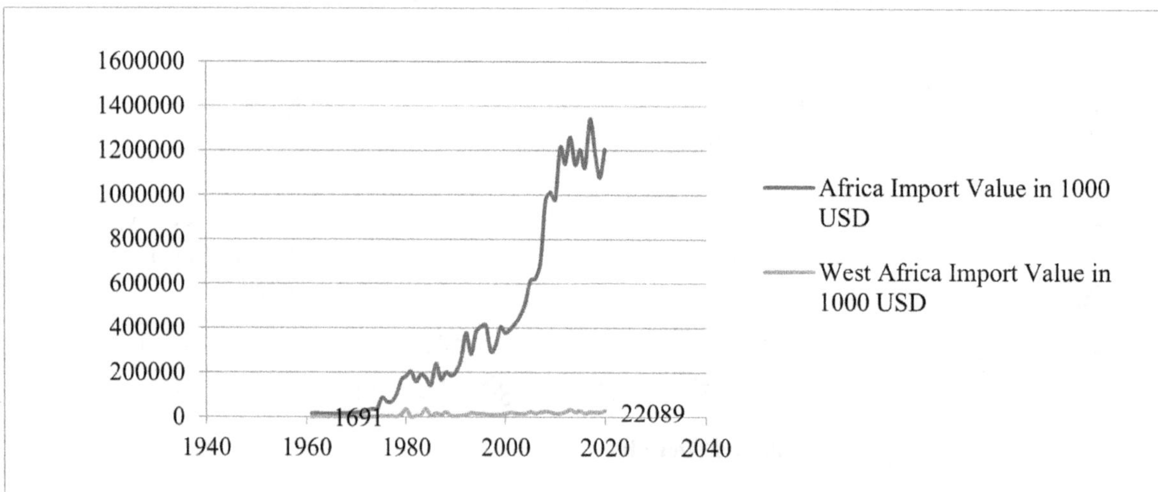

Figure 8. Share of West African Imports (US$) in African pulse total imports (%)

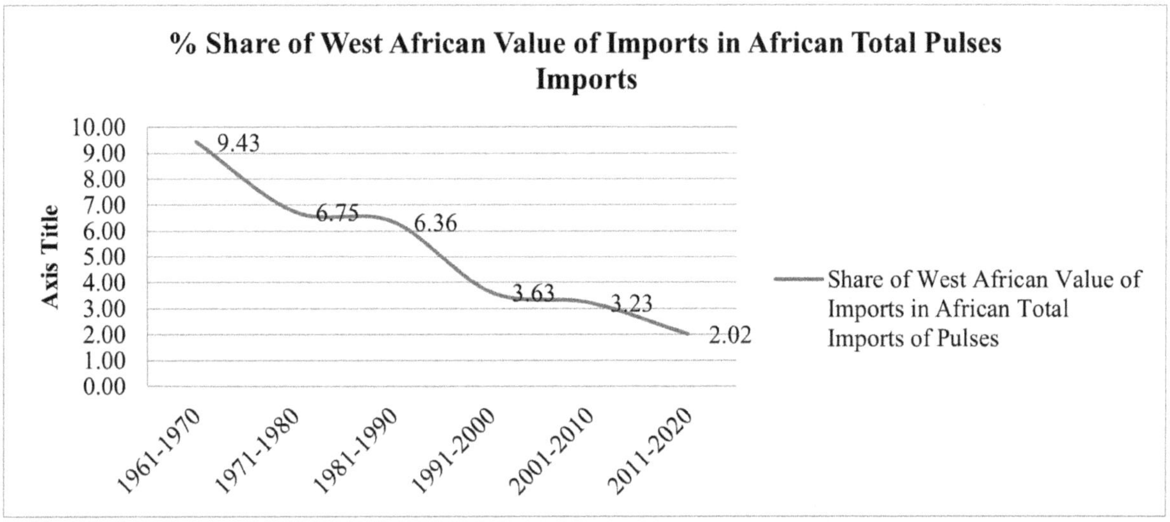

with 24,655 tonnes of pulses exports in 2020. In the African space, it appears that the Africans generally were conscious of the nutritional and ecological benefits of the pulses and this could have buoyed the increase in pulses production and exports, representing another trade and income opportunity, even as West-African agricultural export trade had been very low going by its dismal contribution of exports to African total exports of pulses.

Figure 9. Trend of West African exports (tonnes) in African pulse total exports

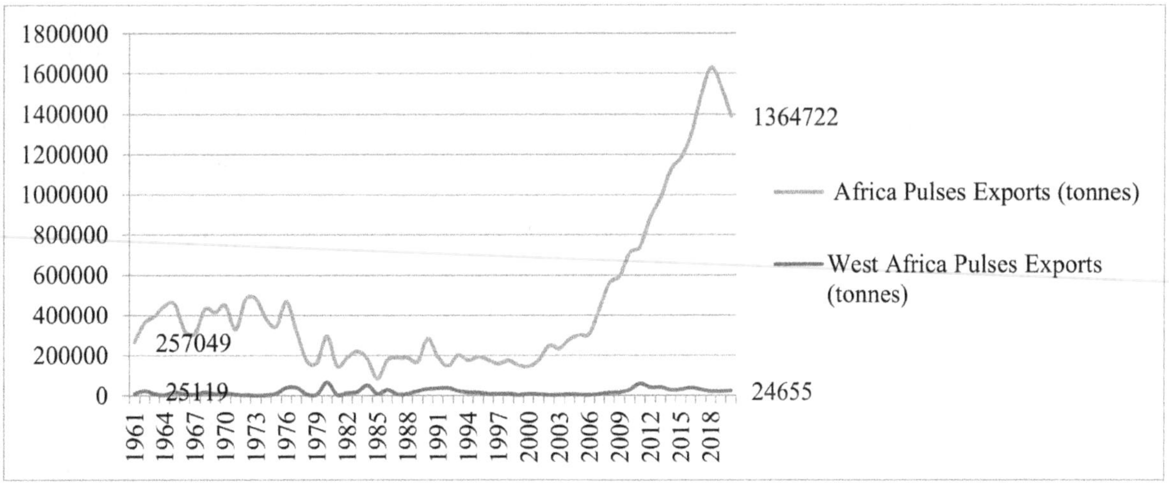

There has been, however, a curious trend in the growth of the proportion of export volumes (i.e. tonnes of pulses exported) of West African pulses in African Total Exports of pulses over the six decades in review (See Figure 10). Figure 10 reveals a fluctuating trend in the export share of exports of pulses for the African continent. This measure exhibits an increasing pattern from 1961-1970 (recording 3.22%)

Figure 10. Trend in % share of West African exports (US$) in African pulse total exports (%)

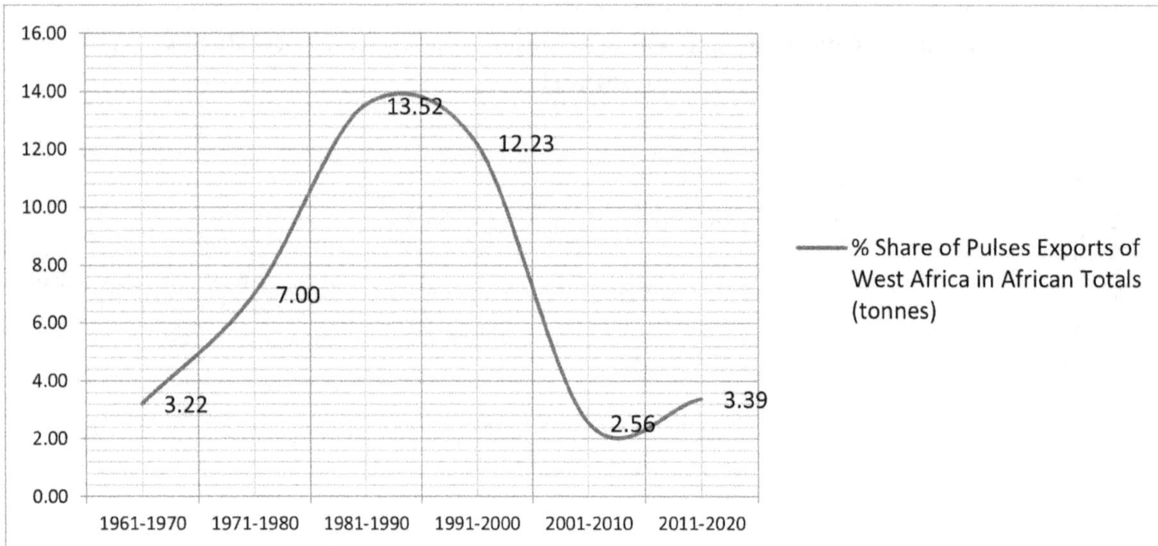

which rose to 13.52 percent by 1981-1990 (decade 3) before dropping continuously till the period 2011-2020 with an improved share of 3.39 percent.

The trend in the growth of the proportion of values of revenues from trade (i.e. export earnings) of West African pulses(US dollars) in African Total Exports of pulses over the six decades in review (See Figure 11) also exhibits a similar pattern. From Figure 11, a fluctuating trend was observed in the share of export revenue earnings from pulses of the African total earnings. This measure as indicated in Figure 11, exhibits an increasing pattern from 1961-1970 (averaging 2.70% per annum) which rose to 14.65 percent by 1981-1990 (decade 3), evidently the golden age for pulses, before dropping continuously till the period 2011-2020 with a decreasing share of 2.17 percent which it maintained from 2001-2010 when the proportion ebbed drastically to 2.37 per cent. The implication of these indicators and trends is that the West African region is losing out in earnings from exports of pulses in Africa even though it stands a chance of being the strongest region with a higher earning capacity of foreign revenue from exports of pulses in the African pulses trade. This requires policies that will improve the market penetration of West African pulse farmers to enable them to regain their pride of place in the African pulse exports market

More evidence on the export performance for the pulses trade in West Africa can be gleaned from the net export volumes and net earnings trend for pulses trade for West Africa as charted in Figures 11 and 12. As shown in these charts, net exports in terms of volumes of pulses traded in international markets as well as the net export earnings from the pulses were positive but had begun to decline between the decades 1961-1970 and 1981-1990 when the net exports became negative. This trend continued in the period from 2011 to 2020. There are obvious yearnings to improve the export earning capacity and total export volumes of pulses in Africa and West Africa. What this means is that the prioritization of pulses as an export crop could hold the key to boosting African participation in the AfCFTA thus increasing the production of food crops, especially those with high nutritional value, in African trade.

Figure 13 presents results for net export earnings growth per decade for the period 1961 to 2020. The results show that net export earnings (in US dollars) of pulses declined on average in West Africa by a whopping 4,210.14 percent over the period of six decades. Similarly, net exports in tonnes of pulses

Figure 11. Trend in % share of West African exports (US$) in African pulse total exports (%)

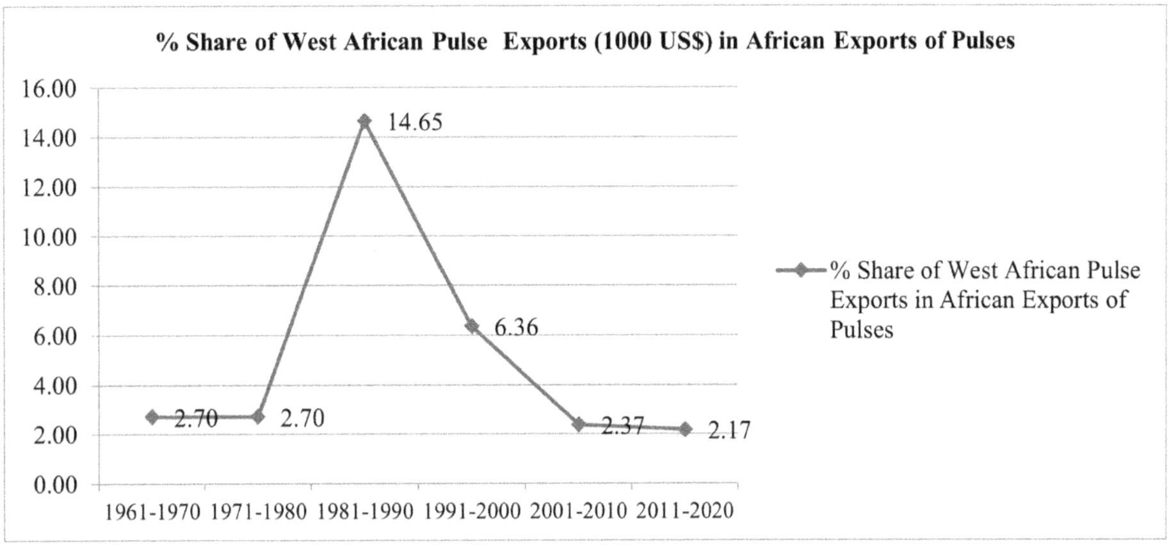

generally decreased at a rate of 447.54 percent. These results have left a huge void in the drive to increase export earnings from pulses and have frustrated the agricultural transformation strategies embarked upon by several West African countries, including the ECOWAS Secretariat itself, to make agriculture a key source of foreign exchange earnings to its member countries, Nevertheless, on the whole, net exports earnings from pulses had been relatively stable for West Africa but not improving in any way. Even at that level, when compared to the net exports trend in Africa generally, the West African region stands a better chance of earning more from pulses trade than the African-wide trend as indicated in Figure 14.

Figure 12. Trend in pulses net Exports in West Africa

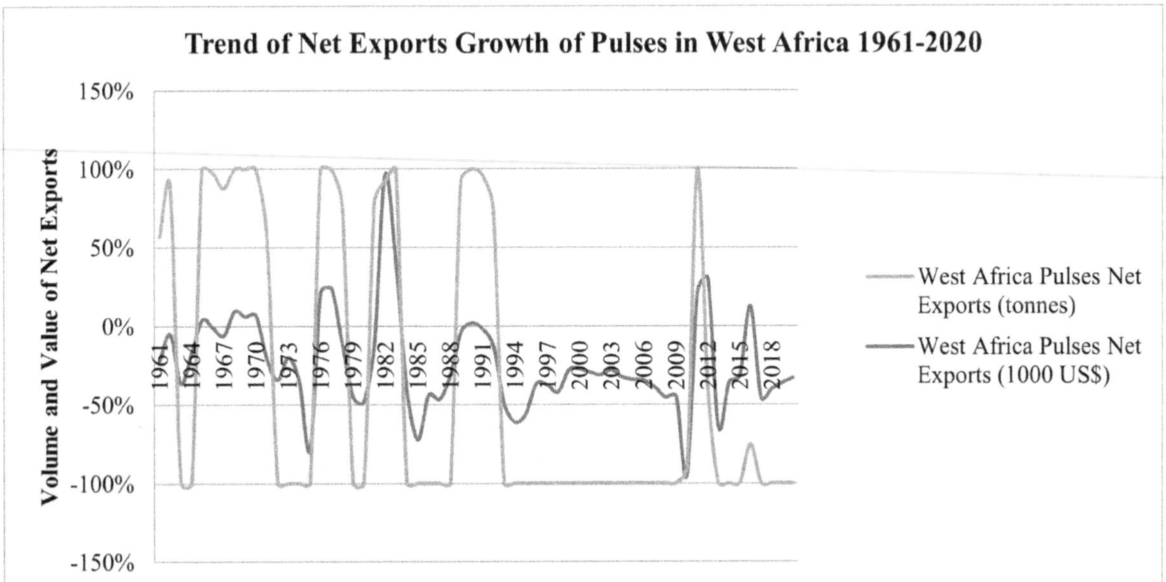

Figure 13. Trend of net export growth in West Africa for six decades (1961-2020)

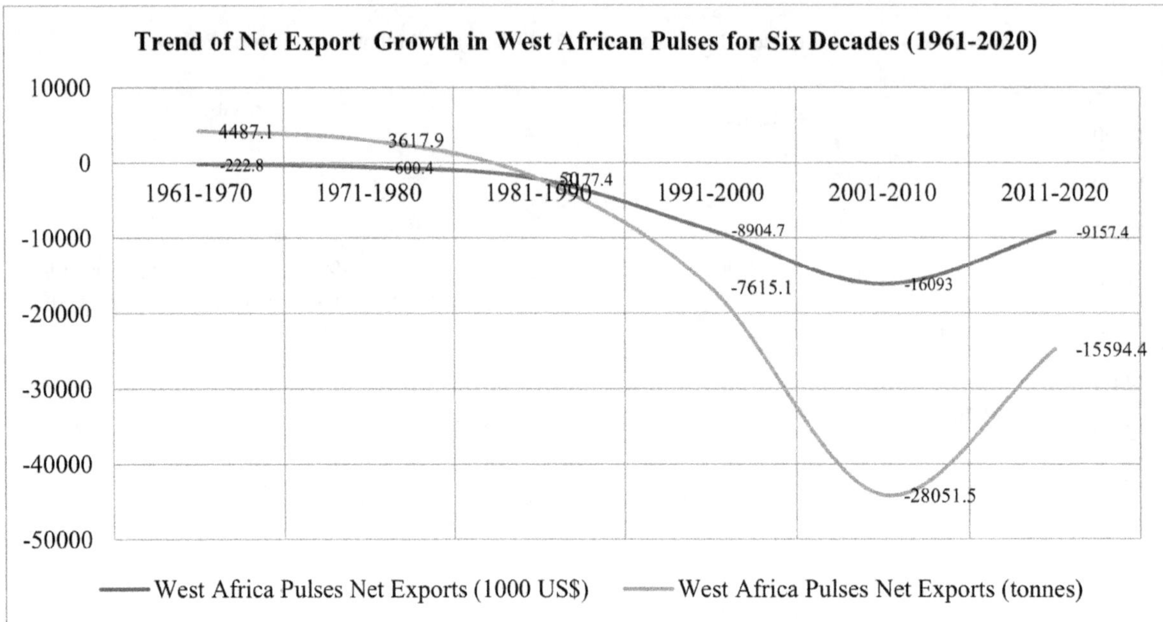

Figure 14. Trend of net export growth in West Africa and Africa for from 1961-2020

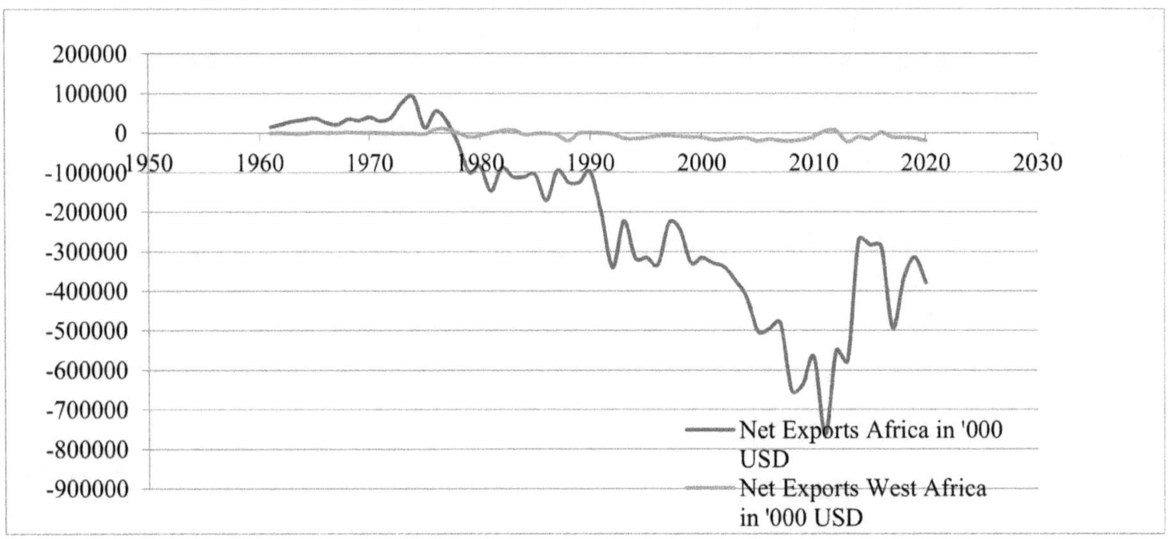

So far, data analysis has proven that African pulses are not being optimized as a foreign exchange earner nor as a key crop to drive the expected opportunities that AfCFTA holds in global trade especially food commodities trade.

Some Key Issues That Will Drive Future Trends of Agricultural Production and Trade in Africa

One issue to consider by AfCFTA is to take note of consumer price behaviour. It is important to note that "customers are willing to pay more for a product when it's difficult to analyze alternatives." (Jones, 2021). This has been confirmed especially in products whose benefits are challenging to know before purchase. Quoting Nagel and Muller, Jones (2021) notes that the difficult-comparison effect can be seen in grocery stores. Thus when suppliers over-package their products, it can make food products seem more valuable than they are. The unusual shapes and sizes of some food brands complicate comparisons. Businesses need to provide information to aid customers who are considering switching to their brand. This could include customer testimonials or quality guarantees. Customers usually want the best deal for their dollar. The lower the price for a quantity of product, the higher the demand.

The growing need to reduce greenhouse gas emissions and consume sustainably are key issues that should also guide choices of commodities to produce and supply into the international market under AfCFTA regime (UNFCC, no date). A report by Pedersen, Kondrup and Børsheim (2013) notes that the future and present demand for food products is likely to be tilted in favour of vegetarian-based diets as epidemiological studies have reported varied associations between dietary proteins and different protein sources and risks of chronic diseases, including cardiovascular disease (CVD), *Type 2* diabetes (T2D), colorectal cancer (CRC), and mortality. Willett et al (2019) and Päivärinta et al (2020) advocate for a transition towards more plant-based diets in order to achieve the UN Sustainable Development Goals and the Paris 2015 Agreement. Päivärinta et al (2020) find that substituting animal protein with plant protein sources in the diet leads to an increase in fibre consumption and improved dietary fat quality as well as blood lipoprotein profile in human beings. It advocates for a Planetary Health Diet that relies on a flexitarian diet composed mainly of plant-based foods which contain only small quantities of animal-based foods (Willett et al, 2019). If the goal of this advocacy is to be met, there is a need for AfCFTA to leverage it in order to achieve a double win of increasing foreign exchange earnings from agricultural (crop products trade) and at the same time transforming from current westernized diets to mainly plant-based diets in Africa. This can be done by further processing the crop products to take advantage of profits inherent in various agricultural value chains, especially of legumes and pulses.

AfCFTA can also benefit from the findings of McDermott et al (2013) who note that the market-oriented smallholder farming approach to agricultural development remains the dominant paradigm to bring about nutrition-sensitive agricultural development. First, smallholder farmers remain the dominant force in countries with a high rate of malnutrition in Sub-Sahara Africa and Asia. Secondly, smallholder farmers will be more market-driven and market-based incentives have the potential of boosting food output, quality and safety. Hence AfCFTA can leverage the diverse value chain approaches, (ranging from informal to supermarket-oriented value chains) to incentivize diversification of food production and further processing which can achieve higher nutritious food availability and access. Adoption of biofortification to increase levels of micronutrients in crops at different value chains is an option. Mc-Dermontt et al (2013) document a model nutrition-sensitive programme called Health and Nutrition Gardens implemented in Mali, West Africa from 2007–2010 that successfully reduced high rates of stunting and micronutrient deficiencies as well as inappropriate feeding practices. The program combined the traditional home gardens approach, such as access to inputs, training in crop production, and postharvest practices, with a number of gender empowerment and nutrition education components. The results of the evaluation of this program can be upscaled by various governments of AfCFTA countries to boost trade in nutritious-sensitive food crops in Africa which suffers from malnutrition currently.

Food Inflation and Food Affordability to Drive the Demand for the Product

The growing trend in global food inflation exacerbated by disruptions in global supply chains, climate change, wars (including Ukraine and insurgency in African countries) and the lingering impact of the Covid-19 pandemic (McKinsey's Agriculture Practice, 2022) can equally serve as an impetus to drive AfCFTA if the African agricultural sectors can respond positively to this cue. This trend is expected to serve as a catalyst for boosting the demand for African agricultural products especially, cereals, legumes and pulses. Unfortunately, as of 2022, it has been indicated (Fusacchia, Balié and Salvatici, 2022) that the current pace of trade growth within Africa still remained slow, a major scenario that prompted the recent commitment by African heads of state to triple intra-continental agricultural trade by 2025. This, Fusacchia et al (2022) notes raises questions about whether it is feasible for Africa to enhance its food security too through its regional supply chains and greater intra-African trade. However, reducing costs of production can help the producers of legumes and pulses to maximize the advantages inherent in rising food prices in the international market thus giving them a marketing edge.

Barriers to Optimization of Pulses as Export-Earning Crops in West Africa

The assertion and findings in this study corroborate earlier research findings of the Japan Association for International Collaboration of Agriculture and Forestry,(JAICAF) (2007) which holds that although it cannot be said that pulse crops are unimportant staples for the people in West Africa, it cannot be denied that its relevance as a cash crop is still virtually low. They held that except groundnut, pulse crops in West Africa (such as cowpeas, kidney beans, bambara groundnuts, soybean, chick peas etc.) do not have the cash-earning ability and are usually planted as mixed crops with staple grain crops (maize, pearl millet and sorghum). Moreover, the farmers are yet to regard the yielding ability of pulses as an important attribute of the crop. Nevertheless, it is obvious that pulse crops have been firmly established as vital components in the West African local diet cultures (JAICAF, 2007). The belief that pulses as crops facilitate the maintenance and preservation of soil fertility coupled with their additional capacity to improve major grain crop productivity makes these crops viable candidates for promotion in the current drive to build African export trades, especially under AfCFTA regime. They are unlike the other staple crops (such as yam, cassava, rice and maize) which mainly supply carbohydrates.

Pulse crops remain very vital as sources of vital vegetable protein. One of the important pulse crops, cowpea, originated in West Africa and it is a major grain food in this region. Besides, the stalks and leaves of cowpea are good sources of forage (animal feeds). Hence, farmers in the countries within the Sahel zone (e.g. Mali and Burkina Faso as well as some parts of Northern Nigeria) where the livestock industry commonly thrives, prefer the prostrate type of cowpea, which is considered as a "non-waste" crop (JAICAF, 2007). Soybean was introduced recently to West Africa for the purpose of improving the nutritional status of the poor and infants. The introduction of soybean was spearheaded by UNICEF and NGOs as foodstuffs for processing soybean milk and soybean flour (*kinako*) etc., and they have become firmly established in some countries and regions. For example, since soybean milk was used in school lunches in some countries, the dietary habits of the young have changed. Soybeans are also considered to be a promising substitute for néré (*Parkia biglobosa*), an ingredient of traditional seasoning. Past trends of production of four major pulse crops (cowpea, groundnut with shell, bambara groundnut and soybean) in West Africa have also been well documented (FAOSTAT, 2021).

Following the review of some literature on challenges faced by pulse production and trade in West and other African states as observed by Rawal and Navarro, eds. (2019), African Development Bank (AfDB) Group (2014) and Nini-Pratt et al (2010) the crucial barriers to pulses production can be summarized as follows:

1. Pest infestation leading to fear of unsafe levels of pesticide residues in some legumes (e.g. cowpea),
2. Production constraints including poor access to key inputs, lack of secure land rights, high labour costs, limited water access under a variable and changing climate,
3. There are issues with poor development of irrigation facilities in many ECOWAS regions. This makes it difficult to optimize the production of pulses in the dry season and during times of drought.,
4. Weak and obsolete production technologies have hampered the productivity of pulses in ECOWAS regions, a common phenomenon in Sub-Saharan African countries,
5. Fragmentation of smallholder producers hinders gaining economies of scale from large-scale commercial farming in the region.,
6. Many farmers in the ECOWAS region are faced with the challenges of limited farm credit access, and technical constraints in processing, This condition retards the possibility of improving value addition to the agricultural products including pulses and legumes.
7. Poor market access by many farmers in the region will hinder the farmers from optimizing the benefits of AfCFTA in the region.
8. Deficient transport/market infrastructure in many countries of the ECOWAS region implies that food loss and waste will increase and very limited finished leguminous crops and /pulses will be able to enter the African market outside the origin of the member states of ECOWAS.

FUTURE RESEARCH DIRECTIONS

The research will be needed in the future to ascertain country-level capacities, policies and community-level initiatives that will drive commercial production and market penetration in all value chains of pulses production in the region and Sub-Sahara Africa as a whole. Assessments of existing programmes of food security in ECOWAS countries which emphasize pulses and leguminous crop production are required in future research which can bring out feasible and best practices for improving production and trade of the commodity under the AfCFTA regime. This may require also mapping of hotspots of legume/pulses production in Sub- Sahara Africa.

CONCLUSION AND RECOMMENDATIONS

Given the numerous nutritional, land restorative capacities and promising role of pulses in boosting African and West African inclusive economic green growth, the potential of pulses remains largely untapped in the regions. For instance, the study finds that monetary values of the net exports of pulses at the African continental levels were positive only in the first two decades from 1960 posting net export earnings of US$14,269,000 in 1961 that fluctuated slightly during the period up to 1977 when it last recorded positive net export earnings of US$28,090,000. Since then (1977) when it first recorded negative net export earnings (US$27,790,000), the negative story of African net exports earnings from pulse crops remained unchanged until as recently as 2020, with the continent still recording dismally negative net export earnings (US$379,004,000). In West Africa, earnings from the pulse trade (net exports) had followed a negative since the beginning of the first decade under review and indeed from 1961 when it posted dismal net export earnings of US$634,000. The largely negative trend in export earnings of pulses continued till 2020 when larger negative values of net export deficits totalling US$19,464,000 were still being recorded from trade in pulses. It is obvious that the African region was consuming more pulses than they were producing as their import bills for this commodity grew geometrically at 8,295.53 percent over a period of six decades.

This underlies the need for investment promotion in pulses production with the aim of closing this demand gap for pulses. West African market trends, especially its import profile leave room to anticipate a potential role for the region as the leader in reducing the import dependency of Africa in pulse trade. This can be seen, for instance, in the fact that, for over the six decades under review, remarkable drops in the share of expenditures on pulses imports of West Africa in the continental total expenditure on pulses imports were recorded. This is an indication that West Africa is gradually improving its portfolio in terms of self-reliance in pulses production and if the regional authorities, including ECOWAS and various national and regional governments, at the country and state levels, as well as investors in the private sector, increase their investment and innovations in this agricultural sub-sector, the potential of this commodity as a driver of growth in agricultural commodities in AfCFTA can be realized given that some major constraints against the production and export performance of these crops are mitigated. The major barriers against the optimization of these crops as export crops and major food output in the region were clearly identified from relevant literature already elaborated in this chapter.

Based on the findings from this research, the following recommendations are made:

1. Regional bodies like ECOWAS, its member countries and national governments as well as organized private sector leaders (small and medium scale Enterprises investors especially) should invest more in pulses production and marketing especially participating in the various value chains of pulses to increase its relevance as a commercial crop for earning foreign exchange and income;
2. Regional bodies like ECOWAS member countries and national governments should make efforts to provide transport infrastructure and enable finance architecture that is inclusive and capable of improving market access to pulses/legume farmers/traders;
3. Profiling of pulses farmers by member countries' governments should be undertaken. Based on such data identified pulses farmers can be properly targeted for the provision of farm production inputs such as fertilizers, credit and improved seed varieties.
4. Irrigation facilities should form a node for agricultural investment by the regional authorities (e.g. national governments, NGOs, private sector leaders/investors, international donor organizations and ECOWAS) to facilitate all-year-round farming for pulses and reduce risks for dry spells inherent in the changing climate;
5. Farmers should be encouraged by national governments and agricultural ministries and NGOs to form cooperatives so as to take advantage of economies of scale accruing from farming on large scale rather than fragmented farm holdings
6. Farmers' processing capacities needs should be well identified and built by governments of the ECOWAS countries where pulses are being produced so as to use as a basis for training them and

equipping them with the appropriate technologies and tools for processing their pulses which should meet up with export standards globally;

7. Agricultural science curricula in secondary and tertiary institutions of learning will need modifications to accommodate more emerging topics such as agronomy, nutritional contents of pulses, best farm management practices, processing of legumes/pulses and other value addition that will help promote the growth and management of pulses in the region.
8. Efforts should be made by governments, private sectors and donor agencies to prioritize the cultivation, processing, packaging and marketing of pulses in national agricultural policies and programmes in West African and SSA regions.

REFERENCES

African Development Bank (AfDB) Group. (2014) Inclusive growth: An imperative for African Agriculture. In Kanu B.S., Salami, A. O. & Namasawa,K. (eds.) Tunisia: African Development Bank (AfDB) Group. (pp 78).

African Union. (2018) Agreement Establishing The African Continental Free Trade Area. African Union: Kigali. 72pp. https://au.int/sites/default/files/treaties/36437-treaty-consolidated_text_on_cfta_-_en.pdf

African Union. (2021) The Comprehensive African Agricultural Development Programme. https://au.int/en/articles/comprehensive-african-agricultural-development-programme

Bjornlund, V., Bjornlund, H. & Van Rooyen, A. F. (2020) Why agricultural production in sub-Saharan Africa remains low compared to the rest of the world – a historical perspective. *International Journal of Water Resources Development, 36*(sup1), S20-S53. doi:10.1080/07900627.2020.1739512

Centre for Disease Control (CDC). (2021). How to prevent chronic diseases. CDC. https://www.cdc.gov

ECOWAS. (2016) *Agriculture and environment.* ECOWAS. https://ecowas.int/ecowas-sectors/agriculture/

Fabinin, A. N. (2022). Pulse, export and staple convenience food: Market analysis in West Africa. *Applied Economics, 54*(7), 764–773. doi:10.1080/00036846.2021.1965534

FAO. (2017) *The future of food and agriculture:Trends and Challenges.* Food and Agriculture Organization of the United Nations: Rome. https://www.fao.org/3/i6583e/i6583e.pdf

FAOSTAT. (2021). *Statistical Yearbook World Food and Agriculture.* Food and Agriculture Organization of the United Nations.

Food and Agriculture Organization. FAO. (1994). *Definition and classification of commodities, 4. Pulses and derived products.* FAO. https://www.fao.org/es/faodef/fdef04e.htm

Food Crisis Prevention Network (RPCA). (2022). Food and Nutrition Crises in The Sahel and Lake Chad Regions: Time to act now and better mobilise in West Africa for the future. https://ec.europa.eu/international-partnerships/system/files/6apr2022_special-event-media-advisory_en-final.pdf

Fusacchia, I., Balié, J., & Salvatici, L. (2022). The AfCFTA impact on agricultural and food trade: A value added perspective *European Review of Agriculture Economics*, *49*(1), 237–284. doi:10.1093/erae/jbab046

ICRISAT. (2021). Legumes for poverty alleviation, nutrition security and sustainable development. *ICRISAT Happenings Newsletter*. International Crop Research Institute for the Semi-Arid Tropics. https://www.icrisat.org/legumes-for-poverty-alleviation-nutrition-security-and-sustainable-development/

Japan Association for International Collaboration of Agriculture and Forestry. (2007) *Production, processing and marketing of pulses in West Africa. Japan Association for International Collaboration of Agriculture and Forestry March 2007 -Considering Inter-Regional Markets and Improving the Living Standards of Local Residents through a Case Study of the Republic of Benin, Alaska.* (164pp). JAICAF. http://jaicaf.or.jp/publications/Production.pdf

Jones, P. (2021) *Price your food product with consumer behavior in mind, part 2*. MSU. https://www.canr.msu.edu/news/price-your-food-product-with-consumer-behavior-in-mind-part-2

Joshi, P. K., & Rao, P. P. (2016). Global and regional pulse economies: Current trends and outlook. *IFPRI Discussion Paper 1544*. Washington, D.C.: International Food Policy Research Institute (IFPRI). http://ebrary.ifpri.org/cdm/ref/collection/p15738coll2/id/130480

Jutzi, S., & Neate, P. J. H. (Eds.). *Proceedings of a workshop, Addis Ababa (Ethiopia).* International Livestock Centre in Africa(ILCA).

Koroma, S., Molina, P. B., Woolfrey, S., Rampa, F., & You, N. (2016). *Promoting regional trade in pulses in the Horn of Africa*. Food and Agriculture Organization of the United Nations.

McDermott, J., Aït-Aïssa, M., Morel, J., & Rapando, N. (2013). Agriculture and household nutrition security—Development practice and research needs. *Food Security*, *5*(5), 667–678. doi:10.100712571-013-0292-6

McKinsey's Agriculture Practice. (2022). *A reflection on global food security challenges amid the war in Ukraine and the early impact of climate change*. McKinsey. https://www.mckinsey.com/industries/agriculture/our-insights/a-reflection-on-global-food-security-challenges-amid-the-war-in-ukraine-and-the-early-impact-of-climate-change

Merga, B.,&Haji, J. (2019). Economic importance of chickpea: production, value and world trade. *Cogent Food and Agriculture*, *5*, 1-12.

Nigeria, O. (2022). Food crisis: Hungry persons in West Africa to reach 38 million by June. Punch NG. https://punchng.com/food-crisis-hungry-persons-in-wafrica-to-reach-38-million-by-june/?utm_source=auto-read-also&utm_medium=web

Nin-Pratt, A., Johnson, M., Magalhaes, E., You, L., Diao, X., & Chamberlin, J. (2010). *Yield gaps and potential agricultural growth in West and Central Africa.* IFPRI Research Monograph.

Obokoh, L. O., & Goldman, G. (2016). Infrastructure deficiency and the performance of small- and medium-sized enterprises in Nigeria's liberalized economy. *ActaCommercii, 16*(1), 1–10.

Päivärinta, E., Itkonen, S. T., Pellinen, T., Lehtovirta, M., Erkkola, M., & Pajari, A. M. (2020). Replacing animal-based proteins with plant-based proteins changes the composition of a whole nordic diet-a randomised clinical trial in healthy Finnish adults. *Nutrients, 12*(4), 943. doi:10.3390/nu12040943 PMID:32231103

Pedersen, A. N., Kondrup, J., & Børsheim, E. (2013). Health effects of protein intake in healthy adults: A systematic literature review. *Food & Nutrition Research, 57*(1), 21245. doi:10.3402/fnr.v57i0.21245 PMID:23908602

Rawal, V., & Navarro, D. K. (2019). The Global Economy of Pulses. Rome, FAO.

Snapp, S., Rahmanian, M., & Batello, C. (2018). *Pulse crops for sustainable farms in sub-Saharan Africa.* Food and Agriculture Organization of the United Nations. doi:10.18356/6795bfaf-en

Teferra, T. F. (2021). Advanced and feasible pulses processing technologies for Ethiopia to achieve better economic and nutritional goals: A review. *Heliyon, 7*(7), 1–9. doi:10.1016/j.heliyon.2021.e07459 PMID:34286131

The Nigerian Economic Summit Group. (2019) Impact assessment study and economy-wide implications of the African Continental Free Trade Area (AfCFTA) on the Nigerian Economy. Abuja.

Tothill, J. C. (1986). *Role of legumes in farming systems of sub-Saharan Africa.* Potentials of forage legumes in farming systems of sub-Saharan Africa, Haque.

UNFCC. (n.d.) Recommendations for reducing greenhouse gas emissions. *Climate Neutral Now* (8pp). https://unfccc.int/sites/default/files/resource/Recommendations%20for%20Reducing%20Greenhouse%20Gas%20Emissions.pdfRawal, .

V. & Navarro, D. K. (2019). *The Global Economy of Pulses.* FAO.

Vidigal, P., Romeiras, M. M., & Monteiro, F. (2019). Crop diversification and the role of orphan legumes to improve the sub-Saharan African farming systems. In M. Hasanuzzaman, M. C. M. T. Filho, M. Fujita, & T. A. R. Nogueira (Eds.), *Sustainable Crop Production* (pp. 1–20). IntechOpen.

World Bank. (2018) *Africa's pulse in five charts: boosting productivity in Sub-Saharan Africa.* World Bank. https://www.worldbank.org/en/region/afr/publication/africas-pulse-in-five-charts-boosting-productivity-in-sub-saharan-af
rica

Chapter 3
Global Certification Systems of Key Importance in the South African Fruit Industry's Access to International Markets

Ndiadivha P. Tempia
https://orcid.org/0000-0002-9873-9895
National Agricultural Marketing Council, South Africa

Moses H. Lubinga
https://orcid.org/0000-0002-6608-8776
National Agricultural Marketing Council, South Africa

ABSTRACT

The chapter focused on assessing the nature of voluntary sustainability standards (VSSs) that affect South Africa's producers of horticultural produce (fruits in particular) destined for markets within Europe and Africa. Eight VSSs affecting South Africa's fruit trade with Europe and within Africa under food safety and traceability themes were identified. The outcomes of global certification as a requirement to participate in global international markets are outlined, with a specific focus on the implication of these standards for smallholder farmers. Given the independent nature of the end goals from the institutions responsible for these VSSs, it becomes problematic and impractical to argue that VSSs that focus on similar thematic areas, e.g., GLOBALG.A.P. Crops and IFOAM, be merged into one standard. Thus, it is prudent that smallholder producers outrightly identify a specific target market and the corresponding standard requirements such that activities are undertaken in accordance with what the market needs.

INTRODUCTION AND BACKGROUND

The last two decades witnessed unprecedented changes in the agro-food sector through the proliferation of private and voluntary sustainability standards (VSSs) which are influencing international

DOI: 10.4018/978-1-6684-4780-2.ch003

agricultural trade. Whereas food safety and traceability-related VSSs are enforced in agro-food markets globally, the implementation of some environmental, sustainability, and labor-related standards is not yet very strict in developing economies, especially in developing countries. Some voluntary standards combine a mixture of food safety, environmental, and social dimensions, while an inherent emphasis is being given to product traceability. Consequently, standards affect not only the safety of final products but also the whole organization of the supply value chain. This significant change raises new opportunities and challenges for small export-oriented farmers in developing countries and has implications for agricultural development programs and policies for local farmers who ought to be benefiting.

It is noteworthy to point out that these certification schemes are mainly driven by private and non-governmental market organizations, which often fill the gaps that governments and international institutions are not willing or unable to regulate (Brandi et al., 2015; DeFries et al., 2017). While the rationale behind certification is clear. Evidence of its benefits is mixed. For example, through certification, producers receive premiums and other benefits for following sustainability criteria set by a certifying entity. A number of agricultural export-oriented tropical crops are of particular interest for certification programs. DeFries et al. (2017) provided reasons behind the establishment of certification programs. First, due to a lack of interest by the government to prioritize sustainable production or an inability to enforce environmental and labor regulations, non-governmental organizations have taken advantage of the gap to develop markets for certified products. Second, while there has been a noticeable reduction in poverty levels, the majority of people live in rural areas. Therefore, improving the livelihoods of farmers and laborers in rural areas while ensuring adequate working conditions is critical in ensuring sustainable supply chains. Third, due to the tradable nature of most tropical crops, certification of these commodities has contributed to the revenues in international supply chains. However, the challenge that remains is to ascertain whether the benefits of VSSs are realized among rural smallholder farmers.

The success of certification programs often rests on the criteria chosen as they relate to environmental, economic, and social sustainability. While empirical evidence indicates that certification may provide benefits in line with sustainable goals, results indicate that certification is not an absolute solution to improve social outcomes and the overall incomes of smallholder farmers. There is growing literature that flags the concerns with VSSs, particularly on their applications to smallholder farmers. Farmers are generally faced with high transaction costs and operate in an environment characterized by information asymmetry that may contribute further to their exclusion from supply value chains which are concentrated in the hands of large and wealthier farmers (Brandi et al., 2015), and non-government organizations (NGOs) and private sector entities from the North that serve their own interests and promote their priorities (Schleifer, 2016).

The implementation of these standards has, in many instances, followed policy-driven or interventionist approaches with the main actors, in this case, North-based multinational corporations and international non-governmental organizations (NGOs) pursuing the "benefits of certifications" without a full understanding of the realities facing the smallholders on the ground. According to Glasbergen (2018):

Most studies on the certification sustainability of these certifications take the objectives of voluntary standard-setting and certifying arrangements as the reference point, thereby implicitly accepting a problem definition that is not necessarily aligned with the needs, interests and preferences of the smallholder producers in a developing context. (p. 243)

This approach has an implication through the working of the market as higher prices paid by the consumers usually trickle down to producers and thus assuring more environmentally friendly and socially responsible production processes (Mitiku et al., 2017; Ansah et al., 2020; Tayleur et al., 2018).

The economic aspect of global certification is also an indication that certification standards have become more comprehensive, in the sense that they expand an original narrow focus (i.e., organic, social, or ecological) towards a broader set of standards that connect social, environmental and economic issues (Vermeulen & Kok, 2012; Glasbergen & Schouten, 2015, in Dompreh et al., 2021). Consequently, standards affect not only the safety of final products but also the whole organization of the supply value chain. This significant change raises new opportunities and challenges for small export-oriented farmers in developing countries and has implications for agricultural development programs and policies.

The pattern of new standard adoption in developing countries has recently received much attention from economists. Consequently, there is conflicting evidence on the intended outcomes of global trade certification. Although not directly articulated by the proponents of VSSs, the adoption of certification standards, can, directly and indirectly, affect farmers' food security in a positive or negative manner (Dompreh et al., 2021). Those pessimistic about standards argue that they may act as a barrier to market access for smallholder farmers due to increased production costs (audits and changes in production practices), which may disrupt access to food security and displacement of food crops to specialize in cash crop production (Fenger et al., 2017; Oosterveer et al., 2014). While this chapter does not necessarily address the relationship between standards and food security, it can be argued that these schemes may be viewed as external imputes to change that interferes with the locally embedded practices and methods of production, and their food crops may be displaced at the expense of livelihoods and food security.

By contrast, some less pessimistic studies find positive effects, arguing that standards can be catalysts for the upgrade by improving farming techniques through the adoption of good agricultural practices and access to extension services. Furthermore, the incomes of the farmers may be enhanced through yield improvement and product quality, thereby allowing smallholder farmers to participate in high-value-added chains (Ansah et al., 2020, Chiputwa, et al., 2016; Gockowski et al., 2011; Fenger et al., 2017)

Given differences in findings, it is becoming generally recognized that evidence is mixed. There are debates over the degree to which compliance processes do indeed act to exclude smallholders and with which conditions smallholder farmers can really comply, pointing out both threshold capital requirements on the one hand, and industry structure and institutional environment on the other hand, that may greatly affect standard adoption by smallholders. It can therefore be argued that while VSSs are recognized and used globally to catalyze more responsible or sustainable practices and have enjoyed impressive growth, uptake by small producers and small and medium enterprises (SMEs) remains both a bottleneck and an opportunity. Unlocking new approaches to meeting producers' needs will allow standards to achieve greater scale and become more relevant to all producers, especially in emerging economies.

The latter is particularly important within the context of the South African dual agricultural sector, composed of commercial and smallholder farmers, in which participation in the mainstream economy by smallholder farmers remains highly restricted. The horticultural industry is the second-largest agricultural sub-sector in South Africa after animal production (R141 billion) and contributed over R92 billion (or 29% of the total agricultural gross value of R312 billion) in gross value during the 2019/20 production year (DALRRD, 2021. Within the horticultural industry, the biggest contributors in terms of gross value during the 2019/20 production year were vegetables (including potatoes) at R26.8 billion or 29%, deciduous fruits at R24.7 billion or 27%, and citrus fruits at R20 billion or 22%.

With the enactment of the largest free trade area on the African continent (the African Continental Free Trade Area—or AfCFTA—agreement), we draw our focus on VSSs affecting South Africa's intra-Africa horticultural trade, and trade with Europe. Horticulture is the most foreign-exchange-generating subsector within the agricultural sector. By the end of 2020, South Africa's horticultural exports registered a 26% increase in value from R48.81 million in 2019. During the same period, intra-Africa trade grew by 5% while the value of exports destined for the European Union rose by 36% (International Trade Centre—ITC, 2022a). Of all South Africa's horticultural exports, the European Union (EU) accounted for a 45.9% share in value in 2021, while intra-Africa trade was estimated at 9.5%. The major export destinations in the EU were the Netherlands, with a 22.6% share in value, followed by Germany, Portugal, and Italy, among others, with share values of 2.5%, 1.7%, and 1.4%, respectively. Within Africa, Botswana's horticultural imports from South Africa contributed a 1.3% share in value, followed by Mozambique (1.1%, and Nigeria (1.0%), among other countries (ITC, 2022b). At the product level, citrus, grapes, berries, and pomes were the most fruit categories exported to the EU in 2021, each generating R8.64 billion, R6.36 billion, R1.50 billion, and R1.48 billion, respectively. Within Africa, horticultural exports generated R6.44 billion for the South African economy, but pomes (i.e., apples, pears, and quinces) exports were valued at R2.64 billion, followed by potatoes (R0.74 billion) and citrus (R0.61 billion).

This chapter investigates the outcomes of global certification as a requirement to participate in global international markets by various market actors with a specific focus on the implication of these standards for smallholder farmers. As a contribution towards market information asymmetry faced by various value chain actors involved in the agricultural sector, most especially in international trade, in this chapter, we characterize the kind of standards to which agricultural commodities are subjected to. The rest of the chapter is presented as follows: the methodology, the findings, a profile of selected standards that affect South Africa's fruit and nuts exports, and the conclusion and recommendations.

METHODOLOGY

The authors focus on VSSs that affect the agricultural sector, with particular emphasis put on South Africa's fruits and nuts Harmonized System (HS) Code 08 given that this subsector generates the highest amount of foreign earnings. Furthermore, tropical fruits have become the focus of certification because of the perceived benefits they may have to farmers. According to the United Nations Forum on Sustainability Standards (UNFSS; 2022), VSSs are specific requirements that are bound to be met by producers, traders, manufacturers, retailers, or service providers in relation to a variety of sustainability measures. Marx and Wouters (2015) contended that VSSs are vital instruments in facilitating the attainment of green economy targets. A study by Chiputwa et al. (2015) analyzed and compared the impacts of three sustainability standards—Fairtrade, Organic, and UTZ—on the livelihoods of smallholder coffee farmers in Uganda. Various VSSs standards relevant to commodities may be developed. However, this chapter focuses on VSSs that involve food safety and traceability themes. The food safety theme entails the following aspects, namely, meeting the required procedures to produce food, risk management, traceability, incident management, non-conformities, meeting Hazard Analysis Critical Control Point (HACCP) requirements, having appropriate monitoring and verification measures as well as recall procedures in case a food item is regarded to be unsafe for human consumption. Complying with these food safety requirements may, in many instances, be a major constraint to smallholder farmers.

On the other hand, the theme for traceability requires that a value chain actor meets the chain of custody and tracking requirements (ITC, 2022c). The chain of custody refers to the documentation of administrative and technical requirements used to track the sourcing of agricultural produce at every step of the supply chain, while ensuring that the product meets the claims and labeling guidelines as per the enforcing authority. A chain of custody proves the integrity of a piece of evidence (Badiye et al., 2022). VSSs were then disaggregated by destination, i.e., Africa and Europe. Europe was purposively chosen given that most of South Africa's fruits are destined for that market, while for Africa, there is a need for South Africa to strategize on how to benefit from the recently enacted AfCFTA. Voegtlin and Scherer (2017) and Inigo et al. (2020) suggest that VSSs are key governance mechanisms through which product developments can easily become commercially accepted. Secondary data extracted from the ITC Standards Map were used. It provides a trusted and neutral source of information on VSSs, audit protocols, codes of conduct, reporting frameworks, as well as company programs on sustainability.

FINDINGS

According to the ITC Standards Map, South Africa's agricultural exports to other African countries are generally subjected to 46 VSSs, while produce destined for the European Union (EU) must comply with 43 VSSs. By focusing on fruits alone, the EU requires that South Africa's products comply with 40 VSSs, yet intra-Africa trade has 43 VSSs that must be complied with. This suggests that South Africa's agricultural produce, specifically fruit, is required to meet more VSSs if destined for Africa, unlike if products are destined for the EU. When the EU was considered as the supplier of fruits to Africa, 48 VSSs were identified, and South Africa requires that 45 standards are complied with. For fruits destined for South Africa from other African countries, 51 VSSs must be met. The high number of VSSs required to be met by African economies to trade with each other is a big concern, most especially because countries are operating under the AfCFTA that aims at smoothening intra-Africa trade. Of the identified VSSs affecting South Africa's trade in fruits with Europe and within Africa, eight of the VSSs do not address food safety and traceability themes and are presented in Table 1. Hence the authors do not report on them further.

Figure 1 shows VSSs that are in force and are based on both the food safety and traceability themes. The figure reveals a number of food safety and traceability criteria requirements that must be complied with for fruits to gain access to markets in Europe and Africa. It is important to note

Table 1. VSS affecting South Africa's trade in fruits with Europe and within Africa

Number	Standard
1	ARSO–Agriculture—Sustainability and eco-labelling
2	Carbon Trust Product Footprint Certification
3	International Labour Organization Labour Standards
4	MPS-ABC
5	Sustainability Initiative of South Africa (SIZA)
6	UN Global Compact
7	UN Guiding Principles on Business and Human Rights
8	Verified Carbon Standard (VCS)

that these criteria are applicable to all countries that ascribe to the standards. The GLOBALG.A.P. Crops standard by far has the highest number of food safety criteria requirements, and the ISCC PLUS collection of standards has the highest number of traceability criteria requirements.

Figure 1. In force, VSS imposing both the food safety and traceability themes

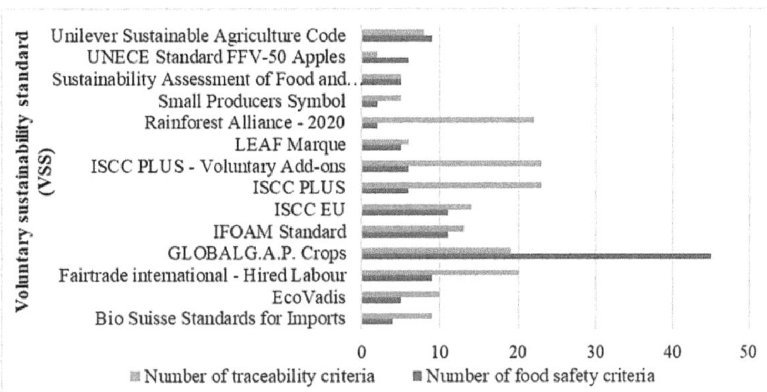

With the exception of the "UNECE Standard FFV-50 Apples" and "Bio Suisse" standards, which are subjected to fruits destined for the EU, all the other VSSs are applicable on the African continent. Furthermore, it is worthwhile to note that "UNECE standard FFV-50 Apples" is only applicable to apples. Conversely, two standards: "Comercio Justo Internacional: Organizaciones de Pequenos Productore" and "Fairtrade International: Small Producers Organizations" are only applicable within the African continent but are not required for fruits destined for Europe. Both standards focus on the traceability theme only. As of December 31, 2021, the Rainforest Alliance (RA 2017) standard expired, hence it is not currently in force. The RA 2017 ranked fourth among VSSs with the highest number (18) of traceability criteria that must be met by exporters of fruits but with relatively fewer food safety criteria requirements (4) as compared to other standards that address the food safety theme.

PROFILING OF A SELECTED VSS AFFECTING SOUTH AFRICA'S FRUIT EXPORTS TO EUROPE AND WITHIN AFRICA

This subsection provides insights into 2–3 selected VSSs with a high proportion of requirements for complying with food safety and traceability criteria. According to Figure 1, GLOBALG.A.P. Crops and ISCC PLUS standards, are discussed. Furthermore, the "UNECE standard FFV-50 Apples" is included, given the importance of South Africa's apple exports in the pome fruit industry. Whereas a detailed discussion on "Comercio Justo Internacional: Organizaciones de Pequenos Productores" and "Fairtrade International: Small Producers Organizations" standards that are only applicable to intra-Africa trade would be of interest, these are limited to the traceability theme. Thus, no further deliberation is made.

GLOBALG.A.P. Crops Standard

GLOBALG.A.P. is an Integrated Farm Assurance (IFA) standard established in the year 1997 and implemented in more than 135 countries. However, over the years, the standard has been revised several times to meet market demands, among other factors. According to Aluzar Systems Consultants (ASC; 2022), the need to comply with the Global Food Safety Initiatives (GFSIs) has been a key driving factor among market-led demands in continuously revising the standard. GLOBALG.A.P.'s flagship standard, IFA V6, is holistic in the sense that it takes into consideration food safety, production processes, animal health and welfare, traceability (legal and management), and environmental sustainability, as well as workers' wellbeing, all constituting responsible farming practices (GLOBALG.A.P., 2022a).

Figure 2. Integrated farm assurance V6 timeline

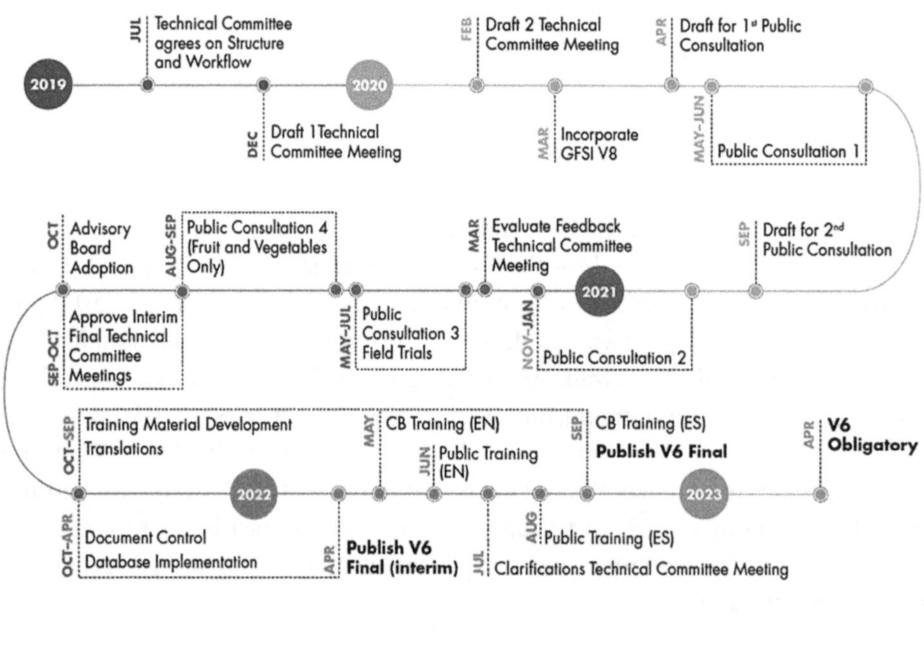

According to Figure 2, the final version of the standard with IFA V6 was published in August 2022. This means that the standard will grant producers an opportunity to make a transition and conversion up until April 2023, when V6 will be obligatory. As soon as IFA V6 becomes obligatory in April 2023, all earlier versions will cease their operation and validity.

Noteworthy, each time the standard is updated, there is also a need for producers to improve upon their production systems in order to comply with new requirements. Whereas this is done in a good spirit, the Department of Agriculture, Forestry, and Fisheries (DAFF, 2016) reckons that the fast-changing voluntary standards requiring producers to meet stringent food safety and quality standards are a major bottleneck limiting smallholder producer access to the commercial markets. This is in part due to the exorbitantly high cost of meeting the requirements not affordable to many smallholder producers. Fiankor et al. (2019) assessed the role of VSSs in bridging institutional gaps between the European Union (EU) and trading partners (South Africa inclusive) on apples, bananas, and grapes destined for the EU between 2010 and 2015. Authors report that the interaction between governance distance and GLOBALG.A.P. certification impedes trade in the selected fruits, especially in countries with non-certified farms as compared with countries with GLOBALG.A.P. certified farms. Policy-wise, the authors recommended producers and firms located in countries with public institutions with low-quality governance but targeting high-value export markets like the EU to acquire certification to a standard(s) acceptable in the target country.

The following are some of the major amendments made in the latest version (version 6) when compared to version 5. First, producers that meet audit criteria will be issued with certificates at the scope level rather than sub-scopes, as was the case. GLOBALG.A.P. exemplifies that "if a producer gets certified for both fruit and vegetables (FV), and flowers and ornamentals (FO), they will receive one certificate for plants—with FV and FO listed as product categories." Second, unlike in version 5, *sub-scopes* are now referred to as *product categories*. Third, 10% of the audits by certification bodies must be unannounced, and this 10% will no longer be considered as additional audits but will constitute the original number of certification audits. Fourth, IFA V6 draws its focus on outcomes achieved instead of the specific measures, thereby providing for more flexibility in the sense that it allows producers to demonstrate compliance that best suits their production practices. Fifth, V6 offers a streamlined risk-based checklist in which principles and criteria that are not applicable based on the producer's risk assessment of the production practices will be labeled as *Not applicable* in the checklist. According to GLOBALG.A.P. (2022a), this implies that, to some extent, checklists

Figure 3. Continuous improvement model at farm level under the new IFA v6

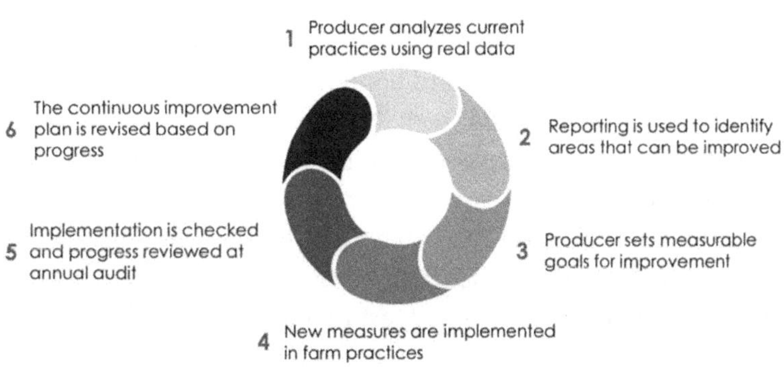

are customizable to suit a producer's specific situation. In general, V6 has reduced the number of principles and criteria applicable to fruits and vegetables by 32, from 222 in V5 to 190. Furthermore, Figure 3 illustrates that IFA V6 presents an opportunity for individual producers to improve upon their production practices continuously. Hence, they are required to have an implementable plan for analyzing current practices, through which hot spots can be identified. Such plans must then show clear measurable goals through which improvement will be attained.

All in all, the new GLOBALG.A.P. IFA V6 is envisioned to offer the following benefits to stakeholders in the horticulture industry: (a) a digital data-driven certification process, (b) certificates issued at the scope level, (c) streamlined, risk-based checklists, (d) continuous improvement plan at individual farm-level, (e) higher level of integrity behind GLOBALG.A.P. certificates, (vi) outcome-oriented approach, and (f) flexibility in the choice of two editions (i.e., Smart and GFS versions).

International Sustainability and Carbon Certification (ISCC) PLUS Standard

The above is a sustainability certification system covering agricultural products, amongst other feedstocks derived from forestry biomass, biogenic wastes and residues, circular materials, and renewables. Globally, the ISCC Plus standard is applicable for all markets and products not regulated under the European Renewable Energy Directive (EU RED) or the Fuel Quality Directive (FQD) covered under the ISCC EU Certification. The standard aims at ensuring that biomass and other raw materials are generated in an environmentally, socially, and economically sustainable manner (International Sustainability and Carbon Certification, 2022; SCS Global Services, n.d.), while putting much emphasis on traceability.

Within the agricultural sector, the ISCC Plus has become an important standard for the fact that there is a growing recovery of wastes and residues as feedstocks for circular plastics and chemical intermediates (SCS Global Services, 2022). According to Bureau Veritas (2022), being certified according to the ISCC Plus scheme is beneficial in that it demonstrates to customers and partners the level of commitment to advanced environmental practices and instills confidence among customers and partners that a firm's activities take cognizance and are part of a circular economy approach in ensuring that business is substantiable. Most importantly, the certificate enables the individual or firm to show that its products conform to international sustainability criteria, thereby fostering market access for its products (Pattison Resource Consulting, 2018).

United Nations Economic Commission for Europe (UNECE) Standard Fresh Fruit and Vegetables (FFV)-50 Apples

Unlike GLOBALG.A.P. and ISCC Plus discussed in the foregoing sub-sections, UNECE is a marketing and commercial quality control standard applicable to apples of varieties (cultivars) grown from *Malus domestica Borkh* to be supplied fresh for consumption purposes. However, the standard does not include apples for industrial processing. It aims at facilitating international trade for apples, fostering high-quality production, improving upon profitability received along the value chain, and protecting consumer interests (UNECE, 2020). The standard is a product of the Working Party on Agricultural Quality Standards of the UNECE. It further stipulates various provisions relating to quality. For instance, maturity requirements suggest that apples should be in a fresh state to enable them to continue their ripening process and to reach a degree of ripeness that meets consumers' satisfaction. Several parameters are used, including firmness, morphological aspects, taste, and the refractometric index, to verify the minimum maturity requirements. The standard also provides clarity on the classification of apples into either "extra" class, Class I or Class II. Classification is mainly based on the textural outlook of apples, i.e., an intact stalk without quality defects such as shape, coloring, and bruises. With regard to sizing, classification is based on either the diameter or weight of the fruit. Other considerations covered by the standard include quality tolerance, how the apples are presented (uniformity and packaging), as well as the markings on the packaging materials.

CONCLUSIONS AND RECOMMENDATIONS

This chapter focused on assessing the nature of VSSs that affect South Africa's producers of horticultural produce (fruits in particular) destined for markets within Europe and Africa. Whereas VSSs has been noted to foster trade in terms of enhancing the quality of fresh produce and enabling producers that comply with the restrictive compliance requirements to access premium markets, many cumbersome compliance issues may be perceived by smallholder producers as a hindrance. Most of the enforced VSSs addressing different themes may render compliance a challenge, most especially for smallholder farmers that are resource constrained. Therefore, given the independent nature of the end goals from the institutions responsible for these VSSs, it becomes problematic and impractical to argue that VSSs focus on similar thematic areas, e.g., GLOBALG.A.P. Crops and IFOAM should be merged into one standard. Thus, it is prudent that smallholder producers outrightly identify a specific target market and the corresponding standard requirements such that all activities are undertaken in accordance with what the market requires.

For instance, for producers targeting a specialty market for organic products in Europe, their priority would be the IFOAM standard. However, to cater for any market-related shocks that might lead to the exclusion of products from a specialty market, it is important for producers to adopt standards that address a broad range of market requirements, e.g., food safety and traceability. In this context, the GLOBALG.A.P. Crops standard and the ISCC PLUS standards become priorities for smallholder producers to focus on, such that when a farm is certified, products can be sold in a wide range of markets. Prioritization of the above standards might not be a sufficient condition to address smallholder farmers' resource constraints. Therefore, there is a need for government and non-government organizations to further support smallholder farmers in adopting and complying with these standards. Support might be through facilitating capacity-building sessions (training), as well as acquiring certification to meet the standards. It is beyond doubt that the high cost of certification is a huge stumbling block to smallholder farmers in South Africa. With the possibility of the AfCTA agreement, it is important that African countries work towards reducing these VSSs to a bare minimum, the basis upon which trade in fruits is bound to flourish. This will also foster the faster distribution of food supplies (fruits in this case) from high-production areas to those where availability is limited, thereby minimizing food insecurity in certain parts of Africa.

REFERENCES

Aluzar Systems Consultants. (2022). *Global GAP IFA V6: What's new?* https://ascconsultants.co.za/global-gap-ifa-v6-whats-new

Ansah, E. O., Kaplowitz, M. D., Lupi, F., & Kerr, J. (2020). Smallholder participation and procedural compliance with sustainable cocoa certification programs. *Agroecology and Sustainable Food Systems*, *44*(1), 54–87. doi:10.1080/21683565.2019.1579776

BadiyeA.KapoorN.MenezesR. (2022). *Chain of custody*. Available at: https://www.ncbi.nlm.nih.gov/books/NBK551677

Brandi, C., Cabani, T., Hosang, C., Schirmbeck, S., Westermann, L., & Wiese, H. (2015). Sustainability Standards for Palm Oil: Challenges for Smallholder Certification under the RSPO. *Journal of Environment & Development*, *24*(3), 292–314. doi:10.1177/1070496515593775

Bureau Veritas. (2022). *ISCC Plus: Biomaterials certification*. https://certification.bureauveritas.com/needs/iscc-plus-biomaterials-certification

Chiputwa, B., & Qaim, M. (2016). Sustainability standards, gender, and nutrition among smallholder farmers in Uganda. *The Journal of Development Studies*, *52*(9), 1241–1257. doi:10.1080/00220388.2016.1156090

Chiputwa, B., Spielman, D. J., & Qaim, R. (2015). Food standards, certification and poverty among coffee farmers in Uganda. *World Development*, *66*(C), 400–412. doi:10.1016/j.worlddev.2014.09.006

DeFries, R. S., Fanzo, J., Mondal, P., Remans, R., & Wood, S. A. (2017). Is voluntary certification of tropical agricultural commodities achieving sustainability goals for small-scale producers? A review of evidence. *Environmental Research Letters*, *12*(3), 033001. doi:10.1088/1748-9326/aa625e

Department of Agriculture, Forestry, and Fisheries. (2016). *SA-GAP Certification Programme for smallholder producers of fresh produce*. Author.

Department of Agriculture Land Reform and Rural development (DALLRD). (2021). *Economic review of the South African agriculture*. http://www.dalrrd.gov.za/phocadownloadpap/Statistics_and_Economic_Analysis/Statistical_Information/Economic%20Review%202021-22.pdf

Dompreh, E. B., Asare, R., & Gasparatos, A. (2021). Sustainable but hungry? Food security outcomes of certification for cocoa and oil palm smallholders in Ghana. *Environmental Research Letters*, *16*(5), 1–18. doi:10.1088/1748-9326/abdf88

Fenger, N. A., Bosselmann, A. S., Asare, R., & de Neergaard, A. (2017). The impact of certification on the natural and financial capitals of Ghanaian cocoa farmers. *Agroecology and Sustainable Food Systems*, *41*(2), 143–166. doi:10.1080/21683565.2016.1258606

Fiankor, D. D. D., Martínez-Zarzoso, I., & Brümmer, B. (2019). Exports and governance: The role of private voluntary agrifood standards. *Agricultural Economics*, *50*(3), 341–352. doi:10.1111/agec.12488

Glasbergen, P. (2018). Smallholders do not eat certiðcates. *Ecological Economics, 147*(C), 243–252. doi:10.1016/j.ecolecon.2018.01.023

Glasbergen, P., & Schouten, G. (2015). Transformative capacities of global private sustainability standards; a reñection on scenarios in the ðeld of agricultural commodities. *Journal of Corporate Citizenship, 58*(58), 11–27.

GLOBALG. A.P. (2022a). *IFA v6. What's new?* https://www.globalgap.org/uk_en/for-producers/globalg.a.p./integrated-farm-assurance-ifa/IFA-V6/

GLOBALG. A.P. (2022b). *Integrated Farm Assurance (IFA) v6 at a glance*. https://www.globalgap.org/uk_en/for-producers/globalg.a.p./integrated-farm-assurance-ifa/IFA-V6/

Gockowski, J., Afari-Sefa, V., Sarpong, D. B., Osei-Asare, Y. B., & Dziwornu, A. K. (2011). Increasing income of Ghanaian cocoa farmers: Is introduction of fine flavour cocoa a viable alternative. *Zeitschrift für Ausländische Landwirtschaft, 75*(2), 175–200.

Inigo, E. A., Garst, J., Blok, V., & Pentaraki, K. M. (2020). Do voluntary standards support responsible innovation implementation and reporting in industry? The case of the European food sector. In E. Yaghmaei & I. van de Poel (Eds.), *Assessment of Responsible Innovation: Methods and Practices* (pp. 14–168). Routledge. doi:10.4324/9780429298998-9

International Sustainability and Carbon Certification. (2022). *ISCC for food and feed*. https://www.iscc-system.org/process/market-applications/iscc-for-food-and-feed/

International Trade Centre. (2022a). *Standard map app*. https://www.standardsmap.org/en/identify?name=&products=Fruits&origin=Europe&destination=South%20Africa&themes=

International Trade Centre. (2022b). *Trade Map: List of importing markets from European Union (EU 28) for a product exported by South Africa* [Data set]. Trademap.org. https://www.trademap.org/Country_SelProductCountry_TS.aspx?nvpm=1%7c710%7c%7c%7c14719%7c08%7c%7c%7c2%7c1%7c1%7c2%7c2%7c1%7c2%7c5%7c1%7c1

Marx, A., & Wouters, J. (2015). Is everybody on board? Voluntary sustainability standards and green restructuring. *Development, 58*(2), 511–520. doi:10.105741301-016-0051-z

Mitiku, F., de Mey, Y., Nyssen, J., & Maertens, M. (2017). Do private sustainability standards contribute to income growth and poverty alleviation? A comparison of different coffee certification schemes in Ethiopia. *Sustainability, 9*(2), 146. doi:10.3390u9020246

Oosterveer, P., Adjei, B. E., Vellema, S., & Slingerland, M. (2014). Global sustainability standards and food security: Exploring unintended effects of voluntary certification in palm oil. *Global Food Security, 3*(3-4), 220–226. doi:10.1016/j.gfs.2014.09.006

Pattison Resource Consulting. (2018). *Agri-sustainability standards, market access, and prairie wetland conservation*. A Research Report for Ducks Unlimited Canada and the Prairie Habitat Joint Venture. https://www.phjv.ca/wp-content/uploads/2019/02/18-03-31-PHJV-Market-Access-Report_FINAL-.pdf

Schleifer, P. (2016). Private governance undermined: India the roundtable on sustainable palm oil. *Global Environmental Politics*, *16*(1), 38–58. doi:10.1162/GLEP_a_00335

SCS Global Services. (2022). *ISCC PLUS certification.* https://www.scsglobalservices.com/services/iscc-plus-certification

SCS Global Services. (n.d.). *ISCC Plus certification for the circular economy and bioeconomy.* https://www.scsglobalservices.com/services/iscc-plus-certification

Tayleur, C., Balmford, A., Buchanan, G. M., Butchart, S. H. M., Corlet Walker, C., Ducharme, H., Green, R. E., Milder, J. C., Sanderson, F. J., Thomas, D. H. L., Tracewski, L., Vickery, J., & Phalan, B. (2018). Where are commodity crops certified, and what does it mean for conservation and poverty alleviation? *Biological Conservation*, *217*, 36–46. doi:10.1016/j.biocon.2017.09.024

United Nations Forum on Sustainability Standards. (2022). *Flagship publication.* https://unfss.org/home/flagship-publication/

Vermeulen, W. J. V., & Kok, M. T. J. (2012). Government interventions in sustainable supply chain governance: Experience in Dutch front-running cases. *Ecological Economics*, *83*, 183–196. doi:10.1016/j.ecolecon.2012.04.006

Voegtlin, C., & Scherer, A. G. (2017). Responsible innovation and the innovation of responsibility: Governing sustainable development in a globalized world. *Journal of Business Ethics*, *143*(2), 227–243. doi:10.100710551-015-2769-z

Chapter 4
Market Attitude Towards Genetically-Modified Food Products:
A Developing Economy Perspective

Yajeshwar Seetul
Heriot-Watt University, Malaysia

Praveen Balakrishnan Nair
https://orcid.org/0000-0002-3617-7049
Heriot-Watt University, Malaysia

ABSTRACT

Public/social behaviour is one of the main elements that determines the success or failure of any commercial product launched in the market, and food product marketing is no exception. There have been several studies conducted to assess attitude, awareness, or acceptance of genetically modified (GM) food products, giving a good insight about the perception towards this product. However, limited attempts have been made to determine the antecedents that influence the attitude towards GM food products in southeast Asian context. This chapter attempts to investigate the influence of perceived benefits, perceived risks, knowledge, general attitude, and trust on attitude towards GM food products in Malaysia, a fast-developing economy. To effectively develop the GM food industry, it is important for the government, policy makers, non-governmental organisations and the industry in general to seriously consider the market's attitude towards GM food products, while formulating commercialisation strategies.

INTRODUCTION

Understanding consumer behavior is central to the success of any consumer product. This article tries to bring together consumer behavior and one of the most debated products in agriculture marketing, genetically modified (GM) food products. Despite numerous benefits of GM products shown by recent

DOI: 10.4018/978-1-6684-4780-2.ch004

studies and acceptance of it as a solution to feed the ever-increasing population worldwide, the public is still skeptical about this technology (Nicolia, Manzo, Veronesi, & Rosellini, 2014). An attempt is made here to understand the attitude of consumers towards GM food products, especially from a developing economy perspective. Genetic modification is a practice whereby molecular genetics tools and methods are used to modify the genetic makeup of an organism to eventually possess features that suit the needs and wants of end users (Anderson, Wachenheim & Lesch, 2005; Macer, Bazer & Gough, 1991). GM food comes from genetically modified organisms and thus, are defined as a change in the genetic makeup of food products by insertion of new genes from other sources and/or deletion of existing gene which lead to characteristics they did not possess naturally (Food and Drug Administration [FDA], 2017; World Health Organization [WHO], 2017).

This expertise created a string of new products in various fields, mainly medicine and agriculture (Salazar and Montenegro, 2009). Continuous large-scale cultivation and commercialization of GM crops in both developing and developed countries in the last three decades have shown numerous benefits like cost effective nourishment for consumers, and sustainable income source for farmers (ISAAA, 2022; Nap, Metz, Escaler, & Conner, 2003; Zika et al. 2007). GM crops are currently cultivated on more than 2.1 billion hectares of land and approved in 45 countries around the world, including developed countries like Australia and developing country like Bangladesh (ISAAA, 2022). Other than economic gains, benefits of GM crops include reduced dependence on insecticides and pesticides and increase in yield. A reduction in pest management and lower quantities of insecticides make agricultural products safer, cheaper and reduces environmental degradation, which benefits farmers, producers, and consumers (Arya, 2015).

Malaysia, which is in the early stages of developing its own GM food industry, was chosen for conducting this study. The country is currently in the research and field-testing stages with no commercialization. In 1960, 44% of Malaysia's Gross Domestic Product (GDP) was based on the agricultural sector but in 2007 this number shrunk to 8% (Frost and Sullivan, 2009). To boost the agricultural sector, National Agriculture Policy plans were developed in 1984 and these policy plans became the precedents to all forthcoming policy developments. The introduction of the Malaysian Biotechnology Policy has shown that Malaysia planned to use biotechnology to drive forward their agricultural sector (Firdaus-Raih et al., 2005). Currently, millions of tons of GM food products, such as corn and soybean, are imported to Malaysia and the Malaysian government recognises the handicap of being heavily dependent on imports. Hence the interest in biotechnological practices to improve food security and develop sustainability. Biotechnology development in Malaysia can be divided into four main phases (Mokhtar & Mahalingam, 2010): The first phase was before 1995 when Malaysia started establishing basic infrastructure to explore biotechnological practices.

During the second phase (1995 – 2000), the national agenda on biotechnology was implemented and further enhanced with the establishment of the National Biotechnology Directorate (NBD). The main objective of the NBD was to promote biotechnology research and development towards commercialization. The third phase (2001 - 2005) saw the development of BioValley Malaysia; a biotechnology cluster of national institutes aimed to accelerate research and commercialization of biotechnological products (Cyranosk, 2005; Mokhtar & Mahalingam, 2010). US$ 550 million was allocated for the development of biotechnology by the government in the 9th Malaysian Plan (2006-2010) (Mokhtar and Mahalingam, 2010). The 9th Malaysian Plan detailed the goals and support in terms of capital and investment to the biotechnology sector, including agro-biotechnology products which are crops and animals (Frost & Sullivan, 2009). Around 2014 Malaysia focused on commercialization efforts and was increasing collaborations in preparation to move in to Phase 4 of the Malaysian Biotechnology Policy (European

Commission, 2014). Marketing activities should be customer centric and agricultural marketing is no exception. The successful marketing of modern biotechnology products in Malaysia depends heavily on consumer awareness and acceptance. Market acceptance depends on the consumer's perception of risks and benefits which may have different facets like taste impact, added nutritional value, convenience, safety, risks, and environmental impact. If consumer acceptance issues are not adequately addressed, then the potential economic and social benefits of modern biotechnology may not be realized. Hence assessment of consumer awareness and acceptance of GM products is crucial. There have been several studies conducted to assess attitude, awareness, or acceptance of GM food products, giving a good insight about the perception towards this product in the mind of the consumers. However, limited attempts have been made to determine the antecedents that influence the attitude towards GM food products in this geographical region and hence this study. The aim of the study is to investigate the consumer attitude towards genetically modified food products in Malaysia. The findings will contribute to the agricultural and the food and beverage industry in Malaysia; bringing more clarification about attitude towards GM food products in the eyes of the consumer and will help to determine the feasibility and viability of such technology in Malaysia. It will help marketers to segregate and identify appropriate market segments and government in the policy decision making as public opinion must be taken into account in a democratic system.

LITERATURE REVIEW

Underpinning Theory

The Fishbein Multi-Attribute Attitude model (Fishbein, 1963) is among the most renowned frameworks regarding the 'bottom-up' formation of consumer attitudes. The theory gives a basis to understand attitude and predict behavior. The 'bottom-up' formation of attitude theory states that the beliefs that make up the attitude towards a product is formed by weighing and evaluating the attributes of the product (Grunert, Bredahl, & Scholderer, 2003). Ajzen and Yee et al. (2008) and Fishbein (1963) defined attitude as a blend of a person`s feeling towards an object, his beliefs about the object and the action tendencies towards the object The socio-scientific field of consumer behavior related to GM food started in the late 1980`s, when it became apparent that modern biotechnologies will be applied in the field of food and agricultural practices (Hamstra, 1998). Since then, several studies have been conducted, using diverse models and Fishbein Multi-Attribute Attitude model was one of the common ones.

Perceived Risk and Benefit

Fishbein (1963) and Frewer, Howard, Hedderley, and Shepherd (1998) concluded that consumer behaviour is guided by their perception of benefits and risks about a product to a point of even ignoring expert opinions (Bredahl, 2001; Chen & Li, 2007; Costa-Font & Gil, 2009; Verdurme & Viaene, 2001). Studies conducted by Gaskell et al. (2004) in Europe classifies the population as optimistic, pessimistic, and undecided while a risk and benefit factor drive their attitude (Costa-Font, Gil, & Traill, 2008). A Spanish study conducted by Rodriguez-Entrena, Salazar-Ordonez, & Sayadi (2013) found that both perceived benefit and perceived risk have an influence on consumer attitude towards GM food. Many other studies in the area found similar observations that more perceived benefit lead to a more positive attitude, while more perceived risk towards GM food led to a negative attitude towards GM foods (Christoph, Bruhn, & Roosen, 2008; Costa-Font & Gil, 2009; Costa-Font et al., 2008; Curtis, McCluskey, an&d Whal, 2004; Kim, 2009; Kim, 2012; Prati, Pietrantoni, & Zani, 2012; Traill et al., 2006).

Furthermore, Moon and Balasubramanian (2004) and Hossain and Onyango (2004) observed that perceived risk weighs more than the perceived benefit when consumers assess agricultural biotechnology. The higher the risk perceived by the consumer towards GM food, the more negative is the attitude (Napier, Tucker, Henry, & Whaley, 2006). On the contrary, the study undertaken by Traill et al. (2006) showed that perceived benefit, rather than risk, is more dominant in the formation of consumer attitude. US consumers fall on both sides of the argument; while they are concerned about the health and environmental risk it poses, they also accept the potential and benefits of GM food (Hossein et al., 2003; Lusk, Daniel, Mark, & Lusk, 2001; Onyango, Ferdaus, Hallman, Schilling, & Adelajan, 2003; Onyango, Govindasamy, & Nayga, 2004). Hence US consumers may accept GM food by weighing the significant benefits and risks (Loureiro & Bugbee, 2005).

The study conducted by Siegrist, Cvetkovich, and Roth (2000) shows that North European consumers` attitude towards a product is not influenced by its perceived benefits. Fortin and Renton (2003) came to similar conclusions when studying the Nordic and New Zealand populations. A study in Taiwan has even shown that there is no relationship between attitude towards GM food and perceived risk (Chen, 2008). Studies about the influence of perceived risk and benefit on attitude towards GM food products in Malaysia are scarce and location specific. A study conducted in Johor Bahru in Malaysia showed that perceived benefit had most impact on the purchase behavior of GM food followed by perceived risk (Ismail et al., 2012a). However, the study by Ismail, Vivishna, Khurram, and Jafri (2012), Amin et al. (2010) and Hassan, John Kua, and Harun (2016) showed that only perceived benefit had an impact on purchase intentions and there was no relationship between perceived risk and purchase intentions. It is observed that the equivocal observations do not allow generalization of the findings from past studies. Attributes of GM food products are generally evaluated in terms of perceived benefits and perceived risks, leading to the first two hypotheses.

Knowledge

Another crucial element in the Fishbein Multi-Attribute Attitude model is knowledge (Fishbein, 1963; Grunert et al., 2003). Knowledge is a measure of consumers' recognition of GM food products which translates to an evaluative judgement (Kim, 2010). Consumer attitude is a dynamic process and knowledge or information about GM food product can be a decisive factor that allows the overall positive evolution of public attitude (Bredahl et al., 1998; Zhu & Xie, 2015). In 2010, 84% of Europeans were aware of GM foods, led by Norway and followed by Germany, Finland, and Netherlands (European Commission, 2010). However, despite the high level of knowledge and awareness about the biotechnological applications in the food industry, the population remained skeptical about their potential benefits. With more knowledge, the public become more critical in their discussions, increasing their skeptical attitude and further knowledge would consolidate their existing negative attitude (Zhang et al., 2010; Hassan et al., 2016). However, A study on German consumers in 2008 found out that there is no significant relationship between knowledge and attitude (Christoph et al., 2008). There are numerous similar studies that found no correlation between attitude and knowledge (Bredahl, 1999; Connor & Siegrist, 2010; Scholderer & Frewer, 2003). Angulo and Gil (2007) even found a negative correlation between acceptance and knowledge. Interesting results were obtained from a study in China which found out that despite the low level of knowledge in the Chinese population, acceptance of GM food product was high showing that knowledge was not one of the factors that impacts positive consumer attitude (Huang, Qiu, Bai, & Pray, 2006). Various other studies showed direct positive relation between knowledge and attitude towards GM food products (Baker & Burnhum, 2002; Boccaletti & Moro, 2000; Chen & Li, 2007; Costa-Font & Mossialos, 2007; Costa-Font et al., 2008; Moerbeek & Casimir, 2005; Moon and Balasubramanian, 2004; Nayga, Fisher, & Onyango, 2006; Vilella-Vila, Costa-Font, & Mossialos, 2005). These findings conclude that more the people have correct knowledge about GM food products (Verdurme & Viaene, 2003). A study conducted on the Japanese consumers, based on Fishbein Multi-Attribute Attitude model, showed that there was a positive relationship between attitude towards GM foods and knowledge (Kim, 2010). The same positive relationship was found in China (Steur et al., 2010).

Malaysian studies primarily measured the level of knowledge and awareness of Malaysians with contradicting results. Amin, Azlan, Hamdan, Samian, and Haron (2011) and Amin and Ibrahim (2011) found that the level of knowledge and awareness of Malaysians on modern biotechnology was from moderate to high. In contrast, a study by Ismail et al. (2012a) showed that Malaysians have low knowledge about GM foods but are nevertheless aware about it (Abu Bakar et al., 2005). Similar results were obtained by Ibrahim et al. (2013) in terms of knowledge and awareness and they also reported a general positive attitude from Chinese consumers despite the low knowledge. Tanius and Seng (2015) evaluated the level of awareness about GM food, and they concluded that awareness is low, and Malaysians have a negative attitude towards GM products There is nonetheless no study conducted to assess the relationship between the attitude towards GM food products and knowledge in Malaysia. The third hypothesis of this study (see below) was based on the statement of Verdurme and Viaene (2003); i.e. perception can be altered by knowledge and accurate knowledge will increase the positive attitude towards GM food products.

General Attitude

Verdurme and Viaene (2001) described general attitudes as consumers` individual attitudes towards technology, food, science, nature, and trust. Bredahl (2001) stated that the attitude towards GM food is grounded in the general attitude of the consumer towards nature and technology. Bredahl further elaborates that "A general attitude is like values in that it is embedded in our minds to influence our thoughts of a specific product" (Chen & Li, 2007). General attitude towards GM food products can be the key factor that determines consumer acceptance or attitude (Onyango et al., 2003). The phenomenon has been studied from various aspects with several studies asserting positively correlated relationships between it and attitude towards GM foods (Chen and Li, 2007; Costa-Font et al., 2008; Frewer et al., 1998; Loureiro and Hine, 2004; Mohr, Harrison, Wilson, Baghurst, & Syrette, 2007; Moon & Balasubramanian, 2004, 2001; Zhu & Xie, 2015). The approach that links attitude and consumer values to product perception was confirmed by Grunert et al. (2001) when examining Swedish, Finish, Norwegian and Danish attitude.

Dreezens, Martijn, Tenbult, Kok, and De Vries (2005) found similar correlation when they observed that consumers who easily accept GM products also accept that "humans should dominate nature." Traill et al. (2004) found that US consumers have a positive attitude towards innovative technology, which contributes to their acceptance of GM products. Grunert, Sondergaard, and Scholderer (2004) study on the European population observed that consumers who support GM food production are the ones who are positive towards novel technology development. Many more studies showed empirical evidence of the relationship between general attitude and attitude towards GM food products in the American and European context (Brant, Tilley, & Mowen, 2004; Bredahl, 2001; Traill et al., 2004). Consumers without knowledge about a product will refer to their general attitude as a guidance to form an attitude towards the product (Grunert et al., 2003; Kajale & Becker, 2015; Zhu & Xie, 2015).

Trust

Trust is a crucial factor in determining credibility of information; it determines whether a person accepts the information or facts provided (Costa-Font et al., 2008; Yee et al., 2008). Consumer skepticism is not based solely on GM foods but all the foundations that are connected to the product. Consumers may no longer assess food quality using only their senses and therefore must count on producers and regulators. Lack of knowledge is complemented with social trust and hence, perception of GM foods may be severely affected with a lack of accurate information and a low level of trust (Kim, 2012; Lobb, Mazzocchi, & Traill, 2007; Siegrist & Cvetkovich, 2000; Zhu & Xie, 2015).

There are various studies showing that customers are more willing to trust scientists, environmental groups and consumer organisations but are pessimistic about the biotechnology industry and the government (Bredahl, Grunert, & Frewer, 1998; Onyango et al., 2003; Savadori et al., 2004; Veeman, Adamowicz, & Hu, 2005). However, studies conducted both in the UK and US showed that the biotech industry and the government are the key pillars in controlling GM technology and therefore, important determinants of attitude towards GM products (Costa-Font et al., 2008; Frewer, Howard, & Shepherd, 1996; Han, 2006; Hossain & Onyango, 2004; Yue, Zhao, Cummings, & Kuzma, 2015). A study done by Moon and Balasubramanian (2001) in the US and UK revealed that consumer acceptance correlates positively with trust in regulatory institutes. Another study by Siegrist (2000) showed a similar positive correlation between trust in research institutions and attitude towards GM products. Similar results were obtained from a study in Taiwan by Chen and Li (2007). González, Johnson, and Qaim (2009) found a positive influence between support for GM food and trust in regulatory authorities. Lack of consumer trust will thus certainly hinder acceptance and attitude toward GM products (Ganiere, Chern, & Hahn, 2006; Kajale & Becker, 2015; Onyango, 2004).

It is however interesting to note that consumers are more willing to trust the negative and risk aspects provided by the institutions rather than the benefits (Traill et al., 2004). Frewer, Scholderer, and Bredahl (2003) and Lusk et al., (2004) observed an inverse relation between attitude and trust by stating that consumers who have a positive attitude towards GM products are more likely to trust institutions that provide beneficial information about the product, while those who have a negative attitude will more likely trust the institutions that provide detrimental information about the product. No previous study in Malaysian context was found, that uses trust as predictor to assess attitude towards GM food products. Trust can be assessed through various dimensions, but this study focuses on the ones most suitable and familiar in Malaysia, trust in government, non-governmental organizations, public media, international organizations, and food producers. The fifth hypothesis (see below) supports the statement of Siegrist (2000) that public trust is among the main drivers of GM food products. Despite huge spending in the elevation of biotechnology and being on the brink of commercialization, there are few studies that have been conducted in Malaysia on GM food products and public perception. Most studies in Malaysia were conducted years back (Amin et al., 2010; Amin, Azlan, Hamdan, Samian, & Haron, 2011; Amin & Ibrahim, 2011; Ismail et al., 2012a.; Ismail, Vivishna, Khurram, & Jafri, 2012); Hassan, John might have changed over time. The Fishbein`s Multi-Attribute Attitude model is the main framework that inspired this study, as it considers perceived benefits, perceived risks, and knowledge as antecedents of attitude towards GM food products. Adaptations were made to study trust and general attitude as additional antecedents to attitude towards GM food products. The conceptual framework of the study is provided in Fig. 1.

Figure 1. Conceptual framework

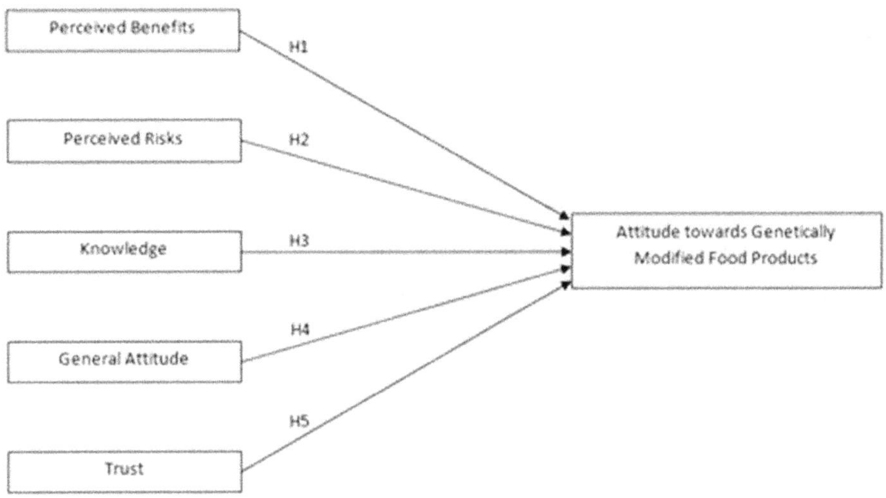

HYPOTHESES

H1: Consumers' positive attitude towards genetically modified food product increases with increasing perceived benefit.

H2: Consumers' positive attitude towards genetically modified food products decreases with increasing perceived risk.

H3: Consumers' positive attitude towards genetically modified food product increases with increasing knowledge.

H4: Consumers' positive attitude towards genetically modified food product increases with a more positive general attitude

H5: Consumers' positive attitude towards genetically modified food product increases with increasing trust.

METHODOLOGY

The aim of this research is to assess the attitude towards GM food products in Malaysia. Since this study is assessing the Malaysian population, the sample frame are respondents that have Malaysian nationality and are of appropriate age to make dietary decisions. A social survey design was deemed most appropriate as a standardized and systematic method to collect quantifiable data. A single indicator, that is a single question, may not be able to yield appropriate response for a construct. Hence, several questions are used to measure each variable, also known as a multiple-indicator measure. Questions for each construct were adapted from earlier research to fit this study. The questions are assessed using metric measure-

Table 1. Constructs, items, and source

Construct	Items	Source
Perceived Benefit	1. Overall, applying gene technology to produce food products will prove beneficial to the environment, me and other people that are important to me. 2. Applying gene technology in food production will improve the food functional features. 3. The world`s food supply will increase through the use of GM. 4. In the long run, a successful genetically modified food industry will be good for the economy. 5. GM technology can reduce the use of pesticide on crops. 6. Genetically modified food products are better quality foodstuffs than other food products. 7. Applying gene technology in food production will reduce the price of food products	Chen and Li, 2007; Rodriguez- Entrena et al., 2013; Yee et al., 2008; Prati et al., 2012
Perceived Risk	1. Overall, applying gene technology to produce food products involves considerable risk to the environment, me and other people that are important to me. 2. Applying gene technology in food production will cause allergy in human beings. 3. Applying gene technology in food production only serves the interest of big companies. 4. The side-effects of eating food produced using GM are largely unknown. 5. Consumption of GM food may give rise to antibiotic resistant diseases. 6. Applying gene technology in food production will cause environmental hazards. 7. Gene technology combines the genetic bases of plants and animal species. Irreversible damage to the environment becomes possible.	Chen and Li, 2007; Rodriguez -Entrena et al., 2013; Yee et al., 2008; Amin et al. 2010; Bredahl, 2001
Knowledge	1. Agricultural crops can be made resistant to certain diseases and plagues by modifying their hereditary material. 2. Animal features can be transferred to plants. 3. GM food contains genes like all conventional food. 4. All source of food contains bacteria which may be harmful or harmless. 5. Soybean contains GM components.	Chen and Li, 2007; Tanius and Seng, 2015.
General Attitude	1. Scientific progress implies social welfare (science). 2. New food products are worth trying (food). 3. New technological inventions and applications make up the driving force of the progress of society. 4. Modifying the environment for human use seldom causes serious problems.	Chen and Li, 2007; Yee et al., 2008; Bredahl, 2001
Trust	1. I trust the government in relation to communication of the risk on GM food. 2. I trust non-governmental organisations in relation to communication of the risk on GM food. 3. I trust public media in relation to communication of the risk on GM food. 4. I trust international organisations such as the Food and Drug Administration (FDA) in relation to communication of the risk on GM food. 5. The authorities can sufficiently monitor whether scientists in the field of gene technology uphold legal regulations and restrictions 6. Food producers take potential damage to the human health and the environmental into account during production.	Rodriguez-Entrena et al., 2013; Prati et al., 2012; Seigrist, 2000; Verdurme and Viaene, 2003
Attitude towards GM food products	1. Applying gene technology in food production is extremely good. 2. Production of GM food improves the current and future society's welfare. 3. Production of GM food is a necessary activity for society. 4. Genetically modified food industry will be good for the economy.	Chen and Li, 2007; Rodriguez -Entrena et al., 2013.

ment scales (5-point Likert scale), which was reckoned to be the most appropriate for a structured and closed-ended survey questionnaire.

A pilot study was conducted with 30 respondents and the questionnaire was reviewed accordingly. Convenience sampling was chosen as a realistic method and the survey was administered through a supervised self-completion questionnaire. In this way, respondents could be assessed for any biasness and a high response rate was guaranteed. Data for this study was collected over a period of one month. To increase the accessibility and positive response of respondents, data was collected around shopping malls in Klang Valley. With a population size of 8 million, 95% confidence interval and 6% margin of error, the sample size should be at least 267. A total of 301 completed questionnaires were collected for this study; 24 responses were discarded as they were either incomplete, answered too quickly (average response time was 4 minutes), respondent being underage or non-Malaysians.

A total of 277 usable responses were available for statistical analysis for which Statistical Package for Social Sciences was used. Cronbach`s alpha was used to ensure the internal reliability of each construct. The variables were constructed through the summation of relevant questions. Extreme outliers were highlighted through boxplots in the constructs and omitted from further analysis to increase the precision and normality of the study. Normality tests were used to find out whether the data were non-parametric or parametric, based on kurtosis, skewness, normal Q-Q plots, and histograms (Bryman and Bell, 2015; Pallant, 2016). Multiple regression analysis was used to analyse the relationships. This analysis allowed the study to assess the significance of the overall model, the relationship of each independent variable with the dependent variable while also displaying which independent variables have the most impact on the dependent variable.

FINDINGS AND DISCUSSION

Out of the 277 surveys considered for statistical analyses, 50.5% of the respondents were males and 49.5% were females. More than 50% of them were between the age range of 20-29 years old. The majority of the sample respondents were Malay, followed by Indian and Chinese. 57.4% of the sample respondents were single and 42.6% of the sample respondents were married with 31% having children. 57% of the respondents had a bachelor's degree and about 3.25% recorded their qualification as 'others', which mainly represented vocational courses. 32.1% were professionals and about 9.03% recorded their occupation as 'others', which mainly represented respondents who have their own businesses, unemployed or retired. The summary of the socio-demographic variables is presented in Table 2.

The reliability results showed that indicators measuring Perceived Benefits, Perceived Risk, Trust, and Attitude towards GM food products are reliable with a Cronbach`s alpha of more than 0.7. However, one question had to be removed from the General Attitude construct to increase the reliability value from 0.44 to 0.63. The Cronbach`s alpha of the indicators of the Knowledge construct was relatively low at 0.50. Variables can be wrongly discarded due to low Cronbach`s alpha values if the concept of reliability is not well understood. Along with internal consistency, unidimentionality is also measured when running Cronbach`s alpha analysis (Tavakol and Dennick, 2011). However, a multidimensional test does not affect the internal consistency of a variable but it still violates the fundamental assumption of the Cronbach`s alpha analysis, hence yielding a low value. Normally an alpha value of 0.7 or above is more suitable in most cases. However, when dealing with multidimensional psychological variables, values below 0.7 should be expected (Field, 2009).

Table 2. Sociodemographic profile of sample

Demographic profile		Sample percentages
Gender	Male	50.5
	Female	49.5
Age	20-29	57.0
	30-39	27.1
	40-49	10.1
	50-59	5.4
	60 and above	0.4
Race	Malay	57.8
	Indian	21.7
	Chinese	16.9
	Others	3.6
Marital Status	Single	57.4
	Married	42.6
Qualification	Secondary	5.8
	Diploma	23.8
	Degree	57.0
	Masters	8.7
	PhD	1.4
	Others	3.3
Occupation	Executive	29.3
	Non-executive	7.2
	Professional	32.1
	Housewife	3.3
	Student	19.1
	Others	9.0
Income	RM 2000 or less	5.1
	RM 2001 – RM 4000	28.5
	RM 4001 – RM 6000	20.2
	RM 6001 – RM 8000	7.6
	RM 8000 or above	12.6
	Not applicable	26.0

The data was checked for multicollinearity through Variance Inflation Factor (VIF), Collinearity Tolerance and Pearson Correlation. They show no significant relationship among the independent variables as the VIF values, Tolerance and Pearson Correlation were less than 10.0, more than 0.10 and below 0.7 respectively. The Analysis of Variance (ANOVA) also shows the statistical significance of the model whereby the p-value was 0.000. The overall significance of the theoretical model represented by the R Square value was 0.521, showing that 52.1% of the dependent variable is explained by the independent

Table 3. Statistical analysis evaluating relationships between dependent and independent variables

	Standardized Coefficient		Collinearity Statistics	
Model	**Beta**	**Significance**	**Tolerance**	**VIF**
Constant	-	.003	-	-
Knowledge	.002	.632	.884	1.184
Perceived Benefits	.532	.000	.707	1.415
Perceived Risks	-.272	.000	.873	1.145
General Attitude	.029	.546	.762	1.313
Trust	.184	.000	.936	1.068

variables. Table 3 shows the analysis that is used to evaluate the contribution of each independent variable to the dependent variable and their respective strengths.

The Beta column under standardized coefficients (Table 3) shows which construct has the strongest contribution to the dependent variable; the lower the value, the weaker the relationship (Pallant, 2016). Perceived benefits (.532) have the most influence on attitude towards GM food products followed by perceived risk (-.272) and trust (.184).

The acceptance/rejection of perceived benefits and attitudes towards GM foods is presented in Table 4.

Table 4. Acceptance / rejection of hypotheses

Hypotheses	*Accept/Reject*
H1: Consumers' positive attitude towards genetically modified food product increases with increasing perceived benefit.	Accept
H2: Consumers' positive attitude towards genetically modified food product decreases with increasing perceived risk.	Accept
H3: Consumers' positive attitude towards genetically modified food product increases with increasing knowledge.	Reject
H4: Consumers' positive attitude towards genetically modified food product increases with a more positive general attitude	Reject
H5: Consumers' positive attitude towards genetically modified food product increases with increasing trust.	Accept

The outcomes are discussed further in the paragraphs below.

Relationship Between Perceived Benefit and Attitude Towards GM Food Products

The empirical results obtained from the statistical analyses accept the first hypothesis, which means that increasing perceived benefits increases consumer`s positive attitude towards genetically modified food products. This result matches the studies of Prati et al. (2012), Rodriguez-Entrena et al. (2013), Hassan et al. (2016) and countless other researches, who came to the same conclusion that the greater the benefits perceived by the consumers, the better their attitude towards GM products (Chen, 2008; Christoph et al., 2008; Costa-Font and Gil, 2009; Costa-Font et al., 2008; Curtis et al., 2004; Kim, 2009, 2012; Traill et al., 2006). However, similar studies conducted by Siegrist et al. (2000) and Fortin and Renton (2003) with North European consumers and the Nordic/New Zealand population respectively, shows no relationship between consumer attitude and perceived benefit. In the study of Fortin and Renton (2003), the perception of benefits might not have been enough to impact a conventionally pessimistic picture of GM products in the Nordic and New Zealand population (Honkanen & Verplanken, 2004). Moon and Balasubramanian (2004) and Hossein and Onyango (2004) clearly contribute to this opinion by concluding that negative attributes outweigh positive attributes in the determination of consumer attitude. The studies from Moon and Balasubramanian (2004) and Hossein and Onyango (2004) assessed the UK and U.S population, where the inhabitants` attitude is already molded in a pessimistic way (Costa-Font et al., 2008; European Commission, 2010). This might be the reason why perceived benefit fell short in the attitude formation process.

Contrasting the aforementioned studies, this study suggests that perceived benefits is the strongest determinant of attitude towards GM food. GM food is a relatively new concept in Malaysia and the level of knowledge is low. Activities promoting the consumption of these products by the government and non-governmental organizations such as Malaysian Biotechnology Information Centre (MABIC) might have made the population more inclined to consider benefits when shaping their attitude ((MABIC, 2017; Tanius and Seng, 2015).

Relationship Between Perceived Risk and Attitude Towards GM Food Products

Despite falling behind perceived benefit, perceived risk is still a significant determinant of attitude towards GM food product. Perceived risk and attitude towards GM food products are inversely related. Hence, the second hypothesis of this study is supported, which states that a consumer`s positive attitude towards genetically modified food product would increase with decreasing perceived risk. This result reflects the results of several other studies. Nappier et al. (2006), Prati et al. (2012) and Rodriguez-Entrena et al. (2013) to name a few, came to the conclusion that the greater the perceived risk, the lower the positive attitude towards GM foods (Christoph et al., 2008; Costa-Font & Gil, 2009; Costa-Font et al., 2008; Curtis et al., 2004; Kim, 2009, 2012; Traill et al., 2006;). Nevertheless, the result refutes the study of Chen (2008) in Taiwan, who concluded that there is no significance between perceived risk and attitude towards GM foods. This might be explained by the highly positive general attitude towards technology and nature of the Taiwanese population. The inclination towards their positive general attitude might have made risk nearly insignificant in shaping their attitude. The importance of both perceived benefit and perceived risk are highlighted in this study. Hence, the Malaysian attitude towards GM food products seem to be guided by weighing both benefits and risks concerning food quality, health and environment; unlike Gaskell et al. (2004) who segregated the European population into optimistic and pessimistic. However, relatable Malaysian studies conducted by Ismail et al. (2012b), Amin et al. (2010) and Hassan et al. (2016) finds no relationship between consumer purchase intentions and perceived risk. The inconsistency in these studies might be due to the low level of awareness recorded, which is a key factor in measuring perceived risk, which should be further explored.

Relationship Between Knowledge and Attitude Towards GM Food Products

The third hypothesis, built around the statement of Verdurme and Viaene (2003), was rejected as the analysis results shows that there is no relationship between the correct/objective knowledge and attitude towards GM food products. The result disputes the works of numerous authors who found a positive relationship between attitude towards GM food and knowledge in China (Steur et al. 2010), Japan (Kim, 2010) and many other geographical locations (Baker & Burnhum, 2002; Boccaletti & Moro, 2000; Chen & Li, 2007; Costa-Font & Mossialos, 2007; Costa-Font et al., 2008; Moerbeek & Casimir, 2005; Moon & Balasubramanian, 2004; Nayga et al., 2006; Vilella-Vila et al., 2005). Likewise, the study also opposes the works of Angulo and Gil (2007), Zhang et al. (2010) and Hassan et al. (2016) who found that there is a negative relationship between attitude towards GM food products and knowledge, meaning that rather than clearing misconceptions, more knowledge promotes a skeptical mindset.

Nevertheless, a study in Germany conducted by Christoph et al. (2008) showed that there is no difference in the knowledge level between opponents and supporters of GM food products, which converge to the same conclusion that knowledge is not a determinant of either negative or positive attitude. Bredahl (1999), Scholderer and Frewer (2003) and Huang et al. (2006) arrived at the same results as Christoph et al. (2008) and this study. Moreover, Conor and Siegrist (2010) concluded that only basic biological and legal knowledge had an impact on attitude while knowledge about gene technology had no relationship with attitude. Since this study covered basic biological knowledge as well as knowledge about GM food products whereby both of them shows no relationship with attitude, one part of the study of Conor and Siegrist (2010) agrees to the conclusion and the other part disputes it. The insignificant relationship between attitude and knowledge might be explained through the low level of awareness on GM food products of the Malaysian population. Their limited knowledge decreases the relevance of the relationship between attitude and knowledge (Christoph et al., 2008; Tanius & Seng, 2015).

Relationship Between General Attitude and Attitude Towards GM Food Products

The fourth hypothesis, based primarily on the meta-analysis of Verdurme and Viaene (2001), was rejected as the results shows that consumer general attitude towards technology, science and food have no influence on the attitude towards genetically modified food. This result challenges the works of Grunert et al. (2004) and Traill et al. (2004) in Europe and America respectively, who concluded that attitude towards novel technology impacts the acceptance and attitude towards GM food. The probable rationale for the difference in results in this study and the aforementioned studies is that the constructs of Grunert et al. (2004) and Traill et al. (2004) were unidimensional, hence increasing the odds of getting significant relationships, while general attitude in this study is multidimensional (Cleveland, Kalamas and Laroche, 2005). The result from this study also challenges the works of Frewer et al. (1998), Mohr et al. (2007) and Christoph et al. (2008) to name a few (Brant et al., 2004; Bredahl, 2001; Chen & Li, 2007; Loureiro & Hine, 2004; Moon & Balasubramanian, 2004; Traill et al., 2004; Zhu & Xie, 2015).

Amin et al. (2006) conducted a similar study to determine the relationship between attitude towards GM soybean and different classes of attitude in Malaysia and found that general attitude had a significant influence on attitude towards GM soybean. This again contradicts the observation from the current study, but it should be noted that the indicators of the construct of study by Amin et al. (2006) were different. The items that measured general attitude were technology optimism, general promise and concern of biotechnology, attachment to religion and custom, predisposition towards science and technology, and nature/materialistic value. The only common items with this study were predisposition towards science, and technology optimism. General promise and concern of biotechnology were measured as perceived benefits and risks and are in line with the study of Amin et al. (2006) as both have significant impact on attitude towards GM food products.

Relationship Between Trust and Attitude Towards GM Food Products

It was hypothesized that an increase in trust would result in a better attitude towards GM food products. The multidimensional construct included trust in relation to the government, non-governmental organizations, international organizations, concerned authorities, food producers and public media. The results show that there is a significant effect of trust in these organizational bodies on the attitude towards GM food products. The results are in line with various studies conducted throughout Europe and America (Costa-Font et al., 2008; Frewer et al., 1996; Han, 2006; Hossain and Onyango, 2004; Yue et al., 2015). The study by Moon and Balasubramanian (2001) also supports the results of this study by concluding that trust in regulatory bodies has a significant influence on acceptance of GM food.

Studies by Siegrist (2000) and Chen and Li (2007) supported this study by documenting a significant relationship between attitude and trust in research institutions. Furthermore, a study by González et al. (2009) showed a positive correlation between attitude towards GM food products and regulatory authorities, which goes along with the results from this study. Ismail et al. (2012a) in their study found that Malaysians' knowledge about GM food products is low. Since they lack information and cannot efficiently assess the quality of the food solely through their senses, the most probable and logical antecedent to form an attitude towards a product would be to trust the concerning bodies about the information given. Trusting the concerned organizations who gives positive feedbacks about GM food product would theoretically enhance the positive attitude towards GM food product, while being sceptic towards these organizations will give rise to misconceptions and doubts.

STUDY IMPLICATIONS

This study adds to the body of literature by concluding that Fishbein`s Multi-Attribute Attitude model is partly supported in the Malaysian context. Perceived benefits, perceived risks and trust are seen to be significantly related to attitude towards GM food products. However, the results show; that the validity of a model varies among regions and populations, as knowledge, which is also part of Fishbein`s model, is not supported in the Malaysian context. Perceived benefit is the strongest factor influencing attitude towards GM food products, followed by perceived risk. Hence, the rational way to have a positive impact on the population is firstly through increased perceived benefits and then reduced perceived risk. Despite knowledge being rejected as a direct antecedent to attitude, it remains a major factor to influence perceived benefits and risks of consumers, as it is documented to be a major antecedent of perceived risk and benefits (Chen & Li, 2007; Costa-Font et al., 2008; Yee et al., 2008). With accurate knowledge from the government organizations, nongovernmental organizations, public media, producers, marketing departments and international organizations, misconceptions can be reduced, which would eventually decrease perceived risk and increase perceived benefits with substantiated information. Added to that, concerned bodies in Malaysia along with the government should promote trustworthiness among their population as it seems to be a significant antecedent to attitude. This would be crucial in the first stages or to the elderly segment of the population as the concept of such a complex process as genetic modification may not be understood by many. The positive impact of transparency and good practice in terms of regulations and frameworks has already been witnessed in the U.S and India (ISAAA, 2016).

CONCLUSION

This study can be considered as a preliminary analysis giving a realistic start for future research. Only 52.1% of the attitude towards GM food products is explained by the antecedents considered in this study and hence, future research is encouraged to determine other factors that significantly influence the attitude towards GM food products. A study segregating the types of food may give a deeper insight and the difference in attitude towards different types of food products. A qualitative study could be a step forward to have a deeper insight of the consumer behavior and attitude towards GM food products. Similarly, a longitudinal study will assess the development and progress of attitude and hence, actions that are being implemented by concerned organizations can be critically analyzed and improved. This study is not free from limitations. It used non-probability sampling method to represent the Malaysian population and concentrates on few antecedents of attitude towards GM food product using Fishbein`s Multi Attribute Attitude Model as the main pillar. It is also important to note that illiterate people and people with low proficiency in English were excluded from this study. These limitations lead to opportunities for further studies in the Malaysian context. Biotechnology has evolved and taken new proportions in Malaysia with the full financial support of the government. However, prior to commercialization, it is important that the consumer behavioral aspect is studied as the success of any product depends on the attitude and acceptance towards that product. This study gives an understanding about the principal aspects where the government, concerned organizations, industry and policy makers should focus to efficiently influence the attitude towards GM food products.

REFERENCES

Abu Bakar, U. K., Pillai, V., Hashim, M., & Daud, H. M. (2005). Sharing Malaysian experience with development of biotechnology-derived food crops. *Food and Nutrition Bulletin*, *26*(4), 317–320. doi:10.1177/15648265050264S312 PMID:16465992

Amin, L., Ahmad, J., Jahi, J. M., Nor, A. R. M., Osman, M., & Mahadi, N. M. (2010). Factors influencing Malaysian public attitude to agro-biotechnology. *Public Understanding of Science (Bristol, England)*, *20*(15), 674–689. PMID:22164706

Amin, L., Azlan, N. A. A., Hamdan, M. F., Samian, A. L., & Haron, M. S. (2011c). Awareness and knowledge on modern biotechnology. *African Journal of Biotechnology*, *10*(58), 12448–12456.

Amin, L., Hashim, H., Sidik, N. M., Zainol, Z. A., & Anuar, N. (2011b). Public attitude towards modern biotechnology. *African Journal of Biotechnology*, *10*(58), 12409–12417.

Amin, L., & Ibrahim, R. (2011). How aware and knowledgeable are Malaysians of modern biotechnology? *Procedia: Social and Behavioral Sciences*, *15*, 3384–3388. doi:10.1016/j.sbspro.2011.04.305

Amin, L., Jahi, J. M., Nor, A. R. M., Osman, M., & Mahadi, N. M. (2006). Uncovering factors influencing Malaysian public attitude towards modern biotechnology. *Asia-Pacific Journal of Molecular Biology and Biotechnology*, *14*(2), 33–39.

Amin, L., Othman, J., Goh, H. L., & Jusoff, K. (2011a). Consumer information and agro-biotechnology: The experience of Malaysia. *American-Eurasian Journal of Agricultural & Environmental Sciences*, *10*(6), 1006–1017.

Anderson, J. C., Wachenheim, C. J., & Lesch, W. C. (2005). Perceptions of genetically modified and organic foods and processes: North Dakota College students. *Agribusiness and Applied Economics Report No. 571,* pp. 1-50.

Angulo, A. M., & Gil, J. M. (2007). Spanish consumers' attitudes and acceptability towards GM food products. *Agricultural Economics Review*, *8*(1), 50–63.

Arya, D. (2015). *Genetically modified foods; Benefits and Risks*. Massachusetts Medical Society.

Baker, G. A., & Burnham, T. A. (2002). The marker for genetically modified food: Consumer characteristics and policy implications. *The International Food and Agribusiness Management Review*, *4*(4), 351–360. doi:10.1016/S1096-7508(02)00063-0

Boccaletti, S., & Moro, D. (2000). Consumer willingness to pay for GM food products in Italy. *AgBioForum*, *3*, 259–267.

Brant, M., Tilley, D. S., & Mowen, J. C. (2004). *Consumer response to genetically modified foods.* Unpublished manuscript, Arkansas: Southern Agricultural Association.

Bredahl, L. (1999). Consumers' cognitions with regard to genetically modified food. Results of a qualitative study in four countries. *Appetite*, *33*(3), 343–360. doi:10.1006/appe.1999.0267 PMID:10625527

Bredahl, L. (2001). Determinants of consumer attitudes and purchase intentions with regard to genetically modified foods – results of a cross-national survey. *Journal of Consumer Policy*, *24*(1), 23–61. doi:10.1023/A:1010950406128

Bredahl, L., Grunert, K. G., & Frewer, L. J. (1998). Consumer attitude and decision-making with regard to genetically engineered food products: A review of the literature and a presentation of models for future research. *Journal of Consumer Policy*, *21*(3), 251–277. doi:10.1023/A:1006940724167

Bryman, A., & Bell, E. (2015). *Business Research Methods* (4th ed.). Oxford University Press.

Chen, M. F. (2008). An integrated research framework to understand consumer attitudes and purchase intentions toward genetically modified foods. *British Food Journal*, *110*(6), 559–579. doi:10.1108/00070700810877889

Chen, M. F., & Li, H. L. (2007). The consumer's attitude toward genetically modified foods in Taiwan. *Food Quality and Preference*, *18*(4), 662–674. doi:10.1016/j.foodqual.2006.10.002

Christoph, I. B., Bruhn, M., & Roosen, J. (2008). Knowledge, attitude towards and acceptability of genetic modification in Germany. *Apetite*, *51*(1), 58–68. doi:10.1016/j.appet.2007.12.001 PMID:18243411

Cleveland, M., Kalamas, M., & Laroche, M. (2005). Shades of green: Linking environmental locus of control and pro-environmental behaviors. *Journal of Consumer Marketing*, *22*(4), 198–212. doi:10.1108/07363760510605317

Connor, M., & Siegrist, M. (2010). Factors influencing people's acceptance of gene technology: The role of knowledge, health expectations, naturalness and social trust. *Science Communication*, *32*(4), 514–538. doi:10.1177/1075547009358919

Costa-Font, J., & Mossialos, E. (2007). Are perceptions of "risks" and "benefits" of genetically modified food (in) dependent? *Food Quality and Preference*, *18*(2), 173–182. doi:10.1016/j.foodqual.2005.09.013

Costa-Font, M., & Gil, J. M. (2009). Structural equation modelling of consumer acceptance of genetically modified (GM) food in the Mediterranean Europe: A cross country study. *Food Quality and Preference*, *20*(6), 399–409. doi:10.1016/j.foodqual.2009.02.011

Costa-Font, M., Gil, J. M., & Traill, W. B. (2008). Consumer acceptance, valuation of and attitudes towards genetically modified food: Review and implications for food policy. *Food Policy*, *33*(2), 99–111. doi:10.1016/j.foodpol.2007.07.002

Curtis, K. R., McCluskey, J. J., & Whal, T. I. (2004). Consumer acceptance of genetically modified food products in the developing world. *AgBioForum*, *7*(1-2), 70–75.

Cyranosk, D. (2005). The valley of ghosts. *Nature*, *436*(7051), 620–621. doi:10.1038/436620a PMID:16079812

Dona, A., & Arvanitoyannis, I. S. (2009). Health Risks of Genetically Modified Foods. *Critical Reviews in Food Science and Nutrition*, *49*(2), 164–175. doi:10.1080/10408390701855993 PMID:18989835

Dreezens, E., Martijn, C., Tenbult, P., Kok, G., & De Vries, N. K. (2005). Food and values: An examination of values underlying attitudes towards genetically modified and organic grown food products. *Appetite*, *44*(1), 115–122. doi:10.1016/j.appet.2004.07.003 PMID:15604038

European Commission. (2010). *Eurobarometer 73.1: Biotechnology*. TNS Opinion and Social.

European Commission. (2014). *International Bioeconomy Profile Malaysia*. Joint Research Centre.

Field, A. (2009). *Discovering statistics using SPSS* (3rd ed.). SAGE.

Firdaus-Raih, M., Senafi, S., Murad, M. A., Sidik, M. N., Lian, K. W., Daud, F., & Mohamed, R. (2005). A nationwide biotechnology outreach and awareness program for Malaysian high schools. *Electronic Journal of Biotechnology*, *8*(1), 9–16.

Fishbein, M. (1963). An investigation of the relationships between beliefs about an object and the attitude toward that object. *Human Relations*, *16*(3), 233–240. doi:10.1177/001872676301600302

Food & Drug Administration. (2017). Consumer Info About Food from Genetically Engineered Plants. FDA. https://www.fda.gov/Food/IngredientsPackagingLabeling/GEPlants/ucm461805.htm

Food and Agriculture Organization of the United Nations. (2015), The State of Food Insecurity in the World 2015. Meeting the 2015 international hunger targets: taking stock of uneven progress. FAO.

Fortin, D. R., & Renton, M. S. (2003). Consumer acceptance of genetically modified foods in New Zealand. *British Food Journal, 105*(1/2), 42–58. doi:10.1108/00070700310467483

Frewer, L. J., Howard, C., Hedderley, D., & Shepherd, R. (1998). Methodological approaches to assessing risk perceptions associates with food-related hazards. *Risk Analysis, 18*(1), 95–102. doi:10.1111/j.1539-6924.1998.tb00919.x PMID:9523447

Frewer, L. J., Howard, C., & Shepherd, R. (1996). The influence of realistic product exposure on attitudes towards genetic engineering of food. *Food Quality and Preference, 7*(1), 61–67. doi:10.1016/0950-3293(95)00017-8

Frewer, L. J., Scholderer, J., & Bredahl, L. (2003). Communicating about the risk and benefits of genetically modified foods: The mediating role of trust. *Risk Analysis, 23*(6), 1117–1133. doi:10.1111/j.0272-4332.2003.00385.x PMID:14641888

Frost, A., & Sullivan, W. (2009). *The Malaysian agricultural biotechnology sector.* Malaysian Biotechnology Corporation.

Ganiere, P., Chern, W. S., & Hahn, D. (2006). A continuum of consumer attitude toward genetically modified foods in the United States. *Journal of Agricultural and Resource Economics, 31*(1), 129–149.

Gaskell, G., Allum, N., Wagner, W., Kronberger, N., Torgersen, H., Hampel, J., & Bardes, J. (2004). GM foods and the misperception of risk perception. *Risk Analysis, 24*(1), 185–194. doi:10.1111/j.0272-4332.2004.00421.x PMID:15028010

González, C., Johnson, N., & Qaim, M. (2009). Consumer acceptance of second generation GM foods: The case of biofortified cassava in the northeast of Brazil. *Journal of Agricultural Economics, 60*(3), 604–624. doi:10.1111/j.1477-9552.2009.00219.x

Grunert, K., Sondergaard, H., & Scholderer, J. (2004). How can we know what we like when we don't understand it? Consumer attitude formation towards complex technical issues. From the *proceedings of the International Conference on Research in Advertising, Aarhus University*. MAPP - Centre for Research on Customer Relations in the Food Sector.

Grunert, K. G., Bredahl, L., & Scholderer, J. (2003). Four questions on European consumers' attitudes toward the use of genetic modification in food production. *Innovative Food Science & Emerging Technologies, 4*(4), 435–445. doi:10.1016/S1466-8564(03)00035-3

Grunert, K. G., Lahteenmaki, L., Nielsen, N. A., Poulsen, J. B., Ueland, O., & Astrom, A. (2001). Consumer perceptions of food products involving genetic modification-results from a qualitative study in four Nordic countries. *Food Quality and Preference, 12*(8), 527–542. doi:10.1016/S0950-3293(01)00049-0

Hamstra, A. (1998). *Public opinion about biotechnology. A survey of surveys.* European Federation of Biotechnology.

Han, J. H. (2006). The Effects of Perceptions on Consumer Acceptance of Genetically Modified (GM) Foods. [Unpublished Doctoral Dissertation, Department of Agricultural Economics and Agribusiness, Louisiana State University].

Hassan, S. H., John Kua, S. B., & Harun, H. (2016). Muslim consumers' perception and purchase intention toward GM food. *International Food Research Journal*, *23*(2), 806–815.

Honkanen, P., & Verplanken, B. (2004). Understanding attitudes towards genetically modified food: The role of values and attitudes strength. *Journal of Consumer Policy*, *27*(4), 401–420. doi:10.100710603-004-2524-9

Hossain, F., & Onyango, B. (2004). Product attributes and consumer acceptance of nutritionally enhanced genetically modified foods. *International Journal of Consumer Studies*, *28*(3), 255–267. doi:10.1111/j.1470-6431.2004.00352.x

Hossain, F., Onyango, B., Schilling, B., Hallman, W., & Adelaja, A. (2003). Product attributes, consumer benefits and public approval of genetically modified foods. *International Journal of Consumer Studies*, *27*(5), 353–365. doi:10.1046/j.1470-6431.2003.00303.x

Huang, J., Qiu, H., Bai, J., & Pray, C. (2006). Awareness, acceptance of and willingness to buy genetically modified foods in urban China. *Appetite*, *46*(2), 144–151. doi:10.1016/j.appet.2005.11.005 PMID:16469414

Ibrahim, B., Rezal, G., Mohamed, Z. A., & Sharifuddin, J. (2013). Determinants of consumer perception towards gentically modified (GM) foods: Malaysian case study. *3rd International Conference on Management Proceeding*, (pp. 488-499).

International Service for the Acquisition of Agri Biotech Applications. (2022) ISAAA. https://www.isaaa.org/gmapprovaldatabase/default.asp. (Accessed 14October, 2022)

Ismail, K., Soehod, K., Vivishna, S., Khurram, W., Jafri, S. K. A., & Ramily, M. K. B. (2012a). Genetically modified food and consumer purchase intentions: A study in Johor Bahru. *International Journal of Business and Social Science*, *3*(5), 197–207.

Ismail, K., Vivishna, S., Khurram, W., & Jafri, S. K. A. (2012b). Evaluating Consumer Purchase Intentions for Genetically Modified Food in Malaysia: A Comparative Study of Muslim and Non-Muslim Consumers. *Research Journal of Applied Sciences, Engineering and Technology*, *4*(5), 466–474.

Kajale, D. B., & Becker, T. C. (2015). Factors Influencing Young Consumers' Acceptance of Genetically Modified Food in India. *Journal of Food Products Marketing*, *21*(5), 461–481. doi:10.1080/10454446.2013.845866

Kim, R.B. (2009). Factors influencing Chinese consumer behavior when buying innovative food products. *Agricultural Economics – Czech,* 55(9), 436-445.

Kim, R.B. (2010). A multiple-attribute model of Japanese consumer`s purchase intention for GM foods. *Agricultural Economics – Czech, 56*(10), 449-459.

Kim, R. B. (2012). Consumer attitude of risk and benefits towards genetically modified foods in South Korea: Implications for food policy. *Inzinerine Ekonomika –. The Engineering Economist*, *23*(2), 189–199.

Lobb, A. E., Mazzocchi, M., & Traill, W. B. (2007). Modelling risk perception and trust in food safety information within the theory of planned behavior. *Food Quality and Preference*, *18*(2), 384–395. doi:10.1016/j.foodqual.2006.04.004

Loureiro, M., & Hine, S. (2004). Preferences and willingness to pay for GM labelling policies. *Food Policy*, *29*(5), 467–483. doi:10.1016/j.foodpol.2004.07.001

Loureiro, M. L., & Bugbee, M. (2005). Enhanced GM Foods: Are Consumer Ready to Pay for the Potential Benefits of Biotechnology? *The Journal of Consumer Affairs*, *39*(1), 52–70. doi:10.1111/j.1745-6606.2005.00003.x

Lusk, J. L., Daniel, M. S., Mark, D. L., & Lusk, C. L. (2001). Alternative Calibration and Auction Institutions for Predicting Consumer Willingness to Pay for Nongenetically Modified Corn Chips. *Journal of Agricultural and Resource Economics*, *26*(1), 40–57.

Lusk, J. L., House, L. O., Valli, C., Jaeger, S. R., Moore, M., Morrow, B., & Traill, W. B. (2004). Effect of information about benefits of biotechnology on consumer acceptance of genetically modified food: Evidence from experimental auctions in the United States, England, and France. *European Review of Agriculture Economics*, *31*(2), 179–204. doi:10.1093/erae/31.2.179

Macer, D., Bazer, H., & Gough, J. (1991). *'Genetic engineering in New Zealand: Science, ethics and public policy', Information paper no.20*. Centre for Resource Management, Lincoln University.

Malaysian Biotechnology Information Centre. (2017). *Homepage*. MBIC. http://www.bic.org.my

Moerbeek, H., & Casimir, G. (2005). Gender differences in consumers' acceptance of genetically modified foods. *International Journal of Consumer Studies*, *29*(4), 308–318. doi:10.1111/j.1470-6431.2005.00441.x

Mohr, P., Harrison, A., Wilson, C., Baghurst, K. I., & Syrette, J. (2007). Attitudes, values, and socio-demographic characteristics that predict acceptance of genetic engineering and applications of new technology in Australia. *Biotechnology Journal*, *2*(9), 1169–1178. doi:10.1002/biot.200700105 PMID:17703489

Mokhtar, K. S., & Mahalingam, R. (2010). The current biotechnology outlook in Malaysia. *Economia. Seria Management*, *13*(1), 11–26.

Moon, W., & Balasubramanian, S. K. (2001). A multiple-attribute model of public acceptance of genetically modified organisms. *American Agricultural Economics Association annual meeting,* (pp. 1-19). AEA.

Moon, W., & Balasubramanian, S. K. (2004). Public attitude toward agrobiotechnology: The mediating role of risk perceptions on the impact of trust. Awareness and outrage. *Review of Agricultural Economics*, *26*(2), 186–208. doi:10.1111/j.1467-9353.2004.00170.x

Nap, J. P., Metz, P. L. J., Escaler, M., & Conner, A. J. (2003). The release of genetically modified crops into the environment. Part I. Overview of the current status and regulations. *The Plant Journal*, *33*(1), 1–18. doi:10.1046/j.0960-7412.2003.01602.x PMID:12943538

Napier, T. L., Tucker, M., Henry, C., & Whaley, S. R. (2006). Consumer attitudes toward GMOs: The Ohio experience. *Journal of Food Science*, *69*(3), 79–76. doi:10.1111/j.1365-2621.2004.tb13344.x

Nayga, R. M., Fisher, M. G., & Onyango, B. (2006). Acceptance of genetically modified food: Comparing consumer perspectives in the United States and South Korea. *Agricultural Economics*, *34*(3), 331–341. doi:10.1111/j.1574-0864.2006.00129.x

Nicolia, A., Manzo, A., Veronesi, F., & Rosellini, D. (2014). An overview of the last 10 years of genetically engineered crop safety research. *Critical Reviews in Biotechnology*, *34*(1), 77–88. doi:10.3109/07388551.2013.823595 PMID:24041244

Onyango, B. (2004). Consumer acceptance of genetically modified foods: The role of product benefits and perceived risks. *Journal of Food Distribution Research*, *35*(1), 154–161.

Onyango, B., Ferdaus, H., Hallman, W., Schilling, B., & Adelajan, A. (2003). Public perception of food biotechnology: Uncovering factors driving consumer acceptance of genetically modified food. *Journal of Food Distribution Research*, *34*(1), 37–42.

Onyango, B., Govindasamy, R., & Nayga, J. R. M. (2004). *Measuring US consumer preferences for genetically modified foods using choice modelling experiments: the role of price, product benefits and technology*. Food Policy Institute.

Pallant, J. (2016). *SPSS Survival Manual: A step by step guide to data analysis using IBM SPSS*. McGraw-Hill.

Prati, G., Pietrantoni, L., & Zani, B. (2012). The prediction of intention to consume genetically modified food: Test of an integrated psychosocial model. *Food Quality and Preference*, *25*(2), 163–170. doi:10.1016/j.foodqual.2012.02.011

Rodriguez-Entrena, M., Salazar-Ordonez, M., & Sayadi, S. (2013). Applying partial least squares to model genetically modified food purchase intentions in southern Spain consumers. *Food Policy*, *40*, 44–53. doi:10.1016/j.foodpol.2013.02.001

Salazar, E., & Montenegro, G. (2009). Genetically modified crops in Chile. *Ciencia e Investigación Agraria*, *36*(3), 353–368. doi:10.4067/S0718-16202009000300003

Savadori, L., Savio, S., Nocotra, E., Rumiati, R., Finucane, M., & Slovic, P. (2004). Expert and public perception of risk from biotechnology. *Risk Analysis*, *24*(5), 1289–1299. doi:10.1111/j.0272-4332.2004.00526.x PMID:15563295

Scholderer, J., & Frewer, L. J. (2003). The biotechnology communication paradox: Experimental evidence and the need for a new strategy. *Journal of Consumer Policy*, *26*(2), 125–157. doi:10.1023/A:1023695519981

Siegrist, M. (2000). The influence of trust and perceptions of risk and benefit on the acceptance of gene technology. *Risk Analysis*, *20*(2), 195–203. doi:10.1111/0272-4332.202020 PMID:10859780

Siegrist, M., & Cvetkovich, G. (2000). Perception of hazards: The role of social trust and knowledge. *Risk Analysis*, *20*(5), 713–719. doi:10.1111/0272-4332.205064 PMID:11110217

Siegrist, M., Cvetkovich, G., & Roth, C. (2000). Salient value similarity, social trust, and risk/benefit perception. *Risk Analysis*, *20*(3), 353–362. doi:10.1111/0272-4332.203034 PMID:10949414

Steur, D. H., Gellynck, X., Storozhenko, S., Liqun, G., Lambert, W., Van Der Straeten, D., & Viaene, J. (2010). Willingness-to-accept and purchase genetically modified rice with high folate content in Shanxi Province, China. *Appetite*, *54*(1), 118–125. doi:10.1016/j.appet.2009.09.017 PMID:19815041

Tanius, E., & Seng, S. W. (2015). Consumer awareness towards genetically modified (GM) foods. *International Journal of Business, Economics and Law*, *6*(2), 17–26.

Tavakol, M., & Dennick, R. (2011). Making sense of Cronbach's alpha. *International Journal of Medical Education*, *2*, 53–55. doi:10.5116/ijme.4dfb.8dfd PMID:28029643

Traill, W. B., Jaeger, S. R., Yee, W. M. S., Valli, C., House, L. O., Lusk, J. L., Moore, M., & Morrow, J. L. (2004). Categories of gm risk-benefit perceptions and their antecedents. *AgBioForum*, *7*, 176–186.

Traill, W. B., Yee, W. M. S., Lusk, J. L., Jaeger, S. R., House, L. O., Morrow, J. L., Valli, C., & Moore, M. (2006). Perception of risks and benefits of GM foods and their influence on willingness to consume. *Food Economics*, *3*(1), 12–19.

U.S. Department of Agriculture. (2016), *Bangladesh Agricultural Biotechnology Annual 2016*. https://www.agcensus.usda.gov/Publications/

Veeman, M., Adamowicz, W., & Hu, W. (2005). *Risk perceptions, social interactions and the influence of information on social attitudes to agricultural biotechnology*. [Unpublished manuscript, Department of Rural Economy, Faculty of Agriculture and Forestry, and Home Economics, University of Alberta, Edmonton, Canada].

Verdurme, A., & Viaene, J. (2001). Consumer attitudes towards GM food: Literature review and recommendations for effective communication. *Journal of International Food & Agribusiness Marketing*, *13*(2-3), 77–98. doi:10.1300/J047v13n02_05

Verdurme, A. and Viaene, J. (2003). Consumer beliefs and attitude towards genetically modified food: Basis for segmentation and implications for communication. *Agribusiness, 19*(1), 91-113.

Vilella-Vila, M., Costa-Font, J., & Mossialos, E. (2005). Consumer involvement and acceptance of biotechnology in the European Union: A specific focus on Spain and the U.K. *International Journal of Consumer Studies*, *29*(2), 108–118. doi:10.1111/j.1470-6431.2004.00425.x

World Health Organization. (2017), *Food, Genetically Modified*. WHO. https://www.who.int/topics/food_genetically_modified/en/ (Accessed: 17 August 2019).

Yee, W. M. S., Traill, W. B., Lusk, J. L., Jaeger, S. R., House, L. O., Moore, M., Morrow, J. L. B., & Valli, C. (2008). Determinants of consumers' willingness to accept GM foods. *International Journal of Biotechnology*, *10*(2/3), 240–259. doi:10.1504/IJBT.2008.018356

Yue, C., Zhao, S., Cummings, C., & Kuzma, J. (2015). Investigating factors influencing consumer willingness to buy GM food and nano-food. *Journal of Nanoparticle Research*, *17*(7), 1–19. doi:10.100711051-015-3084-4

Zhang, X., Huang, J., Qiu, H., & Huang, Z. (2010). A consumer segmentation study with regards to genetically modified food in urban China. *Food Policy*, *35*(5), 456–462. doi:10.1016/j.foodpol.2010.04.008

Zhu, X., & Xie, X. (2015). Effects of knowledge on attitude formation and change towards genetically modified foods. *Risk Analysis*, *35*(5), 790–810. doi:10.1111/risa.12319 PMID:25693867

Zika, E., Papatryfon, I., Wolf, O., Gomez-Barbero, M., Stein, A. J., & Bock, A. K. (2007). *Consequences, opportunities and challenges of modern biotechnology for Europe*. Joint Research Centre of the European Commission.

Chapter 5
Formal and Informal Agricultural Markets in Sub-Saharan Africa

Unity Chipfupa
University of South Africa, South Africa

ABSTRACT

Agricultural marketing determines the extent of smallholder farmer commercialization. However, this needs effective agricultural markets. Evaluating these markets and their interaction with small farmers is critical in harnessing the potential from the rural farming sector. The chapter aims to evaluate formal and informal agricultural markets, highlighting opportunities for smallholder agriculture. The findings reveal that smallholder producers' interaction with formal markets is limited by a lack of knowledge and resources, poor safety and quality standards, and inconsistency in supply. However, they can benefit from opportunities in the formal markets through effective collective action arrangements. Informal markets are critical to smallholders' access to markets, and with legislation that seeks to address infrastructure, safety, and quality standards, they can be a conduit for small producers' access to formal markets. Some informality still exists in the supply chain of formal markets. Thus, informal markets remain important in agriculture marketing in Africa.

INTRODUCTION

The pathway to sustainable rural development and poverty reduction in Sub-Saharan Africa (SSA) is achievable through the commercialization of smallholder agriculture (Ogutu & Qaim, 2019). However, this is not possible without the development of effective agricultural markets (Barrett, 2008). The transformation of subsistence farming to market-oriented production is the backbone of economic development in the smallholder sector (Gc & Hall, 2020; Ogutu & Qaim, 2019; Olwande et al., 2015). The benefits of smallholder participation in the market on income security, food supply, employment creation and poverty reduction are well documented (Carletto et al., 2017; Muricho et al., 2017; Ogutu & Qaim, 2019). The effects of commercialized production also transcends the local economy and agricultural

DOI: 10.4018/978-1-6684-4780-2.ch005

sector, impacting economic growth in other sectors of the economy (Carletto et al., 2017). However, for this to happen it requires, among other things, a sustained active participation of smallholder farmers in lucrative agricultural produce markets (Gc & Hall, 2020).

Agricultural marketing, both produce and input marketing, plays an important role in the transformation of smallholder farming (Olwande et al., 2015). The focus of this paper is on produce markets, although this does not undermine the importance of the input markets. Agricultural produce marketing encompasses all activities involved in the movement of farm produce to the market/consumers, conducted by different players such as traders/hawkers, wholesalers, marketing agencies, exporters, and retailers, among others (Ghosh, 2013). The activities include postharvest operations, assembling, grading, storage, transportation, and distribution. The produce markets are further segmented into formal and informal markets, commonly evident in developing regions such as Sub-Saharan Africa (SSA). Formal agricultural markets are marketing activities that are regulated and monitored, that adhere to taxation and standards as stipulated by the authorities/government (Weng, 2015). The reverse is true for informal markets, they are neither regulated nor monitored by government. However, there is regulation and legitimacy in the informal markets but not by government (Vorley, 2013). The dynamics between these two market systems define the practical possibilities of growing or transforming smallholder agriculture in SSA.

The emergence of informal markets stems from several factors which include the unavailability of formal markets and the failure/unwillingness/inability of the existing formal marketing systems to serve/accommodate the needs of small-scale farmers (Olwande et al., 2015; Vorley et al., 2012). Critically, informal markets also emerge as a survival strategy for small-scale farmers in light of the competitive environment in the formal markets (Vorley, 2013). Smallholder farmers' inability to access formal markets threatens rural livelihoods, making informal markets important for their survival. Linkages do exist between the two marketing sub-systems, i.e., formal, and informal markets, though the relationship is biased in favor of formal markets. There is also competition for space and legitimacy among the two (Vorley et al., 2012). It appears formal produce markets in SSA thrive from the price inefficiencies inherent in the informal markets. In fact, in most cases the informal agricultural market presents as a sub-system of the formal markets, acting as a source of cheap agricultural produce (Vorley, 2013). The natural design of the two sub-systems tends to serve different clientele. Whereas the informal market mainly serves the smallholder farmers, the formal market is focused on large-scale farming and agro-business enterprises. The orientation and relationship between the formal and informal agricultural markets in SSA will determine the future of smallholder farmers in the region.

Several studies argue that without access to lucrative formal agricultural markets, the smallholder farming sector will remain small, unable to compensate the millions of small producers toiling for their survival (Fan & Rue, 2020; Olwande et al., 2015). Hence, governments' attitude towards informal markets is negative due to lack of organization, non-tax paying, un-hygienic and backward nature (Vorley, 2013). There is a drive to transform informal markets so that they can adhere to expected regulations and standards. However, these moves have been criticized as those that ignore or try to regulate informal markets out of existence (Vorley et al., 2012). The reality of the informal sector is often ignored in favor of the formal sector. However, the truth is that smallholder farmers use a combination of both formal and informal markets to sustain their livelihood. This means informal markets serve a specific and critical role in smallholder agriculture and hence should not be undermined. Rather the focus should be on how such markets can be supported to serve their clientele better.

The frictional dynamics between the formal and informal markets of SSA, where the narrative is that they are operating in tangent with each other (Vorley, 2013), might be slowly degrading the avenues of sustainable livelihoods for small-scale rural farmers. A critical analysis of these markets and their interaction with small-scale rural producers is therefore a first step in understanding and developing effective agricultural markets and harnessing the potential from the rural farming sector in Africa. The objective of this chapter is to evaluate formal and informal agricultural markets in SSA, highlighting their organization, linkages, challenges, and opportunities for the smallholder farming sector. Lessons are also drawn from Asian and Latin American countries experiences. In the end, the chapter discusses implications of the findings on smallholder producers' participation in markets and the future of agricultural marketing in SSA.

CONTEXT OF THE PROBLEM

Smallholder agriculture remains the most common and important form of farming in SSA (Yaseen et al., 2018). There are about 33 million small-holder farms in Africa which account for 80% of all farms on the continent (NEPAD, 2013). About 9% of the world's 450-500 million small-holder farms are found in SSA (Conway, 2014; Lowder et al., 2016). Small-holder farms are defined as those that operate not more than 2 hectares of land (Fan & Rue, 2020). About 60% of rural household livelihoods in the region survive on smallholder farming (Siphesihle & Lelethu, 2020). The sector employed over 176 million people in SSA (Proctor & Lucchesi, 2012) and was also responsible for 80% of food produced in the region (Lowder et al., 2014).

The problem is that smallholder agriculture is inherently affected by numerous challenges affecting its growth prospects. Chief among these is the continued limited access and low participation of smallholder farmers in formal produce markets (Rantlo et al., 2020). This can be observed, albeit at different levels, across countries south of the Sahara. Only 1-2% of small-holder farmers are producing commercially, 3-15% regularly sell in the markets and the remainder are subsistence farmers some of who occasionally sell their surplus (Vorley et al., 2012).

Studies suggest that this status quo is a result of weak policies, lack of infrastructure, lack of institutional support, lack of market orientation on the part of farmers, and unwillingness of the farmers to transform, among others (Carletto et al., 2018; Yaseen et al., 2018). Hence, the transaction cost of accessing formal markets is usually steep and beyond the reach of smallholder producers. Consequently, most smallholder farmers' produce is sold at roadsides or to informal traders/hawkers at low prices. Therefore, farmers retain a small portion of the residual value of their produce. The greater part is captured by middlemen and agencies who informally buy from the farmers for sell in the formal markets (Fan & Rue, 2020).

In the past three decades, Sub-Saharan Africa has experienced reformation of agricultural markets. The reforms mostly took the form of market liberalization of agricultural food markets which were predominately controlled by state parastatals. The intention was to open the markets for more involvement of private companies and increase the producer prices to parallel market levels (Sitko & Jayne, 2012). In their book, *Reforming Agricultural Markets in Africa*, Kherallah *et al.* (2002) indicate that the move resulted in more private sector entry into the food market. However, the lack of access to credit and uncertainty in reforms affected the operations of the private sector. In response, in most African countries, the state has continued to intervene in the market leading to uncertainty and speculative behavior by the private sector. Market liberalization also led to an increase in price transmission inefficiencies. Though the urban markets were more integrated, lack of information and poor transport infrastructure meant that rural markets remained segmented. Liberalization also reduced marketing margins due to low consumer prices (Sitko & Jayne, 2012), resulted in high volatility in food market prices (Sarris & Morrison, 2010) and led to the emergence of informal traders whose operations are affected by their small-scale nature, education levels and access to resources. It is within this context that the growth of the informal sector in agricultural marketing has been observed.

METHODOLOGY

The study was qualitative in nature employing a desktop approach to reviewing studies and books from academic and research institutes including those from international organizations with practical experience of supporting agricultural marketing in Africa. A literature search was conducted on Google Scholar, Scopus, ScienceDirect, Web of Science and websites of recognized organizations such as Food and Agriculture Organization of the United Nations, Alliance for a Green Revolution in Africa and International Institute for Environment and Development, with an interest in agricultural marketing issues in Africa. The scope of the search was SSA although inferences were made from experiences in other regions. Scanning was used to identify relevant issues in the literature, which were analyzed through thematic content analysis.

THE MARKETING CONCEPT

The marketing concept in agriculture is best understood within the tenets of existing economic theories such as the spatial models and price theory, among others. These offer the foundation for understanding how markets develop and operate (Bateman, 1976). The marketing concept favors market rather than product orientation (Yeh, 2016). It supports the idea of market segmentation, placing much emphasis on products that meet customer needs. Small-holder farmers have minimum interactions with the final consumers of their products. Much of their interactions are with middlemen or marketing agencies, who themselves introduce inefficiencies in the agricultural value chain (Magesa et al., 2020). To succeed, a farmer should be able to adapt to new markets and/or modify their product for existing markets. In doing so, the farmer should balance the needs of all the players in the value chain. However, the application of the marketing concept in smallholder agriculture is limited by the size of their enterprises and the biological nature of their production (Bateman, 1976). Nevertheless, structural adjustments such as cooperation can make it easier to apply ideas from the marketing concept in the sector.

Farmer marketing decisions are complex because they involve several independent decisions made simultaneously. In these decisions, the farmer should find an optimal mix of the price, product, marketing channel, and promotion that gives them the best result (Bateman, 1976). However, decision making in the real world is often made under uncertainty which affects the ability of humans to adopt the optimal choices/decisions (Simon, 1997). Likewise, smallholder farmers' marketing decisions are boundedly rational because they are often made without the full knowledge of the possible alternatives and consequences of each alternative (Wale & Chipfupa, 2021).

Price mechanisms are key in the transmission of marketing information (Bateman, 1976). Any instability in prices negatively affects this role. Hence, there is a lot of interest in government policy instruments used to address price instability such as price controls, buffer stocks and futures markets, among others. To make profit one ought to sell their produce where marginal costs is equal to marginal revenue

(Colman & Young, 1989). However, information asymmetry and trade-offs between the profit maximization goal and other welfare goals affect the achievement of this optimizing rule by smallholder farmers. Moreover, their marginal costs are often higher than their marginal revenue because, in addition to other inefficiencies brought by poor access to technology and lack of knowledge, they often sell their produce at a lesser price than they ought to sell. Bateman (1976) suggests that low prices in the produce market are often associated with inefficiencies in distribution, low bargaining power and limited value addition.

Cooperation is also critical in agricultural marketing. The plausible conditions for cooperation include opportunities for price bargaining or vertical integration (Ishak et al., 2020; Valentinov, 2007). Spatial challenges involving the physical distribution of agriculture produce across different geographical areas makes cooperation an interesting idea. Smallholder farmers are usually located away from the formal markets or consumers of their products. Cooperation, thus, presents an opportunity for them to access formal markets, bargain for better prices and reduce transaction costs (Francesconi & Wouterse, 2019). The theory of countervailing power suggests that cooperation among farmers can also develop as a countervailing response to oligopsony power on the other side of the market (Galbraith, 1993; Snyder, 1996). Still, this arises from the desire to be competitive and have more negotiating power in the market.

Lastly, marketing decisions are not made in a vacuum. They are made within a given marketing environment, the most important being government policy. Government policy defines the legislative environment in the market within which farmers should operate, i.e., regulations, standards, and permits/licenses, among others. Any failures in the policy affect the growth and development of agricultural markets.

FORMAL AGRICULTURAL MARKETS IN SUB-SAHARAN AFRICA

The formal agricultural produce markets in SSA are often characterized by many players. However, in most cases, the markets resemble an oligopsony, with many sellers and few buyers. Most of these markets are integrated such that they offer access to domestic, regional, and international buyers. The most common are commodity exchange markets, marketing boards, supermarkets, formal urban food markets and online markets.

Commodity Exchange Markets

A commodity exchange is a market, where various commodities are traded between buyers and sellers including agricultural produce, according to the regulations of the exchange (Gomez & Vossenberg, 2018). These markets were established from the desire to reduce marketing costs, increase market transparency, reduce uncertainty, and support price discovery in marketing transactions (Gomez & Vossenberg, 2018; Rashid et al., 2010). The exchanges are also considered as a strategy for empowering farmers. Commodities exchange markets are found in most SSA countries, though the level of operation, functionality and effectiveness differs across countries (Rashid, 2015). Examples include the South African Futures Exchange (SAFEX), Johannesburg Stock Exchange (JSE) Commodity Derivatives Market, Ghana Commodity Exchange (GCX), Ethiopia Commodity Exchange (ECX), Tanzania Mercantile Exchange (TMX), Nigeria Commodity Exchange, and Kenya Agricultural Commodity Exchange, among others. The exchanges are either run by a private company, government, or a combination of the two. The most common agricultural produce traded on these markets include maize (white and yellow), soyabeans, wheat, rice, sorghum, beans, sunflower, beef, lamb, and wool, among others. The products differ by country depending on their comparative advantage in producing the commodity.

Commodity exchange markets are highly regularized, and one should meet a given criteria to be registered in the exchange. Trading involves complex bargaining and brokering; hence, it is mostly conducted by registered brokering agents. A certificate of training in the exchange is required for one to start trading. This training is offered by the exchanges themselves. The JSE in South Africa provides a list of registered brokers on their website that one should contact if they are interested in selling or buying commodities in the exchange (JSE, 2022). However, this comes at a cost. In Ghana, the GCX has different types of membership, i.e., trading member, broker member and associate member. The first two require one to have a registered company, tax/VAT certificate, audit report and net worth between GHS 50,000-80,000 (USD 6,667- 10,667). The Associate member is a free special membership for smallholder farmers and traders which allows them to buy or sell only two non-complex commodities (maize, rice, soya, millet, sorghum). However, the membership is only available for 12 months (GCX, 2022). Under the ECX, a farmer can trade as a client through a member of the exchange subject to meeting the specified requirements. The brokerage fee for such transactions does not exceed 2% of the traded value. One can also trade on this exchange as a non-member direct trader without representation.

The allowed categories are smallholder farmer, supplier, and exporter. However, trade is only for oneself and there are no transferable rights to trade on the exchange (ECX, 2022). According to TMX (2022), the Tanzania Mercantile Exchange operates an online trading system that allows trading between sellers and buyers of agricultural produce. The system does not have a futures exchange but accommodates spot exchange including auctioning. Smallholder farmers are linked to the exchange through the Agricultural Marketing Cooperative Societies (AMCOS) which collect farmers' produce and transport it to licensed warehouses for depositing and quality testing. After the trades, farmers' payments are also implemented through the same cooperatives. However, several studies have indicated the ineffectiveness of agricultural cooperatives in Africa (Francesconi & Wouterse, 2019; Ortmann & King, 2007).

Available evidence shows mixed results in the performance of commodity exchanges in SSA. While emerging economies like South Africa have performed better in their exchanges, other SSA countries such as Malawi, Nigeria, Uganda, Zambia, and Zimbabwe, among others have struggled and even stopped operations (Mezui et al., 2013; Rashid, 2015). Most of those still in operation are supported by the government which is not sustainable. The major challenges have been political interference, weak regulatory systems, unwillingness of financial institutions to support exchange systems and poor viability conditions e.g. very small trade volumes (Emeana, 2017; Rashid, 2015; Sitko & Jayne, 2012). While there has been claims that commodity exchanges such as the ECX have strengthened price relationships, improved smallholder farmers access to global markets and increased export earnings, other studies did not find evidence to that effect (Hernandez et al., 2017; Rashid, 2015). However, there is consensus that there has been some positive impact on smallholders' access to local markets and improved standards/ quality of produce.

Marketing Boards

In many SSA countries there are national marketing boards that are responsible for establishing marketing practices, regulate agreed practices in a market and facilitate credit, risk management and market intelligence (Crawford, 1997). The boards are public entities established as part of government measures to correct market imperfections that disadvantage producers especially poor and small producers (Dlamini et al., 2017). They focus mainly on key or national strategic crops/products such as maize, rice, coffee, tobacco, sugarcane, milk, and beef, among others. The most popular are grain marketing boards (GMB) that continue to play a significant role in the grain market in eastern and southern African in countries.

Most marketing boards have buying operations, often acting as a government-enforced marketing monopoly for a given crop/product (Dumais, 2012). They usually set a fixed price in advance so that farmers know the output price prior to or before planting, the output price, though not always the case. The setting of prices in advance helps farmers make informed farming decisions. There is evidence that the operations of marketing boards can result in expanded crop production, increased access to the market and increased incomes for smallholder farmers (Mason et al., 2015; Mather & Jayne, 2019). However, other studies have shown otherwise or mixed results (Ayinde, 2014; Dlamini et al., 2017). What has affected the impact of marketing boards in SSA has been inefficiencies and corruption from government officials. For example, in Zimbabwe the GMB has been affected by a myriad of inefficiencies and corruption from corrupt government officials. The same concerns are also observed in other SSA countries.

Supermarkets/Food Chain Stores

The past two decades have experienced growth in supermarkets in Africa especially in Eastern and Southern Africa due to increasing incomes, urbanization and diversifying consumption patterns (Balineau et al., 2021; Chikazunga et al., 2007). Supermarkets have rapidly risen to become a major market for agricultural produce in Africa, Asia and Latin America (Balineau et al. 2021; Reardon, Timmer, and Minten 2012). They improve the range and quality of produce, increase price transparency and reduce transaction cost through integration of the supply chain (Balineau et al., 2021). The literature alludes that supermarkets address the needs of the middle-class consumer regarding quality, hygiene and variety, and are emerging in the affluent and wealthy neighborhoods of the main cities in SSA countries (Aderghal et al., 2019; Balineau et al., 2021). However, the Southern Africa experience shows that market segmentation has also seen the emergence of supermarkets designed to serve the lower-class population in high density suburbs or poor communities.

As early as 2003, Weatherspoon and Reardon (2003) wrote that supermarkets in Africa had taken over the food sector and were controlling about 55% of the national food retail in South Africa. An almost similar situation existed in Kenya. The influence extended to international trade where European supermarkets were also procuring raw and fresh cuts of fruits from Africa. Chikazunga et al. (2007), found that in South Africa four supermarket giants (Pick n Pay, Shoprite, Spar and Woolworths) controlled about 94.5% of the food market. The same supermarkets have spread throughout Africa. For example, Shoprite now has presence in 13 African countries (Eastern, Southern and Western Africa) while Pick n Pay is in 7 countries mainly in Southern Africa. However, the footprint of supermarkets in West Africa is limited when compared to other regions in SSA.

Supermarkets have transformed the procurement system of agricultural produce from a decentralized system involving traditional wholesalers to a centralized system through specialized wholesalers or own distribution centers (Balineau et al., 2021). For example, Pick n Pay has the PnP Distribution Centre, Spar uses the Spar Distribution Centre and Shoprite and Checkers use the Fresh mark Distribution Centre. These centers procure at least 90% of their agricultural supply directly from selected growers under contract whilst the remainder is purchased from the Fresh Produce Market (Chikazunga et al., 2007). The system is highly consolidated and demands a high level of quality, safety standards and coordination between producers, retailers and intermediaries (Weatherspoon & Reardon, 2003). For supermarkets such as Woolworths that deal with organically produced products, there is a need for national and international certification, for example, through the South African Organic Sector Organization Standards and the International Federation of Organic Agricultural Movements (IFOAM) standards, respectively. This certification, which exist at different levels (Second and Third Party) depending on the length of the supply chain, comes at fee and is costly (Tung, 2016).

Supermarkets demand a large constant supply of fresh produce. Hence, their distributors are interested in farmers who can constantly supply large volumes of produce. As much as this is an opportunity, it is a challenge for small producers who are limited both in space and other resource capacity. Further, smallholder producers find it difficult to meet the high levels of quality and standards demanded in this market. The knowledge and cost of investing in infrastructure and personnel to improve product quality is beyond most farmers. Thus, they miss out on the rent that they could obtain from supplying supermarkets.

Formal Urban Food Markets

Formal urban food markets are found in urban areas, mainly in big cities and towns. Most are established by legislature, and regulated, and managed as private entities of the local authorities or municipalities. They are designed as a meeting place between producers and consumers of perishable agricultural produce, allowing equal trading opportunities between large-scale and small-scale producers (Louw, 2008; NAMC, 2007). The markets are not driven by profit but by the need to provide a service to farmers (both small-scale and large-scale), industry and consumers at large. Formal urban food markets demand some level of standard and quality in agricultural produce delivered to them. However, they are only found in a few countries in SSA. South Africa is one country with a well-established network of National Fresh Produce Markets, the biggest being the Johannesburg, Durban, Capetown and Pretoria Fresh Produce Markets. Zambia has also recently established the Zambia Fresh Lusaka Market, which is operated by Savana Capital, a private company involved in logistics and supply chain management.

Participation in the formal urban markets is mainly through registered marketing agencies and wholesalers, who through their sales agents, sell produce on behalf of the farmers at a commission of between 5% - 7.5% of the value of the sold product. The markets mainly deal with perishable horticultural produce. As such farmers supplying the market need access to packhouses with cold storage facilities and packaging to preserve the quality of their produce before delivering to the market. This is a challenge for most smallholder farmers who have limited access to such facilities. The situation is made worse by the fact that urban food markets are located far from the rural communities which makes transportation a key component of the value chain. According to Haggblade (2020), the growth in cities across Africa has resulted in long value chains with expanded distributions systems and increased storage, warehousing, processing and packaging. Though this expansion avails opportunities for other entrepreneurs and agribusiness companies it has challenges for poor farmers who cannot absorb the increased transaction cost of accessing the urban markets (Reardon et al. 2019).

The formal urban food markets specialize in bulk buying and supply. The clientele represented by the marketing agents such as supermarkets, government departments (e.g., prisons, hospitals, school feeding programS), exporters and processors, among others, require a continuous bulk supply of different agricultural produce. However, in South Africa, there is legislation that promote equitable access of smallholder producers to these markets. The challenge, however, is that the volumes from the smallholder producers are too low and the quality of their produce is not always up to the required standards. Hence, there are efforts to organize smallholder producers into thriving cooperatives that facilitate easy access to the formal urban markets (Louw, 2008). Farmwise Grain in South Africa is an example of such an entity established by farmers as a procurement and selling agent linking farmers with buyers of their products. There is also evidence that informal fresh produce traders also source their fresh produce from the fresh produce markets (NAMC, 2007).

The transformation in the procurement system of large supermarkets has threatened the very existence of formal urban food markets. In South Africa, the volume of trade through the National Fresh Produce Markets has declined in the past two decades. For example, supermarkets now procure less than 10% of their fresh produce through the Fresh Produce Markets (Chikazunga et al., 2007; Louw, 2008). The steady improvement in the road networks and infrastructure has also made it easier to access other marketing channels (Haggblade, 2020). Still, formal urban food markets play a critical role in the marketing of fresh produce in SSA.

Online Marketing of Agricultural Produce

E-commerce in the marketing of agricultural produce in SSA is a new phenomenon and thus it has not been widely adopted. However, there is so much potential from online platforms that offer a virtual meeting place for buyers and sellers of agricultural produce especially for smallholder farmers (Barends-Jones, 2020). E-marketing reduces transaction costs and marketing time, and removes the middlemen thereby increasing the price that farmers receive for their produce (Joiner & Okeleke, 2019; Magesa et al., 2020). The examples of such spaces include Hellochoice South Africa (https://www.hellochoice.co.za/), OLX Potato Project in Kenya, TruTrade Africa in Kenya and Uganda (http://www.trutradeafrica.net/), Izyshop in Mozambique, among others (Shrader, 2018). Most of the current platforms are privately operated and participation is at a cost unless the platform is donor funded.

Training in the use of online platforms is important so that one does not enter shady deals since there are no face-to-face negotiations. Currently, online agricultural marketing platforms in SSA are mostly used by large-scale commercial farmers and less by smallholder or emerging farmers due to several factors including lack of knowledge, lack of trust in the system, connectivity challenges and high internet data cost in Africa (Barends-Jones, 2020). Other challenges with use of online marketing platforms include security concerns especially with personal information, quality disputes since produce is not physically inspected before purchase, possible late delivery due to unanticipated events and a general unwillingness of the population to change from their traditional marketing practices (Barends-Jones, 2020; Kumar & Timalsina, 2016). Participation in online agricultural markets has several benefits which include access to diverse and larger markets, removal of barriers to entry, reduction in post-harvest losses, more income, potential for value-addition, growth in other services linked to the marketing platform such as infrastructure and transport (Barends-Jones, 2020; Joiner & Okeleke, 2019).

INFORMAL AGRICULTURAL MARKETS IN SUB-SAHARAN AFRICA

Most agricultural produce from smallholder farmers in SSA is marketed through informal agricultural markets. The markets provide platforms upon which those who both cannot afford to participate in the formal system (both buyers and consumers) can meet. Studies conclude that informal markets will remain an important agricultural marketing channel for smallholder farmers in the future (Resnick, 2017; Vorley et al., 2012). This is despite the threat that the sector is facing from market consolidation due to the growth of supermarkets in SSA. Statistics indicate that about 70% of the urban poor in SSA obtain their food from the informal markets (Resnick, 2017). However, this differs from country to country depending on the level of development and incomes. Households in countries such as South Africa purchase their food mostly from the supermarkets while those from countries such as Kenya, Mozambique, Nigeria and Zambia purchase most of their food from traditional markets and street vendors (Hannah et al., 2022; Resnick, 2017; Tschirley et al., 2004).

Informal agricultural markets are defined as those selling legitimately produced goods and services but do not necessarily follow formal processes such as standards regulations, business registration or operational licenses. This status has often put them at loggerheads with authorities who also have concerns regarding food safety in such markets (Resnick, 2017). The examples of informal agricultural markets in SSA include the traditional markets (also known as open air markets), middlemen, roadside vending, local shops ('spaza shops' in South Africa) and hawkers.

Traditional Markets or Open-Air Markets

Traditional markets are found all over across Africa and are a common feature of every community or city, especially in West and Eastern Africa. They are regarded as markets for the poor. In some countries they are more regulated than others in that there is some entity that exercises authority over the market. Most trading is outdoors and in the open space, hence the name open-air market. The infrastructure is mainly makeshift stalls and booths except in some cities where authorities have tried to build permanent stalls. However, even in such cases, due to the large number of traders, there are always some makeshift structures. The size of traditional markets differs depending on the geographical location. Those in the rural areas are small with fewer traders, while those in the cities can be huge with thousands of shops, stalls, and traders. Though many wares are sold in traditional markets there is always a designated area for agricultural produce and products. Examples of some of the traditional markets in SSA include Karatina in Kenya, Merkato in Ethiopia, Kejetia in Ghana, and Mbare Musika in Zimbabwe. A recent study of 18 cities in Kenya and Zambia concluded that open air markets will remain a feature of most urban areas in the future due to the preference and demand from low-middle income consumers (Hannah et al., 2022).

In many instances, participation in traditional markets is open to everyone for a fee that is paid to the managing authority. However, the fee is less compared to rentals one would pay in a formal market. There are no costs of maintaining or securing the facilities borne by the traders. Usually water and sanitation are a challenge in some of the open air markets and hence, the food safety concerns raised by health officials regarding food items sold in such markets (Kang'ethe et al., 2020; Resnick, 2017; Vorley et al., 2012). Likewise, there are no stipulated standards or quality requirements for the produce or products delivered to the market. Anyone can bring their produce for sale, regardless of its quality. Pricing is quite competitive because of the many buyers and sellers.

Middlemen (Also Known As Van Traders)

Middlemen are a key and common feature in the marketing of produce from smallholder farmers in most SSA rural farming communities. They are men and women with bakkies or vans who buy farmers' produce from the farms for resale in markets in towns and cities. They provide a conduit between the smallholder farmers and agricultural markets far away from the rural areas. However, these value chain actors have been accused of a rent seeking behaviour, i.e., benefitting at the expense of farmers whom they pay very low prices compared to what they obtain in the market (Abebe et al., 2016; Magesa et al., 2020). It is said they prey on the farmers' ignorance regarding market information such as prices, products, and supply. However, despite their negatives, middlemen play an important role in agricultural marketing in SSA (Sudrajat et al., 2021). Though they do not add value to the produce, their intermediary role in the distribution of agriculture produce has helped to link farmers to markets that they would otherwise have no opportunity of reaching.

Roadside Vending

Roadside vending is a very important agricultural marketing channel across SSA, especially for poor farmers. It is usually a sign of a lack of access or inability of smallholder farmers to participate in markets far away from their farms. Some vendors construct makeshift stalls along the highways and busy roads while others sell their agricultural produce to motorists and travelers in buses, trucks and carts. Others sell in the open while some have no fixed location, moving up and down the busy roads selling their produce (Munuo, 2013; Wanyama et al., 2019). Roadside vending can be seasonal or all-year depending on the crops being sold. Usually, the vendors are close to their farms and hence others do not display their produce at the roadside but have a signpost directing customers to their farm. The traded commodities include vegetables, fruits, grains, and crafts, among other wares. Some of the produce (e.g., maize and groundnut) is sold raw or processed (boiled/roasted). Since fresh produce is perishable, at times farmers are forced to sell their produce at a give-away price. Like open air markets, food safety is also a challenge with roadside vending. Rarely would one find proper water and sanitation (WASH) facilities around the vending area except when the vendors are near a shopping center, school, clinic or some other public infrastructure or home.

Small Local Grocery Shops (Mini Supermarkets)

Small local grocery shops in rural communities are an alternative marketing channel for fresh produce. The distance from the wholesale markets makes it convenient for the shops to deal directly with local farmers. Smallholder farmers and the shop owners enter into non-binding contracts or agreements to supply vegetables and other horticultural produce. The arrangements are informal and hence are open to abuse. Prices are negotiable on delivery and at that point the farmer has limited or no other alternative for selling their produce. Fresh produce can be rejected on delivery based on quality. The volumes are usually small and hence payments can be immediate or within a week or two. Farmers incur minimum costs of transport compared to those for accessing other markets away from their communities.

Hawkers

Sometimes smallholder farmers resort to hawking as a marketing strategy for selling their fresh produce and other products. They can either be hawkers themselves or they sell their produce to hawkers. The challenge is that the nature of hawking limits the volume of produce that can be sold using this channel. Hawkers move around communities selling their products. They can either be on foot, or bicycle/motorcycle. However, their carrying capacity limits the products they sell. Some of the fresh produce sold by hawkers include vegetables, fruits, green mealies, pulses, milk, and fish, among others.

EXPERIENCES FROM ASIA AND LATIN AMERICA

Productive Alliances as a Form of Collective Action Arrangement

Productive Alliances (PA) is an approach widely implemented in Latin American countries and supported by the World Bank Agriculture Global Practice to link transitional/emerging smallholder farmers to markets. It is an example of how to support smallholder farmers to expand into modern agricultural value chains. Unlike situations where smallholder cooperatives are formed without any support or knowledge on how collective action arrangements work, the PA approach provides technical assistance, productive investments, and business development activities to the smallholder producers. These are the issues that have inherently affected the success of smallholder farmers in SSA. The PA brings together and strengthens the relationship and linkages between producers, buyers, and the public sector. Within the PA project, business plans are developed on promising value chains, evaluated, and financed by public sector grants. A critical element of the PA approach is that smallholder producers are required to contribute about 30% of the grant requirement (in cash or kind) which increased ownership and success of the projects. Several successes were registered in PA projects including 20-60% increase in sales, 30% increase in average net income of producers, stable prices, improved product quality and sustainable long-term vertical alliances between producers and buyers (World Bank, 2016). This is a testimony to the potency of the approach. Another lesson from the Latin American experiences is that the commercial financial sector is reluctant to support smallholder farmers because of their policies, lack of collateral and lack of coverage of financial services in the rural areas. This points to the need to find strategies for leveraging commercial financing for smallholder projects (World Bank, 2016).

Effect of the Growth of Supermarkets on Informal Agricultural Markets

There is consensus in literature that supermarkets have constricted the informal agricultural sector and hence threatened smallholder farmers with no capacity to access the formal markets (Weatherspoon & Reardon, 2003). This skepticism about the growth of supermarkets in SSA have led to a constrained relationship between formal and informal agricultural markets. However, the experience from Asia and Latin America, where the supermarket boom was much earlier than that in SSA, eases such fears. For example, in Colombia, supermarkets have not displaced informal markets but coexist with them. Traditional informal markets continue to serve the thousands of informal shops and stalls, themselves being serviced by small-scale producers. Despite establishing their own procurement systems the supermarkets still obtain about 10% of their supply from traditional markets (Guarín, 2013). Similarly, in Indonesia modern retailers procure between 40-50% of their fresh produce from local specialized suppliers who obtain their supplies from small-holder farmers (Vorley et al., 2012). In China, the modern retail market is serviced by millions of small-holder farms through a coordinated network of small-scale traders (Huang et al., 2009). These examples demonstrate that, with better organization and coordination, the relationship between supermarkets and the informal agricultural markets in SSA can be symbiotic. Smallholder producers can benefit from the supermarket boom in SSA.

CONCLUSION AND POLICY IMPLICATIONS

The main aim of the chapter was to evaluate the formal and informal agricultural markets in SSA and their linkages with the smallholder farming sector. Despite some provisions in some formal agricultural markets to accommodate small-scale farmers, their access to such markets is mainly limited by lack of knowledge, limited access to resources (including technology), poor food safety standards, poor quality of produce and inconsistency in supply. These limitations affect their ability to meet the standards of the clientele serviced by the formal agricultural markets. Opportunities for smallholder farmers to participate in formal markets can be enhanced through effective collective action arrangements. Standards, certification, and cost-effective mechanisms of accessing the formal markets are best achieved through such arrangements. The Latin American countries' experiences with the PA approach could offer lessons to SSA governments on effective strategies for supporting smallholder cooperatives whilst promoting ownership and sustainability.

There is no denying that despite their challenges, informal markets are integral to the survival of the smallholder producers in SSA. Though it is important to promote small-scale farmers access to formal markets, regulating informal markets out of existence will do more harm than good to the smallholder farming sector, especially given the obvious unaddressed challenges of integrating smallholder producers in the formal markets. Thus, policies should not only promote formalization, but legislation should be inclusive of the informal markets to address issues of safety, quality standards and support. In doing so, policymakers will ensure that markets that service smallholder producers are not bypassed by the formal retail sector. This will create more opportunities for smallholder farmers to benefit from the exponential growth in the retail market sector. Deliberate efforts should be made to promote a symbiotic relationship between the formal and informal agricultural markets.

Smallholder farmers' relationship with middlemen is mostly detrimental because farmers lack market information and hence bargaining power in the price negotiations. Without any intervention from the gatekeepers, farmers will continue to lose their hard-earned income to gullible middlemen. Policymakers should put measures that seek to protect farmers such as seasonal price setting. Promotion of e-commerce in the sector might also improve informational capabilities of smallholder farmers. Currently, there are minimum efforts to explore this form of marketing by smallholder farmers in SSA. It is high time that investments were made in online marketing in the smallholder sector. Addressing challenges of internet connectivity and cost of data will open a whole new world of marketing to millions of small-holder farmers in SSA. Going forward, SSA governments should take a more active role in supporting e-commerce in agriculture. They should work with UN agencies such as FAO and other international and local organizations who have projects that are trying to stimulate e-commerce in the sector.

REFERENCES

Abebe, G. K., Bijman, J., & Royer, A. (2016). Are middlemen facilitators or barriers to improve smallholders' welfare in rural economies? Empirical evidence from Ethiopia. *Journal of Rural Studies*, *43*, 203–213. doi:10.1016/j.jrurstud.2015.12.004

Aderghal, M., Lemeilleur, S., & Romagny, B. (2019). *Contribution des systèmes de distribution alimentaire à la sécurité alimentaire des villes: étude de cas sur l'agglomération de Rabat [Contribution of food distribution systems to food security in cities: case study on the agglomeration of Rabat].*

Ayinde, O. (2014). Is Marketing Board a Barrier or a Stimulant of Agricultural Production in West Africa? A Comparative Study of Ghana and Nigeria Cocoa Pricing Eras. *GJDS, 11*(2), 50–66.

Balineau, G., Bauer, A., Kessler, M., & Madariaga, N. (2021). *Food Systems in Africa: Rethinking the Role of Markets*. World Bank Publications. doi:10.1596/978-1-4648-1588-1

Barends-Jones, V. (2020). ECommerce Markets: Opportunities and Challenges for the Western Cape Agricul tural Sector. *Western Cape Department of Agriculture*. https://www.elsenburg.com/wp-content/uploads/2022/03/2020-E-commerce-Markets.pdf

Barrett, C. B. (2008). Smallholder market participation: Concepts and evidence from eastern and southern Africa. *Food Policy, 33*(4), 299–317. doi:10.1016/j.foodpol.2007.10.005

Bateman, D. I. (1976). Agricultural marketing: A review of the literature of marketing theory and of selected applications. *Journal of Agricultural Economics, 27*(2), 171–226. doi:10.1111/j.1477-9552.1976.tb00823.x

Carletto, C., Corral, P., & Guelfi, A. (2017). Agricultural commercialization and nutrition revisited: Empirical evidence from three African countries. *Food Policy, 67*, 106–118. doi:10.1016/j.foodpol.2016.09.020 PMID:28413250

Carletto, C., Corral, P., & Guelfi, A. (2018). Revisiting the Gains from Agricultural Commercialization. In L. Christiaensen & L. Demery (Eds.), *Agriculture in Africa: Telling Myths from Facts* (pp. 39–46). The World Bank.

Chikazunga, D., Joordan, D., Bienabe, E., & Louw, A. (2007). Patterns of restructuring food markets in South Africa: The case of fresh produce supply chains. *AAAE Second International Conference*, (pp. 53–55). AAAE.

Colman, D., & Young, T. (1989). *Principles of agricultural economics: markets and prices in less developed countries*. Cambridge University Press. doi:10.1017/CBO9780511623509

Conway, G. (2014). On being a smallholder. In P. Hazell & A. Rahman (Eds.), *New Directions for Smallholder Agriculture* (pp. 15–35). Oxford University Press. doi:10.1093/acprof:oso/9780199689347.003.0002

Crawford, I. (1997). *Agricultural and Food Marketing Management*. Food and Agriculture Organization of the United Nations.

Dlamini, T., Mohammed, M., & Sacolo, T. (2017). Assessing the Impact of National Marketing Boards on Consumers and Producers' Welfare. *SEPARC Working Paper 006/SZL*. Eswatini Economic Policy Analysis and Research Centre (ESEPARC).

Dumais, M. (2012). *The negative consequences of agricultural marketing boards*. Montreal Economic Institute.

ECX. (2022). *Membership*. ECX. https://www.ecx.com.et/Pages/Membership.aspx#BC

Emeana, K. N. (2017). Global value chain: the role of commodity exchange markets in linking smallholder farmers in Sub-Saharan Africa. *2nd Annual Research Conference of the Centre for African Research on Enterprise and Economic Development (CAREED)*. IEEE.

Fan, S., & Rue, C. (2020). The role of smallholder farms in a changing world. In S. G. y Paloma, L. Riesgo, & K. Louhichi (Eds.), The role of smallholder farms in food and nutrition security (pp. 13–28). Springer, Cham. doi:10.1007/978-3-030-42148-9_2

Francesconi, G. N., & Wouterse, F. (2019). Building the managerial capital of agricultural cooperatives in Africa. *Annals of Public and Cooperative Economics*, *90*(1), 141–159. doi:10.1111/apce.12218

Galbraith, J. K. (1993). *American capitalism: The concept of countervailing power* (1st ed.). Routledge.

Gc, R. K., & Hall, R. P. (2020). The commercialization of smallholder farming—A case study from the rural western middle hills of Nepal. *Agriculture*, *10*(5), 143. doi:10.3390/agriculture10050143

GCX. (2022). *Membership types: Ghana Commodity Exchange*. GCX. https://gcx.com.gh/membership_type/

Ghosh, N. (2013). *India's agricultural marketing: Market Reforms and Emergence of New Channels*. Springer. doi:10.1007/978-81-322-1572-1

Gomez, G., & Vossenberg, S. (2018). Identifying ripple effects from new market institutions to household rules-Malawi's Agricultural Commodity Exchange. *NJAS Wageningen Journal of Life Sciences*, *84*(1), 41–50. doi:10.1016/j.njas.2017.10.003

Guarín, A. (2013). The value of domestic supply chains: Producers, wholesalers, and urban consumers in Colombia. *Development Policy Review*, *31*(5), 511–530. doi:10.1111/dpr.12023

Haggblade, S. (2020). Overview. In AGRA (Ed.), Feeding Africa's Cities: Opportunities, Challenges, and Policies for Linking African Farmers with Growing Urban Food Markets (pp. 15–24). Alliance for a Green Revolution in Africa.

Hannah, C., Davies, J., Green, R., Zimmer, A., Anderson, P., Battersby, J., Baylis, K., Joshi, N., & Evans, T. P. (2022). Persistence of open-air markets in the food systems of Africa's secondary cities. *Cities (London, England)*, *124*, 103608. doi:10.1016/j.cities.2022.103608

Hernandez, M. A., Rashid, S., Lemma, S., & Kuma, T. (2017). Market institutions and price relationships: The case of coffee in the Ethiopian commodity exchange. *American Journal of Agricultural Economics*, *99*(3), 683–704. doi:10.1093/ajae/aaw101

Huang, J., Zhi, H., Huang, Z., Jia, X., & Rozelle, S. (2009). Smallholder incomes, vegetable marketing and food safety: evidence from China. *2009 Conference of the International Association of Agricultural Economists*. AAE.

Ishak, S., Omar, A. R. C., Sum, S. M., Othman, A. S., & Jaafar, J. (2020). Smallholder Agriculture Co-operatives' Performance: What is in the Minds of Management? *Journal of Co-Operative Organization and Management*, *8*(2), 100110. doi:10.1016/j.jcom.2020.100110

Joiner, J., & Okeleke, K. (2019). E-commerce in agriculture: new business models for smallholders' inclusion into the formal economy. *GSMA*. https://www.gsma.com/mobilefordevelopment/resources/e-commerce-in-agriculture-new-business-models-for-smallholders-inclu
sion-into-the-formal-economy/

JSE. (2022). *Agricultural derivatives*. JSE. https://www.jse.co.za/trade/derivitive-market/interest-rate-derivatives-market/agricultural-derivitives

Kang'ethe, E. K., Grace, D., Alonso, S., Lindahl, J. F., Mutua, F., & Haggblade, S. (2020). Food safety and public health implications of growing urban food markets. In AGRA (Ed.), Africa Agriculture Status Report. Feeding Africa's Cities: Opportunities, Challenges, and Policies for Linking African Farmers with Growing Urban Food Markets (Issue 8) (pp. 101–119). Alliance for a Green Revolution in Africa (AGRA).

Kherallah, M., Delgado, C. L., Gabre-Madhin, E. Z., Minot, N., & Johnson, M. (2002). *Reforming agricultural markets in Africa: Achievements and challenges*. IFPRI.

Kumar, S. K., & Timalsina, A. K. (2016). Challenges for Adopting E-Commerce in Agriculture in Nepalese Context—A Case Study of Kathmandu Valley. *Proceedings of IOE Graduate Conference*, 305–312.

Louw, A. (2008). The role of Fresh Produce Markets in South Africa. Regoverning Markets. [University of Pretoria.]. *Policy Brief*, 4.

Lowder, S. K., Skoet, J., & Raney, T. (2016). The number, size, and distribution of farms, smallholder farms, and family farms worldwide. *World Development*, *87*, 16–29. doi:10.1016/j.worlddev.2015.10.041

Lowder, S. K., Skoet, J., & Singh, S. (2014). *What do we really know about the number and distribution of farms and family farms in the world?* Background paper for The State of Food and Agriculture 2014 (*ESA Working Paper* No. 14-02).

Magesa, M. M., Michael, K., & Ko, J. (2020). Access and use of agricultural market information by smallholder farmers: Measuring informational capabilities. *The Electronic Journal on Information Systems in Developing Countries*, *86*(6), e12134. doi:10.1002/isd2.12134

Mason, N. M., Jayne, T. S., & Myers, R. J. (2015). Smallholder supply response to marketing board activities in a dual channel marketing system: The case of Zambia. *Journal of Agricultural Economics*, *66*(1), 36–65. doi:10.1111/1477-9552.12066

Mather, D., & Jayne, T. S. (2019). The impact of state marketing board operations on smallholder behavior and incomes: The case of Kenya. *Gates Open Research*, *3*(993), 993.

Mezui, C. A. M., Rutten, L., Sekioua, S., Zhang, J., N'Diaye, M. M., Kabanyane, N., Arvanitis, Y., Duru, U., & Nekati, B. (2013). *Guidebook on African commodity and derivatives exchanges*. African Development Bank.

Munuo, E. R. (2013). *Roadside vending as a business strategy and marketing option for agricultural crops: a case study of Moshi-Arusha highway*. Mzumbe University.

Muricho, G., Manda, D., Sule, F., & Kassie, M. (2017). Smallholder agricultural commercialization and poverty: empirical evidence of panel data from Kenya. *91st Annual Conference*. AgEcon. 10.22004/ag.econ.258618

NAMC. (2007). *Investigation: National Fresh Produce Markets*. The National Agricultural Marketing Council.

NEPAD. (2013). *African agriculture, transformation and outlook*. NEPAD Agency for the African Union.

Ogutu, S. O., & Qaim, M. (2019). Commercialization of the small farm sector and multidimensional poverty. *World Development*, *114*, 281–293. doi:10.1016/j.worlddev.2018.10.012

Olwande, J., Smale, M., Mathenge, M. K., Place, F., & Mithöfer, D. (2015). Agricultural marketing by smallholders in Kenya: A comparison of maize, kale and dairy. *Food Policy*, *52*, 22–32. doi:10.1016/j.foodpol.2015.02.002

Ortmann, G. F., & King, R. P. (2007). Agricultural cooperatives II: Can they facilitate access of small-scale farmers in South Africa to input and product markets? *Agrekon*, *46*(2), 219–244. doi:10.1080/03031853.2007.9523769

Proctor, F., & Lucchesi, V. (2012). *Small-scale farming and youth in an era of rapid rural change*. IIED/HIVOS.

Rantlo, A. M., Tsoako, M., & Muroyiwa, B. (2020). Institutional Factors Influencing Diary Farmers Participation in Formal and Informal Milk Markets in Maseru Urban, Lesotho, South Africa. *Journal of Agricultural Extension*, *24*(2), 48–59. doi:10.4314/jae.v24i2.6

Rashid, S. (2015). Commodity Exchanges and Market Development: What Have we Learned? *Agriculture in an Interconnected World, International Conference on Agricultural Economist 8-14 August 2015*. AAE.

Rashid, S., Winter-Nelson, A., & Garcia, P. (2010). Purpose and potential for commodity exchanges in African economies. *Washington DC: International Food Policy Research Institute Discussion Paper, 1035*.

Reardon, T., Timmer, C. P., & Minten, B. (2012). Supermarket revolution in Asia and emerging development strategies to include small-holder farmers. *Proceedings of the National Academy of Sciences*, *109*(31), 12332–12337.

Reardon, T., Awokuse, T., Haggblade, S., Kapuya, T., Liverpool-Tasie, S., Meyer, F., Minten, B., Nyange, D., Rusike, J., & Tschirley, D. (2019). The quiet revolution in agri-food distribution (wholesale, logistics, retail) in Sub-Saharan Africa. In AGRA (Ed.), Africa Agriculture Status Report: The Hidden Middle: A Quiet Revolution in the Private Sector Driving Agricultural Transformation (pp. 13–28). Alliance for a Green Revolution in Africa.

Resnick, D. (2017). Informal Food Markets in Africa's Cities. In Institute International Food Policy Research (Ed.), Global Food Policy Report 2017 (pp. 50–56). International Food Policy Research Institute.

Sarris, A., & Morrison, J. (Eds.). (2010). *Food security in Africa: Market and trade policy for staple foods in Eastern and Southern Africa*. Edward Elgar Publishing. doi:10.4337/9781849806367

Shrader, L. (2018). *Benchmarking E-Commerce Models for Africa's Smallholders: Mercy Corps Agrifin Accelerate Program*. MercyCorps. www.mercycorpsafa.org

Simon, H. A. (1997). *Models of bounded rationality: Empirically grounded economic reason* (Vol. 3). MIT press. doi:10.7551/mitpress/4711.001.0001

Siphesihle, Q., & Lelethu, M. (2020). Factors affecting subsistence farming in rural areas of yandeni local municipality in the Eastern Cape Province. *South African Journal of Agricultural Extension*, *48*(2), 92–105. doi:10.17159/2413-3221/2020/v48n2a540

Sitko, N. J., & Jayne, T. S. (2012). Why are African commodity exchanges languishing? A case study of the Zambian Agricultural Commodity Exchange. *Food Policy*, *37*(3), 275–282. doi:10.1016/j.foodpol.2012.02.015

Snyder, C. M. (1996). A dynamic theory of countervailing power. *The RAND Journal of Economics*, *27*(4), 747–769. doi:10.2307/2555880

Sudrajat, J., Isytar, I., & Arifin, N. (2021). Farmers' perception and engagement with the role of middlemen: A case study of the vegetable farmers. *Makara Human Behavior Studies in Asia*, *25*(1), 45–54. doi:10.7454/hubs.asia.1070220

TMX. (2022). *How Tanzania Mercantile Exchange works*. TMX.. https://www.tmx.co.tz/trading.php

Tschirley, D. L., Ayieko, M., Mathenge, M., & Weber, M. T. (2004). *Where Do Consumers in Nairobi Purchase their Food and Why Does this Matter? The Need for Investment to Improve Kenya's "Traditional" Food Marketing System*. Tegemeo Institute for Agricultural Policy Development. https://www.tegemeo.org/

Tung, O. J. L. (2016). Organic food certification in South Africa: A private sector mechanism in need of state regulation? *Potchefstroom Electronic Law Journal/Potchefstroomse Elektroniese Regsblad, 19*(1).

Valentinov, V. (2007). Why are cooperatives important in agriculture? An organizational economics perspective. *Journal of Institutional Economics*, *3*(1), 55–69. doi:10.1017/S1744137406000555

Vorley, B. (2013). *Meeting small-scale farmers in their markets: understanding and improving the institutions and governance of informal agrifood trade*. International Institute for Environment and Development/HIVOS/Mainumby.

Vorley, B., Del Pozo-Vergnes, E., & Barnett, A. (2012). *Small producer agency in the globalised market: Making choices in a changing world*. IIED/HIVOS.

Wale, E., & Chipfupa, U. (2021). Entrepreneurship concepts/theories and smallholder agriculture: Insights from the literature with empirical evidence from KwaZulu- Natal, South Africa. *Transactions of the Royal Society of South Africa*, *76*(1), 67–79. doi:10.1080/0035919X.2020.1861122

Wanyama, R., Gödecke, T., Chege, C. G. K., & Qaim, M. (2019). How important are supermarkets for the diets of the urban poor in Africa? *Food Security*, *11*(6), 1339–1353. doi:10.100712571-019-00974-3

Weatherspoon, D. D., & Reardon, T. (2003). The rise of supermarkets in Africa: Implications for agrifood systems and the rural poor. *Development Policy Review*, *21*(3), 333–355. doi:10.1111/1467-7679.00214

Weng, X. (2015). *The rural informal economy: Understanding drivers and livelihood impacts in agriculture, timber and mining* (IIED Working Paper).

World Bank. (2016). *Linking Farmers to Markets through Productive Alliances: An Assessment of the World Bank Experience in Latin America*. The World Bank.

Yaseen, A., Bryceson, K., & Mungai, A. N. (2018). Commercialization behaviour in production agriculture: The overlooked role of market orientation. *Journal of Agribusiness in Developing and Emerging Economies*, *8*(3), 579–602. doi:10.1108/JADEE-07-2017-0072

Yeh, Y.-P. (2016). Market orientation and service innovation on customer perceived value: The case of supermarket retailers. *Management Research Review*, *39*(4), 449–467. doi:10.1108/MRR-08-2014-0205

Chapter 6
Images of Organic Food Products, Consumers, Makers, and Distributors: An Image Congruence Study

Bình Nghiêm-Phú
University of Hyogo, Japan

Jillian Rae Suter
https://orcid.org/0000-0003-1289-0760
Shizuoka University, Japan

ABSTRACT

Compared to the other fields of business, research on image congruence in the food business sector, particularly organic food, is relatively scarce. Previous studies could only identify the congruence of certain psychological elements of perceived image. Other functional elements, on the other hand, are mostly overlooked. Considering these issues, this study aimed to examine customer perceptions of themselves as organic food consumers, of organic food products, and of organic food makers and distributors. Using data gathered from nine informants, this study found that organic food consumers held certain images concerning organic food products, their makers and distributors, and themselves. Some images of organic food consumers and organic food products shared the same nature. In addition, some images of organic food consumers, makers and distributors had the same essence. Implications of these findings, together with the limitations of this study and directions for future research, were discussed.

INTRODUCTION

In recent years, organic food has become a new trend of consumption. Worldwide sales revenue of organic food was 167167.85 billion USD in 2020 (Report Linker, 2021). In the US alone, the revenue of this market might have reached 56.4 billion US dollars in 2020 (Statista, 2021). Organic food has transformed

DOI: 10.4018/978-1-6684-4780-2.ch006

itself into a mainstream sector in this country. Facilitating organic food purchases will even strengthen and expand this sector. Nevertheless, research on organic food seems to prefer a general and theoretical concept to real or existing products (Arumugam, Govindasamy, Vellangany, & Gohil, 2017; Konuk, 2018; Richetin, Mattavelli, & Perugini, 2016; Tandon, Dhir, Kaur, Kushwah, & Salo, 2020; Tandon, Jabeen, Talwar, Sakashita, & Dhir, 2021). In addition, research on customer perceptions of organic food makers or distributors or their perceived images is largely missing, although some studies examined both consumers' and makers' attitudes toward organic food consumption and production (McEachern & Willock, 2004; Storstad & Bjørkhaug, 2003), and several emphasized the origin of organic food products (Sirieix, Kledal, & Sulitang, 2011; Thøgersen, Pedersen, Paternoga, Schwendel, & Aschemann-Witze, 2017). In order to provide more practical implications for makers and distributors to improve the production and distribution processes of organic food, such a gap in the literature needs to be filled.

The consumption of organic food is facilitated by customers' perceived self-image and perceived product image (Gianluigi Guido, Peluso, Maloumby-Baka, & Buffa, 2010; Hwang, 2016; Soyez, Francis, & Smirnova, 2012; Sultan, Wong, & Sigala, 2018). The congruence between self-image and product image, in particular, and other types of image congruence (e.g., self-image vs. endorser image and endorser image vs. product image), in general, is essential with customers' purchase intentions toward and satisfaction with products and their providers (Pool, Khodadadi, & Asadi, 2018; Pradhan, Duraipandian, & Sethi, 2016; Wallace, Buil, & de Chernatony, 2017), including food products (Chien, Wu, & Huang, 2018; de Droog, Buijzen, & Valkenburg, 2014; Septianto, Kemper, & Paramita, 2019). However, compared to the other fields of business, research on image congruence in the food business sector, in general, and organic food, in particular, is relatively scarce. In addition, by applying the structured methods (pre-determined scales), previous studies could only identify the congruence of certain psychological elements of perceived image, such as personality and gender (Beldad, Hegner, & Hoppen, 2016; Malodia, Singh, Goyal, & Sengupta, 2017; Pradhan, Duraipandian, & Sethi, 2016). On the other hand, other functional elements (e.g., quality and value) are mostly overlooked since each image bearer (e.g., customer, endorser, product, and brand) is attributed to different characteristics due to their specific natures. Unfortunately, a lack of understanding of the functional elements may cause difficulties in the projection and delivery of the psychological elements.

The purpose of this study, therefore, is to examine customer perceptions of themselves as organic food consumers, organic food products, and organic food makers and distributors to fill in the gap concerning organic food and related stakeholders' images. In addition, adopting the image congruence approach, this study identifies the differences or similarities among these types of images to support future management and marketing practices involving organic food products. Nevertheless, unlike the majority of previous research, this study chooses a qualitative method to reveal the true and authentic perceptions of organic food customers and avoid the limitation of the conventional structured quantitative method. The US market is specifically selected as the context of the study, given that it already is an established market. Customer perception, thus, has been more solidly formed. The findings of this study will enrich the literature on image congruence and provide practical implications for organic food management and marketing.

LITERATURE REVIEW

An image is generally regarded as a representation of a product or service. On the one hand, it might be created by a business owner, maker, or distributor to promote the selling and consumption of the product or service (i.e., the projected image). On the other hand, the image is held by the customers as a result of encountering various information sources and experiencing said product and service (i.e., the perceived image). The perceived image, or the perception, can be an overall impression or an aggregation of many attributes, characteristics, features, and (de)merits. The projected image can also consist of the latter components. The following subsections describe the images of organic food products, consumers, makers, and distributors. They also explain the formations and impacts of these images as documented by the existing literature.

Organic Food Products

According to the US Department of Agriculture, organic food products are those made from plants "grown on soil that had no prohibited substances applied for three years prior to harvest" and animals "raised in living conditions accommodating their natural behaviors (like the ability to graze on pasture), fed 100% organic feed and forage, and not administered antibiotics or hormones" (McEvoy, 2019). Processed and packaged organic food products must contain "at least 70% organically produced ingredients" and are free of "genetically modified organisms" and "artificial preservatives, colors, or flavors" (McEvoy, 2019). Organic food is often regarded as a general concept. Marketers can categorize specific organic food typologies, but customers may see these types under the organic umbrella (Baudry, et al., 2016). As per Schleenbecker and Hamm's (2013) review, customers care about four significant attributes of organic food: product quality, labelling, packaging, and product range. With product quality, customers especially emphasize organic food's nutrition and safety or healthiness (Aschemann-Witzel, Maroscheck, & Hamm, 2013; Rana & Paul, 2020; Stolz, Stolze, Hamm, Janssen, & Ruto, 2011). They seem to emphasize less on the products' sensory attributes, such as taste (Asioli, et al., 2014; Gilsenan, Burke, & Barry-Ryan, 2012).

Organic food consumers are also heavily affected by product labels and packaging and presentation methods. Specifically, organic-labelled products can encourage customers to purchase (Aitken, Watkins, Williams, & Kean, 2020; Lee, Shimizu, Kniffin, & Wansink, 2013). However, well-known and government-certified organic logos are more persuasive than other types of logos (Janssen & Hamm, 2012; Sønderskov & Daugbjerg, 2011). Moreover, customer intentions may differ among product categories (van Doorn & Verhoef, 2011). Specifically, customers are less willing to pay for vice organic products (e.g., products containing a high sugar content) compared to virtue organic products (e.g., products containing vitamins). In addition, price is another attribute that organic food consumers consider (Aschemann-Witzel & Zielke, 2017; Marian, Chrysochou, Krystallis, & Thøgersen, 2014). Having a high price has been a prominent feature of organic food products. The price element, together with the abovementioned attributes, helps project the "superiority" image of organic food products (Vega-Zamora, Torres-Ruiz, Murgado-Armenteros, & Parras-Rosa, 2014).

Organic Food Consumers

As previous research has identified, organic food customers are relatively older and have a high education level (Ngobo, 2011; van Herpen, van Nierop, & Sloot, 2012; Xie, Wang, Yang, Wang, & Zhang, 2015). Consequently, they have an adequate budget for these products. However, there potentially is a difference between male and female customers (Fatha & Ayoubi, 2021; Grubor & Djokic, 2016; Ureña, Bernabéu, & Olmeda, 2008). Specifically, men may be more willing to pay a higher price for organic food, while women may have a more positive attitude toward buying and eating these products. The primary reason is that men tend to associate organic food with environmental protection, and women seem to be affected more by their peers (Ureña, Bernabéu, & Olmeda, 2008).

From an intrinsic perspective, organic food consumers are health and safety conscious, both for themselves and for their offspring (Chen, Lobo, & Rajendran, 2014; Eisinger-Watzl, Wittig, Heuer, & Hoffmann, 2015; Kareklas, Carlson, & Muehling, 2014; Rana & Paul, 2020). This characteristic is directly related to their healthy lifestyle (i.e., an egoistic value). From an extrinsic standpoint, organic food consumers are concerned about the environment (Nasir & Karakaya, 2014; Shin, Moon, Jung, & Severt, 2017; Tandon, Dhir, Kaur, Kushwah, & Salo, 2020). This feature is directly related to the altruistic meaning that these consumers value.

Organic Food Makers and Distributors

Consumers of organic food highly regard the specific merits of producing these products. For example, consumers believe that organic production can help guarantee and improve animal welfare (Harper & Makatouni, 2002; Storstad & Bjørkhaug, 2003). In addition, properly using pesticides and other chemicals will help protect the environment (Cranfield, Deaton, & Shellikeri, 2009; Storstad & Bjørkhaug, 2003). The origin of the products, whether local or imported, seems not to be an essential characteristic (Thøgersen, Pedersen, Paternoga, Schwendel, & Aschemann-Witze, 2017). Thus, as Sirieix, Kledal, and Sulitang (2011) observed, although consumers appreciate several altruistic and ethical attributes of organic food production, these attributes may be only some "emerging" issues.

Regarding organic food distribution, customers especially care about the social responsibility aspect of the retailers (Hanks & Mattila, 2016; Perrini, Castaldo, Misani, & Tencati, 2010). A more positive perception of the retailers' social responsibility will lead to a higher level of trust and, consequently, more willingness to pay a premium price and loyalty. Moreover, the image of the retailers' stores can also impact customers' evaluations. Specifically, store image is a significant antecedent of perceived quality and trust (Konuk, 2018). Store image can also influence customers' demand for organic food products (Ngobo & Jean, 2012). Furthermore, the perceived fit of the retailers with their products may mediate the correlation between customers' attitudes toward organic food purchases and customers' choice of a retailer (Hwang & Chung, 2019). With restaurants, an organic menu is an essential supporter of a more favourable image, both in terms of external social responsibility and internal product quality (Lu & Gursoy, 2017; Namkung & Jang, 2013).

Organic Food Products, Consumers, Makers, and Distributors' Image Congruence

Generally, image congruence is the matching of different types of images. In the business sectors, the critical congruences are those between customer self-image and product or brand image (Klabi, 2020; Wallace, Buil, & de Chernatony, 2017) and those among customer self-image, endorser image and product or brand image (Pradhan, Duraipandian, & Sethi, 2016). Image congruence is an essential antecedent of customer purchase intention and satisfaction (Pool, Khodadadi, & Asadi, 2018; Pradhan, Duraipandian, & Sethi, 2016; Wallace, Buil, & de Chernatony, 2017). The prior review of the literature reveals that there generally is a considerable lack of studies directly investigating the images of organic food products, consumers, makers, and distributors. More importantly, studies on the congruence of these images are largely missing. The only exception is the work of Hwang and Chung (2019), in which the researchers examined the congruence between product and distributor images. These remaining issues are the focus of this chapter.

METHODOLOGY

Given its exploratory nature, this study opted to collect and descriptively analyze qualitative data to identify American organic food customers' perceptions of themselves, organic food products, and organic food makers and distributors. First, the research group implemented direct interviews (Fujii, 2018) to gather first-hand information about American customers' opinions. To do this, the second researcher of the research group, an American citizen, initially called on their personal social media (i.e., Facebook – almost 500 friends and followers and Instagram – about 200 friends and followers) to recruit the interviewees. Six female acquaintances responded to the call, but only three interviewees were chosen. The other recruits were excluded because two recruits did not strictly consider themselves organic consumers although they sometimes consumed organic foods, and one recruit did not consider herself American because she did not grow up in the United States. The recruitment of the interviewees was a purposeful sampling process (Palinkas et al., 2015), in which the researcher's acquaintanceship served as the first and internal selection criterion and the respondents' mindset as the second and external selection principle. The two researchers of the research group then conducted online semi-structured interviews via Zoom (https://zoom.us/) in March and April 2021. Each interview lasted between 30 to 60 minutes and was recorded with the participant's permission. In the interviews, the participants were asked about their history of organic food consumption and their opinions.

Second, the two researchers explored public social media platforms to collect more user-generated information (Kim, 2012). By chance, the researchers found many videos posted on YouTube describing shopping experiences at Whole Foods, an American multinational supermarket chain. After that, they used the keywords "organic" and "Whole Foods" to specify the relevant videos for this study. They sorted the videos by their upload dates and watched more than 200 programs longer than four minutes. From this pool, three videos uploaded by three American YouTubers between mid and late 2020 were explicitly targeted for two reasons: (1) the experiences retold in the videos were directly related to organic food purchases even though Whole Foods is not an exclusive organic product store and (2) the information provided was the latest. These videos ranged from 7 to 28 minutes long.

Table 1. Sources of data

Source	Code	Details
Direct interview	Informant 1	Female individual, 30s
	Informant 2	Female individual, 30s
	Informant 3	Female individual, 30s
YouTube videos	Informant 4	Male individual, 30s https://www.youtube.com/watch?v=aDd2kwWnip0
	Informant 5	Male individual, 40s https://www.youtube.com/watch?v=MQmLD34erEg
	Informant 6	Male individual, 30s https://www.youtube.com/watch?v=pznYAAR9ZQg
Blogs	Informant 7	Organization https://www.organicauthority.com/
	Informant 8	Organization https://www.organic-center.org/
	Informant 9	Organization https://www.organicconsumers.org/

Third, the two researchers searched for blogs where writers describe their experiences in the written language (Hellmann, Griesbaum, & Mandl, 2010). To do this, the researchers went through a blogging agglomeration website (https://www.feedspot.com/) that they found after an extensive search using Google. Fortunately, FeedSpot has already compiled a list of the "25 Best Organic Food Blogs and Websites," and the researchers opted to use this list as the initial sample. In a theory-based qualitative study, 25 information sources were a relevant initial sample (Francis, et al., 2010). Again, three particular blogs were critically chosen because they were ranked high on the list of blogs about organic food and verified as being based in the US. Specifically, on organicauthority.com, the bloggers write more about the consumer side of organic foods, including recipes and wellness guides. On organic-center.org, however, the bloggers focus more on the science and farming facet of organic foods. On the third one, organicconsumer.org, the writers use advocacy and education campaigns to promote organic foods on behalf of organic consumers. The diversity of the blogs' content helps provide more information about organic consumers' opinions from a professional and collective standpoint, which is different from the first and second sources described earlier. Combining the three sources provides a valid and reliable compilation of information (Kassarjian, 1977) about the research topic (Table 1).

After the sources of information were identified and the data was collected, the researchers deductively analyzed them under four categories (deductive content analysis) (Elo & Kyngäs, 2008):): (1) characteristics of organic food consumers, (2) characteristics of organic food products, (3) characteristics of organic food makers, and (4) characteristics of organic food distributors. The raw interviews, videos, and blog materials were used in the analysis to maintain the relation between the content and the context (Wellard & McKenna, 2002), which could easily be eliminated if only the former was emphasized. In the analysis process, the second researcher initially reviewed all the contents and noted the answers for each of the four categories in a Microsoft Excel file. Later, the first researcher analyzed these contents again and identified the missing or differences in the answers. After that, the two researchers discussed with each other to address the gaps and reached the final result. The total agreement between the researchers served as the reliability criterion of the analysis (Kassarjian, 1977).

Given the exploratory and descriptive nature of the study and the availability of the resources, the research group did not intend to reach the point where the data could be considered saturated (Saumure & Given, 2008). Instead, the researchers compared and combined their findings with those of previous studies to triangulate the information and generalize it to the American context (Oppermann, 2000). They did this under the guidance of the creative inquiry literature review method (Montuori, 2005).

FINDINGS

Images of Organic Food Consumers

In total, the information gathered from nine informants was analyzed. It was revealed that only one informant had held specific images of organic food customers who were health- and welfare-wise conscious.

Informant 7: "Today's consumers are smart; they care about the food they eat, how it is grown, how it is processed. Their commitment goes beyond fresh. They want clean food, clean beauty, wellness, and home products. They want the truth."

However, several informants also mentioned their family tradition concerning organic food production and consumption as one reason they use organic products. For example, some informants talked about the influence their parents' gardening had on them growing up and how it affects how they see organic choices today. Despite these past and present exercises, most informants were not strictly organic consumers. In other words, they could be considered flexible buyers and eaters of food products, including organic ones.

Informant 1: "I try to buy organic foods my daughter is eating for the most part. Organic milk, avocado, bananas."

Informant 2: "We try to keep all of our vegetables and our fruits to be organic… I know that the meat industry here in Texas is not necessarily always organic."

Informant 3: "I [use] just whatever vegetables I need for my usual dinners."

Images of Organic Food Products

Regarding the images of the organic food products, while several informants mentioned specific labels (e.g., Egg Bites, Keto Cups), others named some essential qualities, including better nutrition and better sensory properties. In addition, some of them also said that organic foods are expensive and may be small in quantity.

Informant 1: "[Organic] produce just last longer, looked better, fresher. I feel like you notice a bigger difference with certain things like raspberries and strawberries… [Moreover] I do think the flavours are better. I think you notice it a lot with tomatoes [and carrots], in particular."

Informant 2: "When I think of organic foods, I always think of produce… Sometimes I notice the organic food may be a little smaller in size."

Informant 5: "This is clean, but that price is not something I would want."

Informant 6: "I love the idea they have garbanzo beans. Super low glycemic. Pumpkin seeds, flax seeds, so we have a nice omega-3 profile there. Plus chia seeds. So, we are really rounding out that omega-3."

Images of Organic Food Makers

Concerning the images of the organic food makers, most informants believed that these were the responsible ones, particularly regarding the natural environment and the welfare of the human labourers and the animals. When they heard something terrible about a maker, especially those from abroad or other locations, they would try to avoid them. However, more often than not, there was not much that the informants could actively do to improve the benefits or decrease the costs to the environment, the labor, and the animals.

Informant 9: "The phase-out of factory farms, including the elimination of GMO [Genetically Modified Organisms] animal feed, toxic pesticides, antibiotics and other drugs in meat, poultry and dairy production; the end of dangerous and unjust factory farm labor practices; the end of unfair contracts that enrich huge corporations and put small family farms at risk of bankruptcy; and the end of inhumane treatment of animals… support for farmers transitioning to an organic regenerative paradigm that improves public health, revives strong local economies, renews biodiversity, reduces environmental pollution and restores climate stability."

Informant 2: "I am okay with certain foods coming from other states. I do not think it is that bad. Other countries I tend to shy away from, because I might not know the best way [that] they are prepared. Because it is going over a border, there might mean additional costs. If they are not preserving it correctly, maybe I am not getting it the way it is intended to be."

Images of Organic Food Distributors

Finally, concerning the images of the organic food distributors, the impressions held by the informants varied according to the local options that they had. In most cases, the distributors' images were actually those of the stores or outlets where the purchases were made. They might include such elements as product types, product categories, and prices.

Informant 1: "Sprouts is actually really close to us, and I like to go there. I think their produce is a lot fresher… Sprouts does not have most of the main nationally distributed brands. They have a lot of their own brands."

Informant 2: "And there is a Whole Foods as well. But I honestly do not like Whole Foods. I just feel like I am paying extra for every single thing."

Informant 3: "Sprouts and Whole Foods are quite similar. Sprouts is a little bit cheaper. It is more like family friendly. Whole Foods I feel like is a little bit fancier… [In addition] I found that Albertsons, that is just a regular grocery chain, their food, in general, is more expensive than any other place, including Sprouts."

Informant 4: "Every time I come to Whole Foods, there are always new products."

Informant 5: "You guys know I love this collage flavored water from vital proteins, but if you are a Costco shopper, it is 1.24 per bottle at Costco, it is 3.49 a bottle here at Whole Foods."

Informant 6: "I do not always get the best vibes from some Whole Foods, but when I go in there, I usually do find some interesting, new keto things… There is a lot of super deceptive stuff at Whole Foods, too. Just because it has a Whole Foods name, or is at Whole Foods, does not mean that it is healthy."

Figure 1. The main images held of organic consumers, makers, distributors, and products

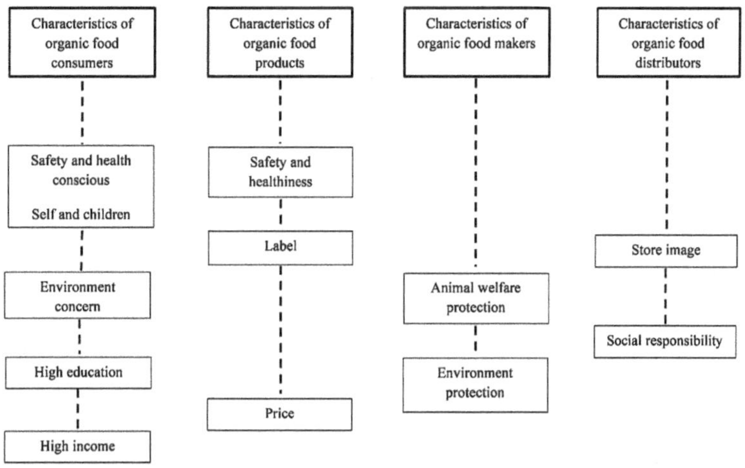

DISCUSSION

The information collected from the interviewees, the YouTubers, and the bloggers have helped reveal some images of organic food consumers, makers, distributors, and organic food products, particularly in the US. Figure 1 combines this information with previous studies' findings to summarize the primary observations (Figure 1).

Firstly, organic food consumers are those who are conscious about their own and their children's health and safety and are concerned about environmental issues to some degree (Chen, Lobo, & Rajendran, 2014; Eisinger-Watzl, Wittig, Heuer, & Hoffmann, 2015; Kareklas, Carlson, & Muehling, 2014; Nasir & Karakaya, 2014; Shin, Moon, Jung, & Severt, 2017). They can also have high education and high-income characteristics (Ngobo, 2011; van Herpen, van Nierop, & Sloot, 2012; Xie, Wang, Yang, Wang, & Zhang, 2015). However, most individuals only consider themselves as casual consumers, those who sometimes make organic choices, not fully organic consumers.

Secondly, organic food products are regarded as safe and healthy (Aschemann-Witzel, Maroscheck, & Hamm, 2013; Stolz, Stolze, Hamm, Janssen, & Ruto, 2011). Product label and price are mainly considered (Aschemann-Witzel & Zielke, 2017; Lee, Shimizu, Kniffin, & Wansink, 2013; Marian, Chrysochou, Krystallis, & Thøgersen, 2014). The origins of the products, whether local or imported, may also be taken into account. However, this observation is different from previous studies (Thøgersen, Pedersen, Paternoga, Schwendel, & Aschemann-Witze, 2017).

Thirdly, organic food makers are often seen as responsible businesses that care about and protect the animals' welfare and the environment's sustainability (Cranfield, Deaton, & Shellikeri, 2009; Storstad & Bjørkhaug, 2003; van Herpen, van Nierop, & Sloot, 2012). In addition, they may also be responsible for the benefits of the human labor they use to produce these products.

Finally, organic food distributors also are attached to the image of socially responsible business units (Hanks & Mattila, 2016; Perrini, Castaldo, Misani, & Tencati, 2010). However, customers' purchase decision is often affected by the images of the stores (Lu & Gursoy, 2017; Namkung & Jang, 2013), including product range and price (Aschemann-Witzel & Zielke, 2017; Marian, Chrysochou, Krystallis,

& Thøgersen, 2014; van Doorn & Verhoef, 2011). The findings of this study, thus, are consistent with those of the previous research.

The existing observations have also shown that there potentially are some congruences in the images of organic food consumers and organic food products and those of organic food consumers, makers, and distributors. In the former sense, organic food products' safety and healthiness characteristics also are those of organic food consumers and vice versa. In the latter sense, social and environmental responsibility is a trait shared by all the involved human stakeholders, including the consumers, the makers, and the distributors of organic food products. These findings have helped extend the existing knowledge about perceived psychological image congruence (e.g., personality and gender traits) (Beldad, Hegner, & Hoppen, 2016; Malodia, Singh, Goyal, & Sengupta, 2017; Pradhan, Duraipandian, & Sethi, 2016) to that concerning the functional images, such as safety, healthiness, and social and environmental responsibility.

PRACTICAL IMPLICATIONS

Image congruence is significant in consumers' purchase intention and satisfaction (Chien, Wu, & Huang, 2018; de Droog, Buijzen, & Valkenburg, 2014; Hwang & Chung, 2019; Septianto, Kemper, & Paramita, 2019). Fortunately, such congruences have already existed in the market, particularly the US market. It is necessary to provide correct information about the safety and healthiness of the products and the commitments to ensure the quality of the makers and distributors to facilitate organic food purchases. In addition, the responsibilities toward the environment, the animals, and the labor of associated with the latter during the production and distribution processes should also be emphasized.

Based on this information, the makers and distributors of organic food could try to advocate the price element of these products further. Both theoretically and practically, high-quality and ethical products cannot be cheaply produced and distributed, and customers must be appropriately informed and persuaded about this. Moreover, as many consumers in the market are casual, increasing the variety (production) and availability (distribution) of organic products would help increase organic consumption. The makers and distributors could do more to cultivate a new consumption culture to create a healthier, safer, and more ethically responsible future for organic food products in their own country.

CONCLUDING REMARKS

More and more people are consuming organic food. They, of course, hold specific images concerning organic food products, their makers and distributors, and themselves. Some images of organic food consumers and organic food products share the exact nature, such as safety and healthiness. In addition, some images of organic food consumers, makers, and distributors have the same essence, such as social and environmental responsibility. Image congruence in the organic food business subsector, thus, has existed. Moreover, thanks to the qualitative method adopted by this study, the congruence is also found in the functional image category, not only in the psychological image category as the literature has suggested (Beldad, Hegner, & Hoppen, 2016; Malodia, Singh, Goyal, & Sengupta, 2017; Pradhan, Duraipandian, & Sethi, 2016). With the projection and promotion of proper information, image congruence can be improved to further facilitate future organic food product consumption. This study, however, could not avoid some limitations. First, the number of informants was limited, although they were found through different channels. Consequently, the results of this study might not be able to be generalized to other developed markets. In addition, the comparison of the images held by different groups of customers, such as age, income, and biological sex (Fatha & Ayoubi, 2021; Grubor & Djokic, 2016; Ngobo, 2011; Ureña, Bernabéu, & Olmeda, 2008; van Herpen, van Nierop, & Sloot, 2012; Xie, Wang, Yang, Wang, & Zhang, 2015), could not be done. Future studies could generate a larger sample and apply a structured method to quantify the images and the congruences to address this issue. Second, due to the periodical nature of the research, congruence was identified, but the change of congruence with the inputs of additional information could not be explored. Future researchers may want to carry out experiments or longitudinal studies to clarify this issue. The abovementioned efforts will help expand the literature on organic food and the involved businesses in the upcoming periods. This has further implications for the farmers, agricultural marketing and organic food regulatory authorities as this indicates what current and future consumers expect from the organic food market.

REFERENCES

Ahmed, N., Li, C., Khan, A., Qalati, S. A., Naz, S., & Rana, F. (2021). Purchase intention toward organic food among young consumers using theory of planned behavior: Role of environmental concerns and environmental awareness. *Journal of Environmental Planning and Management*, *64*(5), 796–822. doi:10.1080/09640568.2020.1785404

Aitken, R., Watkins, L., Williams, J., & Kean, A. (2020). The positive role of labelling on consumers' perceived behavioural control and intention to purchase organic food. *Journal of Cleaner Production*, *255*, 120334. doi:10.1016/j.jclepro.2020.120334

Arumugam, S., Govindasamy, R., Vellangany, I., & Gohil, H. (2017). An analysis of organic fresh produce choice: A consumer preference study in the Mid-Atlantic USA. *Journal of Food Distribution Research*, *48*(1), 101–102. doi:10.22004/ag.econ.274576

Aschemann-Witzel, J., Maroscheck, N., & Hamm, U. (2013). Are organic consumers preferring or avoiding foods with nutrition and health claims? *Food Quality and Preference*, *30*(1), 68–76. doi:10.1016/j.foodqual.2013.04.011

Aschemann-Witzel, J., & Zielke, S. (2017). Can't buy me green? A review of consumer perceptions of and behavior toward the price of organic food. *The Journal of Consumer Affairs*, *51*(1), 211–251. doi:10.1111/joca.12092

Asioli, D., Canavari, M., Pignatti, E., Obermowe, T., Sidali, K. L., Vogt, C., & Spiller, A. (2014). Sensory experiences and expectations of Italian and German organic consumers. *Journal of International Food & Agribusiness Marketing*, *26*(1), 13–27. doi:10.1080/08974438.2012.755718

Baudry, J., Touvier, M., Allès, B., Péneau, S., Méjean, C., Galan, P., Hercberg, S., Lairon, D., & Kesse-Guyot, E. (2016). Typology of eaters based on conventional and organic food consumption: Results from the NutriNet-Santé cohort study. *British Journal of Nutrition*, *116*(4), 700–709. doi:10.1017/S0007114516002427 PMID:27311793

Beldad, A., Hegner, S., & Hoppen, J. (2016). The effect of virtual sales agent (VSA) gender – product gender congruence on product advice credibility, trust in VSA and online vendor, and purchase intention. *Computers in Human Behavior*, *60*, 62–72. doi:10.1016/j.chb.2016.02.046

Britwum, K., Bernard, J. C., & Albrecht, S. E. (2021). Does importance influence confidence in organic food attributes? *Food Quality and Preference*, *87*, 104056. doi:10.1016/j.foodqual.2020.104056

Chen, J., Lobo, A., & Rajendran, N. (2014). Drivers of organic food purchase intentions in mainland China – Evaluating potential customers' attitudes, demographics and segmentation. *International Journal of Consumer Studies*, *38*(4), 346–356. doi:10.1111/ijcs.12095

Chien, S.-H., Wu, J.-J., & Huang, C.-Y. (2018). "We made, we trust": Coproduction and image congruence in the food-tourism factories. *Asia Pacific Management Review*, *23*(4), 310–317. doi:10.1016/j.apmrv.2018.01.002

Cranfield, J., Deaton, B. J., & Shellikeri, S. (2009). Evaluating consumer preferences for organic food production standards. *Canadian Journal of Agricultural Economics*, *57*(1), 99–117. doi:10.1111/j.1744-7976.2008.01140.x

De Canio, F., & Martinelli, E. (2021). EU quality label vs organic food products: A multigroup structural equation modeling to assess consumers' intention to buy in light of sustainable motives. *Food Research International*, *139*, 109846. doi:10.1016/j.foodres.2020.109846 PMID:33509469

de Droog, S. M., Buijzen, M., & Valkenburg, P. M. (2014). Enhancing children's vegetable consumption using vegetable-promoting picture books. The impact of interactive shared reading and character–product congruence. *Appetite*, *73*, 73–80. doi:10.1016/j.appet.2013.10.018 PMID:24216486

Eisinger-Watzl, M., Wittig, F., Heuer, T., & Hoffmann, I. (2015). Customers purchasing organic food - Do they live healthier? Results of the German national nutrition survey II. *European Journal of Nutrition & Food Safety*, *5*(1), 59–71. doi:10.9734/EJNFS/2015/12734

Elo, S., & Kyngäs, H. (2008). The qualitative content analysis process. *Journal of Advanced Nursing*, *62*(1), 107–115. doi:10.1111/j.1365-2648.2007.04569.x PMID:18352969

Fatha, L., & Ayoubi, R. (2021). A revisit to the role of gender, age, subjective and objective knowledge in consumers' attitudes towards organic food. *Journal of Strategic Marketing*, 1–17. Advance online publication. doi:10.1080/0965254X.2021.1939405

Francis, J. J., Johnston, M., Robertson, C., Dlidewell, L., Entwistle, V., Eccles, M. P., & Grimshaw, J. M. (2010). What is an adequate sample size? Operationalising data saturation for theory-based interview studies. *Psychology & Health*, *25*(10), 1229–1245. doi:10.1080/08870440903194015 PMID:20204937

Fujii, L. A. (2018). *Interviewing in Social Science Research: A Relational Approach*. Routledge.

Gianluigi Guido, M. I., Peluso, A. M., Maloumby-Baka, R. C., & Buffa, C. (2010). The role of ethics and product personality in the intention to purchase organic food products: A structural equation modeling approach. *International Review of Economics*, *57*(1), 79–102. doi:10.100712232-009-0086-5

Gilsenan, C., Burke, R. M., & Barry-Ryan, C. (2012). Do organic cherry vine tomatoes taste better than conventional cherry vine tomatoes? A sensory and instrumental comparative study from Ireland. *Journal of Culinary Science & Technology*, *10*(2), 154–167. doi:10.1080/15428052.2012.679232

Grubor, A., & Djokic, N. (2016). Organic food consumer profile in the Republic of Serbia. *British Food Journal*, *118*(1), 164–182. doi:10.1108/BFJ-06-2015-0225

Hanks, L., & Mattila, A. S. (2016). Consumer response to organic food in restaurants: A serial mediation analysis. *Journal of Foodservice Business Research*, *19*(1), 109–121. doi:10.1080/15378020.2016.1129228

Harper, G. C., & Makatouni, A. (2002). Consumer perception of organic food production and farm animal welfare. *British Food Journal*, *104*(3-5), 287–299. doi:10.1108/00070700210425723

Hellmann, R., Griesbaum, J., & Mandl, T. (2010). Quality in blogs: How to find the best user generated content. In W. Abramowicz, & R. Tolksdorf, *Business Information Systems. BIS 2010.* [Berlin, Heidelberg: Springer.]. *Lecture Notes in Business Information Processing*, *47*, 47–58. doi:10.1007/978-3-642-12814-1_5

Hwang, J. (2016). Organic food as self-presentation: The role of psychological motivation in older consumers' purchase intention of organic food. *Journal of Retailing and Consumer Services*, *28*, 281–287. doi:10.1016/j.jretconser.2015.01.007

Hwang, J., & Chung, J.-E. (2019). What drives consumers to certain retailers for organic food purchase: The role of fit for consumers' retail store preference. *Journal of Retailing and Consumer Services*, *47*, 293–306. doi:10.1016/j.jretconser.2018.12.005

Janssen, M., & Hamm, U. (2012). Product labelling in the market for organic food: Consumer preferences and willingness-to-pay for different organic certification logos. *Food Quality and Preference*, *25*(1), 9–22. doi:10.1016/j.foodqual.2011.12.004

Kareklas, I., Carlson, J. R., & Muehling, D. D. (2014). "I eat organic for my benefit and yours": Egoistic and altruistic considerations for purchasing organic food and their implications for advertising strategists. *Journal of Advertising*, *43*(1), 18–32. doi:10.1080/00913367.2013.799450

Kassarjian, H. H. (1977). Content analysis in consumer research. *The Journal of Consumer Research*, *4*(1), 8–18. doi:10.1086/208674

Kim, J. (2012). The institutionalization of YouTube: From user-generated content to professionally generated content. *Media Culture & Society*, *34*(1), 53–67. doi:10.1177/0163443711427199

Klabi, F. (2020). Self-image congruity affecting perceived quality and the moderation of brand experience: The case of local and international brands in the Kingdom of Saudi Arabia. *Journal of Global Marketing*, *33*(2), 69–83. doi:10.1080/08911762.2019.1614242

Konuk, F. A. (2018). The role of store image, perceived quality, trust and perceived value in predicting consumers' purchase intentions towards organic private label food. *Journal of Retailing and Consumer Services*, *43*, 304–310. doi:10.1016/j.jretconser.2018.04.011

Lee, W.-C. J., Shimizu, M., Kniffin, K. M., & Wansink, B. (2013). You taste what you see: Do organic labels bias taste perceptions? *Food Quality and Preference*, *29*(1), 33–39. doi:10.1016/j.foodqual.2013.01.010

Liu, C., Zheng, Y., & Cao, D. (2021). An analysis of factors affecting selection of organic food: Perception of consumers in China regarding weak signal. *Appetite*, *161*, 105145. Advance online publication. doi:10.1016/j.appet.2021.105145 PMID:33515620

Lu, L., & Gursoy, D. (2017). Does offering an organic food menu help restaurants excel in competition? An examination of diners' decision-making. *International Journal of Hospitality Management*, *63*, 72–81. doi:10.1016/j.ijhm.2017.03.004

Ma, Z., Chen, J., Tian, G., Guo, B., & Cheng, F. (2021). Regulations on the corporate social irresponsibility in the supply chain under the multiparty game: Taking China's organic food supply chain as an example. *Journal of Cleaner Production*, *317*, 128459. doi:10.1016/j.jclepro.2021.128459

Malodia, S., Singh, P., Goyal, V., & Sengupta, A. (2017). Measuring the impact of brand-celebrity personality congruence on purchase intention. *Journal of Marketing Communications*, *23*(5), 493–512. doi:10.1080/13527266.2017.1322125

Marian, L., Chrysochou, P., Krystallis, A., & Thøgersen, J. (2014). The role of price as a product attribute in the organic food context: An exploration based on actual purchase data. *Food Quality and Preference*, *37*, 52–60. doi:10.1016/j.foodqual.2014.05.001

McEachern, M. G., & Willock, J. (2004). Producers and consumers of organic meat: A focus on attitudes and motivations. *British Food Journal*, *106*(7), 534–552. doi:10.1108/00070700410545737

McEvoy, M. (2019, March 13). *Organic 101: What the USDA organic label means*. US Department of Agriculture.

Montuori, A. (2005). Literature review as creative inquiry: Reframing scholarship as a creative process. *Journal of Transformative Education*, *3*(4), 374–393. doi:10.1177/1541344605279381

Nagaraj, S. (2021). Role of consumer health consciousness, food safety & attitude on organic food purchase in emerging market: A serial mediation model. *Journal of Retailing and Consumer Services*, *59*, 102423. doi:10.1016/j.jretconser.2020.102423

Namkung, Y., & Jang, S. (2013). Effects of restaurant green practices on brand equity formation: Do green practices really matter? *International Journal of Hospitality Management*, *33*, 85–95. doi:10.1016/j.ijhm.2012.06.006

Nasir, V. A., & Karakaya, F. (2014). Consumer segments in organic foods market. *Journal of Consumer Marketing*, *31*(4), 263–277. doi:10.1108/JCM-01-2014-0845

Ngobo, P. V. (2011). What drives household choice of organic products in grocery stores? *Journal of Retailing*, *87*(1), 90–100. doi:10.1016/j.jretai.2010.08.001

Ngobo, P.-V., & Jean, S. (2012). Does store image influence demand for organic store brands? *Journal of Retailing and Consumer Services*, *19*(6), 621–628. doi:10.1016/j.jretconser.2012.08.003

Oppermann, M. (2000). Triangulation - A methodological discussion. *International Journal of Tourism Research*, *2*(2), 141–145. doi:10.1002/(SICI)1522-1970(200003/04)2:2<141::AID-JTR217>3.0.CO;2-U

Palinkas, L. A., Horwitz, S. M., Green, C. A., Wisdom, J. P., Duan, N., & Hoagwood, K. (2015). Purposeful sampling for qualitative data collection and analysis in mixed method implementation research. *Administration and Policy in Mental Health, 42*(5), 533–544. doi:10.100710488-013-0528-y PMID:24193818

Perrini, F., Castaldo, S., Misani, N., & Tencati, A. (2010). The impact of corporate social responsibility associations on trust in organic products marketed by mainstream retailers: A study of Italian consumers. *Business Strategy and the Environment, 19*(8), 512–526. doi:10.1002/bse.660

Pool, J. K., Khodadadi, M., & Asadi, A. (2018). The impact of congruence between self-concept and destination personality on behavioural intentions for visiting an Islamic-historical destination. *Tourism and Hospitality Research, 18*(3), 378–387. doi:10.1177/1467358416663820

Pradhan, D., Duraipandian, I., & Sethi, D. (2016). Celebrity endorsement: How celebrity–brand–user personality congruence affects brand attitude and purchase intention. *Journal of Marketing Communications, 22*(5), 456–473. doi:10.1080/13527266.2014.914561

Rana, J., & Paul, J. (2020). Health motive and the purchase of organic food: A meta-analytic review. *International Journal of Consumer Studies, 44*(2), 162–171. doi:10.1111/ijcs.12556

Richetin, J., Mattavelli, S., & Perugini, M. (2016). Increasing implicit and explicit attitudes toward an organic food brand by referencing to oneself. *Journal of Economic Psychology, 55*, 96–108. doi:10.1016/j.joep.2016.01.006

Richetin, J., Mattavelli, S., & Perugini, M.ReportLinker. (2021). *Global Organic Food Market, By Product Type, By Distribution Channel Type (Hypermarket/Supermarket, Speciality Stores, Departmental Stores, Online and Others), By Region, Competition, Forecast Opportunities, 2026*. Report Linker. doi:10.1016/j.joep.2016.01.006

Saumure, K., & Given, L. M. (2008). Data saturation. In L. M. Given (Ed.), *The SAGE Encyclopedia of Qualitative Research Methods* (pp. 195–196). SAGE.

Schleenbecker, R., & Hamm, U. (2013). Consumers' perception of organic product characteristics. A review. *Appetite, 71*, 420–429. doi:10.1016/j.appet.2013.08.020 PMID:24012637

Septianto, F., Kemper, J., & Paramita, W. (2019). The role of imagery in promoting organic food. *Journal of Business Research, 101*, 104–115. doi:10.1016/j.jbusres.2019.04.016

Shin, Y. H., Moon, H., Jung, S. E., & Severt, K. (2017). The effect of environmental values and attitudes on consumer willingness to pay more for organic menus: A value-attitude-behavior approach. *Journal of Hospitality and Tourism Management, 33*, 113–121. doi:10.1016/j.jhtm.2017.10.010

Sirieix, L., Kledal, P. R., & Sulitang, T. (2011). Organic food consumers' trade-offs between local or imported, conventional or organic products: A qualitative study in Shanghai. *International Journal of Consumer Studies, 35*(6), 670–678. doi:10.1111/j.1470-6431.2010.00960.x

Sønderskov, K. M., & Daugbjerg, C. (2011). The state and consumer confidence in eco-labeling: Organic labeling in Denmark, Sweden, The United Kingdom and The United States. *Agriculture and Human Values, 28*(4), 507–517. doi:10.100710460-010-9295-5

Soyez, K., Francis, J. N.-P., & Smirnova, M. M. (2012). How individual, product and situational determinants affect the intention to buy and organic food buying behavior: A cross-national comparison in five nations. *der markt, 51*, 27-35. doi:10.1007/s12642-011-0073-8

Statista. (2021). *Organic food sales in the United States from 2005 to 2020*. Satista: https://www.statista.com/statistics/196952/organic-food-sales-in-the-us-since-2000/

Stolz, H., Stolze, M., Hamm, U., Janssen, M., & Ruto, E. (2011). Consumer attitudes towards organic versus conventional food with specific quality attributes. *NJAS Wageningen Journal of Life Sciences, 58*(3-4), 67–72. doi:10.1016/j.njas.2010.10.002

Storstad, O., & Bjørkhaug, H. (2003). Foundations of production and consumption of organic food in Norway: Common attitudes among farmers and consumers? *Agriculture and Human Values, 20*(2), 151–163. doi:10.1023/A:1024069627349

Sultan, P., Wong, H. Y., & Sigala, M. (2018). Segmenting the Australian organic food consumer market. *Asia Pacific Journal of Marketing and Logistics, 30*(1), 163–181. doi:10.1108/APJML-10-2016-0211

Talwar, S., Jabeen, F., Tandon, A., Sakashita, M., & Dhir, A. (2021). What drives willingness to purchase and stated buying behavior toward organic food? A Stimulus–Organism–Behavior–Consequence (SOBC) perspective. *Journal of Cleaner Production, 293*, 125882. doi:10.1016/j.jclepro.2021.125882

Tandon, A., Dhir, A., Kaur, P., Kushwah, S., & Salo, J. (2020). Why do people buy organic food? The moderating role of environmental concerns and trust. *Journal of Retailing and Consumer Services, 57*, 102247. doi:10.1016/j.jretconser.2020.102247

Tandon, A., Jabeen, F., Talwar, S., Sakashita, M., & Dhir, A. (2021). Facilitators and inhibitors of organic food buying behavior. *Food Quality and Preference, 88*, 104077. doi:10.1016/j.foodqual.2020.104077

Thøgersen, J., Pedersen, S., Paternoga, M., Schwendel, E., & Aschemann-Witze, J. (2017). How important is country-of-origin for organic food consumers? A review of the literature and suggestions for future research. *British Food Journal, 119*(3), 542–557. doi:10.1108/BFJ-09-2016-0406

Truong, V. A., Lang, B., & Conroy, D. M. (2021). Are trust and consumption values important for buyers of organic food? A comparison of regular buyers, occasional buyers, and non-buyers. *Appetite, 161*, 105123. doi:10.1016/j.appet.2021.105123 PMID:33493610

Ureña, F., Bernabéu, R., & Olmeda, M. (2008). Women, men and organic food: Differences in their attitudes and willingness to pay. A Spanish case study. *International Journal of Consumer Studies, 32*(1), 18–26. doi:10.1111/j.1470-6431.2007.00637.x

van Doorn, J., & Verhoef, P. C. (2011). Willingness to pay for organic products: Differences between virtue and vice foods. *International Journal of Research in Marketing, 28*(3), 167–180. doi:10.1016/j.ijresmar.2011.02.005

van Herpen, E., van Nierop, E., & Sloot, L. (2012). The relationship between in-store marketing and observed sales for organic versus fair trade products. *Marketing Letters, 23*(1), 293–308. doi:10.100711002-011-9154-1

Vega-Zamora, M., Torres-Ruiz, F. J., Murgado-Armenteros, E. M., & Parras-Rosa, M. (2014). Organic as a heuristic cue: What Spanish consumers mean by organic foods. *Psychology and Marketing*, *31*(5), 349–359. doi:10.1002/mar.20699

Wallace, E., Buil, I., & de Chernatony, L. (2017). Consumers' self-congruence with a "Liked" brand: Cognitive network influence and brand outcomes. *European Journal of Marketing*, *51*(2), 367–390. doi:10.1108/EJM-07-2015-0442

Wellard, S. J., & McKenna, L. (2002). Turning tapes into text: Issues surrounding the transcription of interviews. *Contemporary Nurse*, *11*(2/3), 180–186. doi:10.5172/conu.11.2-3.180 PMID:11924614

Wu, X., Xiong, J., Yan, J., & Wang, Y. (2021). Perceived quality of traceability information and its effect on purchase intention towards organic food. *Journal of Marketing Management*, *37*(13/14), 1267–1286. doi:10.1080/0267257X.2021.1910328

Xie, B., Wang, L., Yang, H., Wang, Y., & Zhang, M. (2015). Consumer perceptions and attitudes of organic food products in Eastern China. *British Food Journal*, *117*(3), 1105–1121. doi:10.1108/BFJ-09-2013-0255

Yu, W., Han, X., Ding, L., & He, M. (2021). Organic food corporate image and customer co-developing behavior: The mediating role of consumer trust and purchase intention. *Journal of Retailing and Consumer Services*, *59*, 102377. doi:10.1016/j.jretconser.2020.102377

ADDITIONAL READING

Ahmed, N., Li, C., Khan, A., Qalati, S. A., Naz, S., & Rana, F. (2021). Purchase intention toward organic food among young consumers using theory of planned behavior: Role of environmental concerns and environmental awareness. *Journal of Environmental Planning and Management*, *64*(5), 796–822. doi:10.1080/09640568.2020.1785404

Aitken, R., Watkins, L., Williams, J., & Kean, A. (2020). The positive role of labelling on consumers' perceived behavioural control and intention to purchase organic food. *Journal of Cleaner Production*, *255*, 120334. doi:10.1016/j.jclepro.2020.120334

Arumugam, S., Govindasamy, R., Vellangany, I., & Gohil, H. (2017). An analysis of organic fresh produce choice: A consumer preference study in the Mid-Atlantic USA. *Journal of Food Distribution Research*, *48*(1), 101–102. doi:10.22004/ag.econ.274576

Aschemann-Witzel, J., Maroscheck, N., & Hamm, U. (2013). Are organic consumers preferring or avoiding foods with nutrition and health claims? *Food Quality and Preference*, *30*(1), 68–76. doi:10.1016/j.foodqual.2013.04.011

Aschemann-Witzel, J., & Zielke, S. (2017). Can't buy me green? A review of consumer perceptions of and behavior toward the price of organic food. *The Journal of Consumer Affairs*, *51*(1), 211–251. doi:10.1111/joca.12092

Asioli, D., Canavari, M., Pignatti, E., Obermowe, T., Sidali, K. L., Vogt, C., & Spiller, A. (2014). Sensory experiences and expectations of Italian and German organic consumers. *Journal of International Food & Agribusiness Marketing*, *26*(1), 13–27. doi:10.1080/08974438.2012.755718

Baudry, J., Touvier, M., Allès, B., Péneau, S., Méjean, C., Galan, P., Hercberg, S., Lairon, D., & Kesse-Guyot, E. (2016). Typology of eaters based on conventional and organic food consumption: Results from the NutriNet-Santé cohort study. *British Journal of Nutrition*, *116*(4), 700–709. doi:10.1017/S0007114516002427 PMID:27311793

Beldad, A., Hegner, S., & Hoppen, J. (2016). The effect of virtual sales agent (VSA) gender – product gender congruence on product advice credibility, trust in VSA and online vendor, and purchase intention. *Computers in Human Behavior*, *60*, 62–72. doi:10.1016/j.chb.2016.02.046

Britwum, K., Bernard, J. C., & Albrecht, S. E. (2021). Does importance influence confidence in organic food attributes? *Food Quality and Preference*, *87*, 104056. doi:10.1016/j.foodqual.2020.104056

Chen, J., Lobo, A., & Rajendran, N. (2014). Drivers of organic food purchase intentions in mainland China – Evaluating potential customers' attitudes, demographics and segmentation. *International Journal of Consumer Studies*, *38*(4), 346–356. doi:10.1111/ijcs.12095

Chien, S.-H., Wu, J.-J., & Huang, C.-Y. (2018). "We made, we trust": Coproduction and image congruence in the food-tourism factories. *Asia Pacific Management Review*, *23*(4), 310–317. doi:10.1016/j.apmrv.2018.01.002

Cranfield, J., Deaton, B. J., & Shellikeri, S. (2009). Evaluating consumer preferences for organic food production standards. *Canadian Journal of Agricultural Economics*, *57*(1), 99–117. doi:10.1111/j.1744-7976.2008.01140.x

De Canio, F., & Martinelli, E. (2021). EU quality label vs organic food products: A multigroup structural equation modeling to assess consumers' intention to buy in light of sustainable motives. *Food Research International*, *139*, 109846. doi:10.1016/j.foodres.2020.109846 PMID:33509469

de Droog, S. M., Buijzen, M., & Valkenburg, P. M. (2014). Enhancing children's vegetable consumption using vegetable-promoting picture books. The impact of interactive shared reading and character–product congruence. *Appetite*, *73*, 73–80. doi:10.1016/j.appet.2013.10.018 PMID:24216486

Eisinger-Watzl, M., Wittig, F., Heuer, T., & Hoffmann, I. (2015). Customers purchasing organic food - Do they live healthier? Results of the German national nutrition survey II. *European Journal of Nutrition & Food Safety*, *5*(1), 59–71. doi:10.9734/EJNFS/2015/12734

Elo, S., & Kyngäs, H. (2008). The qualitative content analysis process. *Journal of Advanced Nursing*, *62*(1), 107–115. doi:10.1111/j.1365-2648.2007.04569.x PMID:18352969

Fatha, L., & Ayoubi, R. (2021). A revisit to the role of gender, age, subjective and objective knowledge in consumers' attitudes towards organic food. *Journal of Strategic Marketing*, 1–17. Advance online publication. doi:10.1080/0965254X.2021.1939405

Francis, J. J., Johnston, M., Robertson, C., Dlidewell, L., Entwistle, V., Eccles, M. P., & Grimshaw, J. M. (2010). What is an adequate sample size? Operationalising data saturation for theory-based interview studies. *Psychology & Health, 25*(10), 1229–1245. doi:10.1080/08870440903194015 PMID:20204937

Fujii, L. A. (2018). *Interviewing in Social Science Research: A Relational Approach.* Routledge.

Gianluigi Guido, M. I., Peluso, A. M., Maloumby-Baka, R. C., & Buffa, C. (2010). The role of ethics and product personality in the intention to purchase organic food products: A structural equation modeling approach. *International Review of Economics, 57*(1), 79–102. doi:10.100712232-009-0086-5

Gilsenan, C., Burke, R. M., & Barry-Ryan, C. (2012). Do organic cherry vine tomatoes taste better than conventional cherry vine tomatoes? A sensory and instrumental comparative study from Ireland. *Journal of Culinary Science & Technology, 10*(2), 154–167. doi:10.1080/15428052.2012.679232

Grubor, A., & Djokic, N. (2016). Organic food consumer profile in the Republic of Serbia. *British Food Journal, 118*(1), 164–182. doi:10.1108/BFJ-06-2015-0225

Hanks, L., & Mattila, A. S. (2016). Consumer response to organic food in restaurants: A serial mediation analysis. *Journal of Foodservice Business Research, 19*(1), 109–121. doi:10.1080/15378020.2016.1129228

Harper, G. C., & Makatouni, A. (2002). Consumer perception of organic food production and farm animal welfare. *British Food Journal, 104*(3-5), 287–299. doi:10.1108/00070700210425723

Hellmann, R., Griesbaum, J., & Mandl, T. (2010). Quality in blogs: How to find the best user generated content. In W. Abramowicz, & R. Tolksdorf, *Business Information Systems. BIS 2010.* [Berlin, Heidelberg: Springer.]. *Lecture Notes in Business Information Processing, 47*, 47–58. doi:10.1007/978-3-642-12814-1_5

Hwang, J. (2016). Organic food as self-presentation: The role of psychological motivation in older consumers' purchase intention of organic food. *Journal of Retailing and Consumer Services, 28*, 281–287. doi:10.1016/j.jretconser.2015.01.007

Hwang, J., & Chung, J.-E. (2019). What drives consumers to certain retailers for organic food purchase: The role of fit for consumers' retail store preference. *Journal of Retailing and Consumer Services, 47*, 293–306. doi:10.1016/j.jretconser.2018.12.005

Janssen, M., & Hamm, U. (2012). Product labelling in the market for organic food: Consumer preferences and willingness-to-pay for different organic certification logos. *Food Quality and Preference, 25*(1), 9–22. doi:10.1016/j.foodqual.2011.12.004

Kareklas, I., Carlson, J. R., & Muehling, D. D. (2014). "I eat organic for my benefit and yours": Egoistic and altruistic considerations for purchasing organic food and their implications for advertising strategists. *Journal of Advertising, 43*(1), 18–32. doi:10.1080/00913367.2013.799450

Kassarjian, H. H. (1977). Content analysis in consumer research. *The Journal of Consumer Research, 4*(1), 8–18. doi:10.1086/208674

Kim, J. (2012). The institutionalization of YouTube: From user-generated content to professionally generated content. *Media Culture & Society, 34*(1), 53–67. doi:10.1177/0163443711427199

Klabi, F. (2020). Self-image congruity affecting perceived quality and the moderation of brand experience: The case of local and international brands in the Kingdom of Saudi Arabia. *Journal of Global Marketing*, *33*(2), 69–83. doi:10.1080/08911762.2019.1614242

Konuk, F. A. (2018). The role of store image, perceived quality, trust and perceived value in predicting consumers' purchase intentions towards organic private label food. *Journal of Retailing and Consumer Services*, *43*, 304–310. doi:10.1016/j.jretconser.2018.04.011

Lee, W.-C. J., Shimizu, M., Kniffin, K. M., & Wansink, B. (2013). You taste what you see: Do organic labels bias taste perceptions? *Food Quality and Preference*, *29*(1), 33–39. doi:10.1016/j.foodqual.2013.01.010

Liu, C., Zheng, Y., & Cao, D. (2021). An analysis of factors affecting selection of organic food: Perception of consumers in China regarding weak signal. *Appetite*, *161*, 105145. doi:10.1016/j.appet.2021.105145 PMID:33515620

Lu, L., & Gursoy, D. (2017). Does offering an organic food menu help restaurants excel in competition? An examination of diners' decision-making. *International Journal of Hospitality Management*, *63*, 72–81. doi:10.1016/j.ijhm.2017.03.004

Ma, Z., Chen, J., Tian, G., Guo, B., & Cheng, F. (2021). Regulations on the corporate social irresponsibility in the supply chain under the multiparty game: Taking China's organic food supply chain as an example. *Journal of Cleaner Production*, *317*, 128459. Advance online publication. doi:10.1016/j.jclepro.2021.128459

Malodia, S., Singh, P., Goyal, V., & Sengupta, A. (2017). Measuring the impact of brand-celebrity personality congruence on purchase intention. *Journal of Marketing Communications*, *23*(5), 493–512. doi:10.1080/13527266.2017.1322125

Marian, L., Chrysochou, P., Krystallis, A., & Thøgersen, J. (2014). The role of price as a product attribute in the organic food context: An exploration based on actual purchase data. *Food Quality and Preference*, *37*, 52–60. doi:10.1016/j.foodqual.2014.05.001

McEachern, M. G., & Willock, J. (2004). Producers and consumers of organic meat: A focus on attitudes and motivations. *British Food Journal*, *106*(7), 534–552. doi:10.1108/00070700410545737

McEvoy, M. (2019, March 13). *Organic 101: What the USDA organic label means*. US Department of Agriculture.

Montuori, A. (2005). Literature review as creative inquiry: Reframing scholarship as a creative process. *Journal of Transformative Education*, *3*(4), 374–393. doi:10.1177/1541344605279381

Nagaraj, S. (2021). Role of consumer health consciousness, food safety & attitude on organic food purchase in emerging market: A serial mediation model. *Journal of Retailing and Consumer Services*, *59*, 102423. Advance online publication. doi:10.1016/j.jretconser.2020.102423

Namkung, Y., & Jang, S. (2013). Effects of restaurant green practices on brand equity formation: Do green practices really matter? *International Journal of Hospitality Management*, *33*, 85–95. doi:10.1016/j.ijhm.2012.06.006

Nasir, V. A., & Karakaya, F. (2014). Consumer segments in organic foods market. *Journal of Consumer Marketing, 31*(4), 263–277. doi:10.1108/JCM-01-2014-0845

Ngobo, P. V. (2011). What drives household choice of organic products in grocery stores? *Journal of Retailing, 87*(1), 90–100. doi:10.1016/j.jretai.2010.08.001

Ngobo, P.-V., & Jean, S. (2012). Does store image influence demand for organic store brands? *Journal of Retailing and Consumer Services, 19*(6), 621–628. doi:10.1016/j.jretconser.2012.08.003

Oppermann, M. (2000). Triangulation - A methodological discussion. *International Journal of Tourism Research, 2*(2), 141–145. doi:10.1002/(SICI)1522-1970(200003/04)2:2<141::AID-JTR217>3.0.CO;2-U

Palinkas, L. A., Horwitz, S. M., Green, C. A., Wisdom, J. P., Duan, N., & Hoagwood, K. (2015). Purposeful sampling for qualitative data collection and analysis in mixed method implementation research. *Administration and Policy in Mental Health, 42*(5), 533–544. doi:10.100710488-013-0528-y PMID:24193818

Perrini, F., Castaldo, S., Misani, N., & Tencati, A. (2010). The impact of corporate social responsibility associations on trust in organic products marketed by mainstream retailers: A study of Italian consumers. *Business Strategy and the Environment, 19*(8), 512–526. doi:10.1002/bse.660

Pool, J. K., Khodadadi, M., & Asadi, A. (2018). The impact of congruence between self-concept and destination personality on behavioural intentions for visiting an Islamic-historical destination. *Tourism and Hospitality Research, 18*(3), 378–387. doi:10.1177/1467358416663820

Pradhan, D., Duraipandian, I., & Sethi, D. (2016). Celebrity endorsement: How celebrity–brand–user personality congruence affects brand attitude and purchase intention. *Journal of Marketing Communications, 22*(5), 456–473. doi:10.1080/13527266.2014.914561

Rana, J., & Paul, J. (2020). Health motive and the purchase of organic food: A meta-analytic review. *International Journal of Consumer Studies, 44*(2), 162–171. doi:10.1111/ijcs.12556

Richetin, J., Mattavelli, S., & Perugini, M.ReportLinker. (2021). *Global Organic Food Market, By Product Type, By Distribution Channel Type (Hypermarket/Supermarket, Speciality Stores, Departmental Stores, Online and Others, By Region, Competition, Forecast Opportunities, 2026.* Lyon: ReportLinker.
Richetin, J., Mattavelli, S., & Perugini, M. (2016). Increasing implicit and explicit attitudes toward an organic food brand by referencing to oneself. *Journal of Economic Psychology, 55*, 96–108. doi:10.1016/j.joep.2016.01.006

Saumure, K., & Given, L. M. (2008). Data saturation. In L. M. Given (Ed.), *The SAGE Encyclopedia of Qualitative Research Methods* (pp. 195–196). SAGE.

Schleenbecker, R., & Hamm, U. (2013). Consumers' perception of organic product characteristics. A review. *Appetite, 71*, 420–429. doi:10.1016/j.appet.2013.08.020 PMID:24012637

Septianto, F., Kemper, J., & Paramita, W. (2019). The role of imagery in promoting organic food. *Journal of Business Research, 101*, 104–115. doi:10.1016/j.jbusres.2019.04.016

Shin, Y. H., Moon, H., Jung, S. E., & Severt, K. (2017). The effect of environmental values and attitudes on consumer willingness to pay more for organic menus: A value-attitude-behavior approach. *Journal of Hospitality and Tourism Management, 33*, 113–121. doi:10.1016/j.jhtm.2017.10.010

Sirieix, L., Kledal, P. R., & Sulitang, T. (2011). Organic food consumers' trade-offs between local or imported, conventional or organic products: A qualitative study in Shanghai. *International Journal of Consumer Studies*, *35*(6), 670–678. doi:10.1111/j.1470-6431.2010.00960.x

Sønderskov, K. M., & Daugbjerg, C. (2011). The state and consumer confidence in eco-labeling: Organic labeling in Denmark, Sweden, The United Kingdom and The United States. *Agriculture and Human Values*, *28*(4), 507–517. doi:10.100710460-010-9295-5

Soyez, K., Francis, J. N.-P., & Smirnova, M. M. (2012). How individual, product and situational determinants affect the intention to buy and organic food buying behavior: A cross-national comparison in five nations. *der markt, 51*, 27-35. doi:10.1007/s12642-011-0073-8

Statista. (2021). *Organic food sales in the United States from 2005 to 2020*. Satista: https://www.statista.com/statistics/196952/organic-food-sales-in-the-us-since-2000/

Stolz, H., Stolze, M., Hamm, U., Janssen, M., & Ruto, E. (2011). Consumer attitudes towards organic versus conventional food with specific quality attributes. *NJAS Wageningen Journal of Life Sciences*, *58*(3-4), 67–72. doi:10.1016/j.njas.2010.10.002

Storstad, O., & Bjørkhaug, H. (2003). Foundations of production and consumption of organic food in Norway: Common attitudes among farmers and consumers? *Agriculture and Human Values*, *20*(2), 151–163. doi:10.1023/A:1024069627349

Sultan, P., Wong, H. Y., & Sigala, M. (2018). Segmenting the Australian organic food consumer market. *Asia Pacific Journal of Marketing and Logistics*, *30*(1), 163–181. doi:10.1108/APJML-10-2016-0211

Talwar, S., Jabeen, F., Tandon, A., Sakashita, M., & Dhir, A. (2021). What drives willingness to purchase and stated buying behavior toward organic food? A Stimulus–Organism–Behavior–Consequence (SOBC) perspective. *Journal of Cleaner Production*, *293*, 125882. doi:10.1016/j.jclepro.2021.125882

Tandon, A., Dhir, A., Kaur, P., Kushwah, S., & Salo, J. (2020). Why do people buy organic food? The moderating role of environmental concerns and trust. *Journal of Retailing and Consumer Services*, *57*, 102247. Advance online publication. doi:10.1016/j.jretconser.2020.102247

Tandon, A., Jabeen, F., Talwar, S., Sakashita, M., & Dhir, A. (2021). Facilitators and inhibitors of organic food buying behavior. *Food Quality and Preference*, *88*, 104077. doi:10.1016/j.foodqual.2020.104077

Thøgersen, J., Pedersen, S., Paternoga, M., Schwendel, E., & Aschemann-Witze, J. (2017). How important is country-of-origin for organic food consumers? A review of the literature and suggestions for future research. *British Food Journal*, *119*(3), 542–557. doi:10.1108/BFJ-09-2016-0406

Truong, V. A., Lang, B., & Conroy, D. M. (2021). Are trust and consumption values important for buyers of organic food? A comparison of regular buyers, occasional buyers, and non-buyers. *Appetite*, *161*, 105123. doi:10.1016/j.appet.2021.105123 PMID:33493610

Ureña, F., Bernabéu, R., & Olmeda, M. (2008). Women, men and organic food: Differences in their attitudes and willingness to pay. A Spanish case study. *International Journal of Consumer Studies*, *32*(1), 18–26. doi:10.1111/j.1470-6431.2007.00637.x van Doorn, J., & Verhoef, P. C. (2011). Willingness to pay for organic products: Differences between virtue and vice foods. *International Journal of Research in Marketing*, *28*(3), 167–180. doi:10.1016/j.ijresmar.2011.02.005

van Herpen, E., van Nierop, E., & Sloot, L. (2012). The relationship between in-store marketing and observed sales for organic versus fair trade products. *Marketing Letters*, *23*(1), 293–308. doi:10.100711002-011-9154-1

Vega-Zamora, M., Torres-Ruiz, F. J., Murgado-Armenteros, E. M., & Parras-Rosa, M. (2014). Organic as a heuristic cue: What Spanish consumers mean by organic foods. *Psychology and Marketing*, *31*(5), 349–359. doi:10.1002/mar.20699

Wallace, E., Buil, I., & de Chernatony, L. (2017). Consumers' self-congruence with a "Liked" brand: Cognitive network influence and brand outcomes. *European Journal of Marketing*, *51*(2), 367–390. doi:10.1108/EJM-07-2015-0442

Wellard, S. J., & McKenna, L. (2002). Turning tapes into text: Issues surrounding the transcription of interviews. *Contemporary Nurse*, *11*(2/3), 180–186. doi:10.5172/conu.11.2-3.180 PMID:11924614

Wu, X., Xiong, J., Yan, J., & Wang, Y. (2021). Perceived quality of traceability information and its effect on purchase intention towards organic food. *Journal of Marketing Management*, *37*(13/14), 1267–1286. doi:10.1080/0267257X.2021.1910328

Xie, B., Wang, L., Yang, H., Wang, Y., & Zhang, M. (2015). Consumer perceptions and attitudes of organic food products in Eastern China. *British Food Journal*, *117*(3), 1105–1121. doi:10.1108/BFJ-09-2013-0255

Yu, W., Han, X., Ding, L., & He, M. (2021). Organic food corporate image and customer co-developing behavior: The mediating role of consumer trust and purchase intention. *Journal of Retailing and Consumer Services*, *59*, 102377. doi:10.1016/j.jretconser.2020.102377

KEY TERMS AND DEFINITIONS

Functional Image: The ideas held of a product or brand about its practicality and usefulness.

Image Congruence: The image of an item, particularly a brand, matches the self-image held by the consumer.

Organic Consumers: People who purchase and eat foods that are labelled organic, either exclusively or selectively.

Organic Distributers: People or companies responsible for processing organic food and making them available to consumers.

Organic Food Products: Edible items that use little or no manufactured chemicals during production.

Organic Makers: People or companies who grow or raise plants and animals with little or no manufactured chemicals.

Psychological Image: The ideas held of a product or brand about its emotional and non-physical values.

Sustainability: To make, distribute, and consume food products in a way that does not damage the environment.

Chapter 7
Fresh Produce Market Challenges and Opportunities:
A Case for the Johannesburg Municipal Fresh Produce Market

Ndiadivha P. Tempia
https://orcid.org/0000-0002-9873-9895
National Agricultural Marketing Council, South Africa

Elvis Nakana
https://orcid.org/0000-0002-4688-7511
National Agricultural Marketing Council, South Africa

Makhanana Malungane
University of Pretoria, South Africa

Marc Wegerif
University of Pretoria, South Africa

ABSTRACT

This chapter examines the functioning of the national fresh produce markets (NFPMs) with specific reference to three themes: 1) challenges in the regulatory environment and the importance of social relation, 2) networks in shaping the functioning of the NFPMs alongside the narrow economic relations that exist among the market participants in the NFPMs, and 3) the importance of the informal sector for the NFPMs and the distribution of fresh produce. Specific reference is made to the Johannesburg Fresh case study for the Johannesburg Fresh Produce Market (JFPM). The chapter proposes solutions to the mentioned challenges.

DOI: 10.4018/978-1-6684-4780-2.ch007

INTRODUCTION

Fresh produce markets (FPMs) are vital to food security in South Africa. They are key nodes in the flow of fresh produce from primary producers to the end-users. As such, they contribute to determining the viability of farmers and the accessibility of fresh produce across the country. Despite the critical role played by FPMs, there are competing views as to the future of these markets and the most effective means of distributing fresh produce across the country, as well as regionally and globally.

Rapid urbanization across the African continent and rising levels of food insecurity, including urban food insecurity, make effective urban food markets even more important (Baeumler et al., 2021; Balineau et al., 2021). With the double burden of malnutrition and high levels of obesity—increasing especially in cities—access to fresh fruits and vegetables is essential for balanced diets and people's health (Food and Agriculture Organization et al., 2021; Harvard School of Public Health, 2011; Willett et al., 2019). It is also widely argued that increasing the production and consumption of fresh produce is essential for reducing the drivers of climate change and other negative environmental impacts of agriculture (Benton & Bailey, 2019; Benton et al., 2021; Willett et al., 2019). Rising urban incomes and changing lifestyles are also increasing the demand for a greater variety of foods including fresh produce (Balineau et al., 2021). The National Development Plan (NDP) in South Africa has identified fresh produce farming as an area with high potential for new and emerging black farmers and high job creation potential (National Planning Commission, 2011). The nature of markets is one factor that will determine if these challenges are met and opportunities are realized.

It is in this context that this chapter focuses on the Municipal Fresh Produce Markets (MFPMs) in South Africa with specific reference to the Johannesburg Fresh Produce Market (JFPM) as a case study and establish how the functioning markets can contribute to the realization of the NDP goals. Municipal markets have been the mainstay of fresh produce trade in South Africa for decades and have their roots in different forms of markets that emerged more than a century ago to meet the food needs of the new and expanding cities in South Africa (Beavon, 2001; Cripps, 2012). These markets are, however, facing multiple pressures such as shifts in supermarket procurement practices, demands for transformation in what remains a white-dominated agri-food sector, and well-publicized municipal management challenges (Pieterse, 2021; Smit & Madubela, 2021; Staff, 2021). It is essential, given such pressures, that a well-informed discussion takes place about the role of these markets and how they function.

The word "market" has multiple meanings. One is the demand for and the possibility to sell goods, that is, there is a market for them. A market is also a particular place, a building, where goods are sold and bought. The Municipal Fresh Produce Markets (MFPMs) create a physical market space where buyers and sellers come together to trade. Through this platform, they also represent and help to create access to a particular market in the form of a demand for products—in this case, fresh produce. It is important that both sellers and buyers have knowledge of the physical site of these markets and confidence that the will be enough supply and demand for the fresh produce. Also important is that the MFPMs do not only face competition from the creation of other physical market spaces where buyers and sellers can meet but also from the creation of other market channels through which buyers can obtain the produce they want from producers, notably supermarket supply chains. Given the scale of the horticultural industry in South Africa, it is not surprising that participation in this industry is challenging.

The horticultural industry is the second-largest agricultural subsector in South Africa after animal production (R141 billion) and contributed over R92 billion—29% of the total agricultural gross value of R312 billion—in gross value during the 2019-2020 production year (Department of Agriculture Land

Reform and Rural Development [DALRRD], 2021a). Within the horticultural industry, the biggest contributors in terms of gross value during the 2019-2020 production year were vegetables, including potatoes, at R26.8 billion or 29%, deciduous fruits at R24.7 billion or 27%, and citrus fruits at R20 billion or 22%. This chapter is structured into four sections. The next section provides the methodology of the JFPM case study, followed by section three, which gives an overview of the operations of the National Fresh Produce Market (NFPMs) to the actual day-to-day operations with specific reference to JFPM and the regulatory environment. Reflection on experiences and debates on the challenges facing smallholder producers, which will also inform the discussion at the end of this chapter about options for the future of Fresh Produce Markets (FPMs) in South Africa, will also be discussed in section three. Conclusions and recommendations are discussed in section four around three fundamental issues: 1) the challenges in the regulatory environment, 2) the importance of social relations and networks in shaping the actual functioning of the NFPMs alongside the narrow economic relations, and 3) the importance of the informal sector for the NFPMs and for the distribution of fresh produce.

METHODOLOGY

A mixed-method approach was used in the research that informs this chapter. This included 1) primary research, especially at the Johannesburg Fresh Produce Market (JFPM) case study; 2) the use of secondary sources and data such as existing research reports and market statistics; and 3) information and lessons from seminars, webinars, and other discussions with market stakeholders.

Primary Data Collection

The aim of the study was to provide a qualitative description of the JFPM as a case study with reference to 1) the functioning and challenges in the regulatory environment, 2) the importance of social relations and networks in shaping the actual functioning of the NFPMs alongside the narrow economic relations, and 3) the importance of the informal sector for the JFPM and the distribution of fresh produce. Due to the descriptive nature of the study, no sample size was calculated. The study population covered the full range of market actors including market agents and traders of vegetables, fruit, potatoes, and onions and the Mandela and Unity markets at the JFPM, market management, government officials, porters, farmers, and buyers. The buyers who were interviewed ranged from small-scale informal traders to wholesalers and exporters. Primary data collection involved ethnographic research including conducting interviews, gathering life histories of market actors, and making observations. Due to the diverse nature of the economic agents at the JFPM, both semi-structured and structured interviews were conducted.

A survey was conducted in 2019 using a structured questionnaire in which 34 market agents from the JFPM participated. Additionally, 93 structured interviews and 40 semi-structured interviews were conducted in 2021. Life histories gathered from some of the research participants, through the semi-structured interviews, gave insights into how people came to work in the market as well as the social and economic relations that underpin the functioning of the market. These life histories were crucial to understanding how the macro factors in the market are interlinked with the micro activities of the actors. This allowed for the investigation of sociological themes and historical patterns that could explain the functioning of the market (Goodson, 2012; Ojermark, 2007).

Extensive observation at the JFPM throughout the research period helped reveal the spatial organization of the market; identify all actors involved, including some that could easily have been overlooked; understand the daily, weekly, and seasonal routines of the market and networks among the actors.

Secondary Data Collection

Journal articles and other research and academic publications were used to contextualize the current FPMs in South Africa and to gather information on national and international food market debates and experiences. Research was gathered online using search engines such as Google Scholar as well as from the University of Pretoria library. Newspaper articles and photographs were gathered from the Johannesburg library archives; online sources were used to fill in both historical and current information gaps about the JFPM and other markets. Policy documents and reports from the Department of Agriculture Land Reform and Rural Development (DALRRD) as well as the National Agriculture Marketing Council (NAMC) and the South African Competition Commissions were gathered. JFPM reports and information found on their websites was also studied. Statistical data studies collected included information from the JFPM, South African Union of Fresh Produce Markets (SAUFPM), and NAMC.

MARKETING OF FRESH PRODUCE IN SOUTH AFRICA

In South Africa, fruits and vegetables are marketed through sales in the NFPMs, direct sales to intermediaries such as supermarkets, export sales, purchases for processing, and selling directly to buyers at a local level. Within the NFPMs, informal traders play an integral role, although there is a lack of data to back this claim. The NFPMs in South Africa operate on commission-based systems and are governed by the Rules for Fresh Produce Agents issued in terms of the Agricultural Produce Agents Act, 1992 (Act No. 12 of 1992). The market authorities are entitled to 5% of the total sales at the markets, and market agents charge between 5% and 7.5% commission on the sales made.

There are currently twenty-two NFPMs, the majority of which are directly owned and managed by municipalities as departments or business units. There are four categories ownership and management of NFPMs: 1) municipally owned and operated in the same way as other municipal departments (e.g., Durban), 2) corporatized municipal entities with a semi-independent status but still fully owned by municipalities that also appoint board members (e.g., Johannesburg), 3) those where the sites and infrastructure are municipally owned but the markets are privately managed (e.g., Cape Town), and 4) private entities under private ownership and management (e.g. Moeketsi). Referenced in this chapter is data from DALRRD collected from NFPMs over the past twenty years.

NFPMs have remained an important part of the price-setting mechanism, distribution, and marketing of fresh produce in South Africa. However, these markets have witnessed limited growth in the volume of fresh produce traded during the post deregulation period. This can be attributed to the expansion of new markets, particularly retail market outlets and supermarkets, which increased the availability of fresh produce on their shelves (Chikazunga et al., 2008; DALRRD, 2021a; National Agricultural Marketing Council, 2021; NAMC, 1998). According to Chikazunga et al. (2008), the South African NFPM system started as a government initiative to provide regulated meeting places where trade could take place between producers and consumers with equal opportunities for large-scale commercial producers and smallholder farmers. NFPMs were created through regulation and are bound by law not to discriminate based on race, color, or operation size. Smallholder farmers, including informal traders, face economic and regulatory barriers to entering the NFPMs, particularly since they can only trade through registered agents. Most fresh produce supplied to the NFPMs is still dominated by large-scale commercial producers that account for 80% to 90% of the value of produce delivered to the NFPMs (NAMC, 2000a, 2005b).

Figures 1 and 2 show the performance of NFPMs over the past two decades and indicate that they have witnessed limited growth in the volumes of fresh produce traded despite the overall production of fresh produce having increased. Volumes of important vegetables sold on the major NFPMs during the period 2001 to 2020 are depicted in Figure 1. A total of 2.45 million tons of vegetables were sold on the NFPMs in 2020, and this was 25% higher than the volume sold on the same platform 20 years ago (1.96 million tons in 2001). The largest annual volume of vegetables sold in the NFPMs during the 20 years under review was achieved in 2018 at 2.63 million tons, while the lowest sales volume was achieved in 2002 at 1.86 million tons. The major vegetables sold on the NFPMs during 2020 were potatoes (1.17 million tons or 48%), followed by onions (0.39 million tons or 16%), and tomatoes (0.29 million tons or 6%). These top three vegetables usually account for over 70% of the total annual vegetable sales on the NFPMs. Over the past 20 years (2001 to 2020), the volume of potatoes sold on the NFPMs grew by 30%, while the sales of onions and tomatoes grew by 63% and 26%, respectively.

Figure 1. Quantities of important vegetables sold on major fresh produce markets, 2001 to 2020

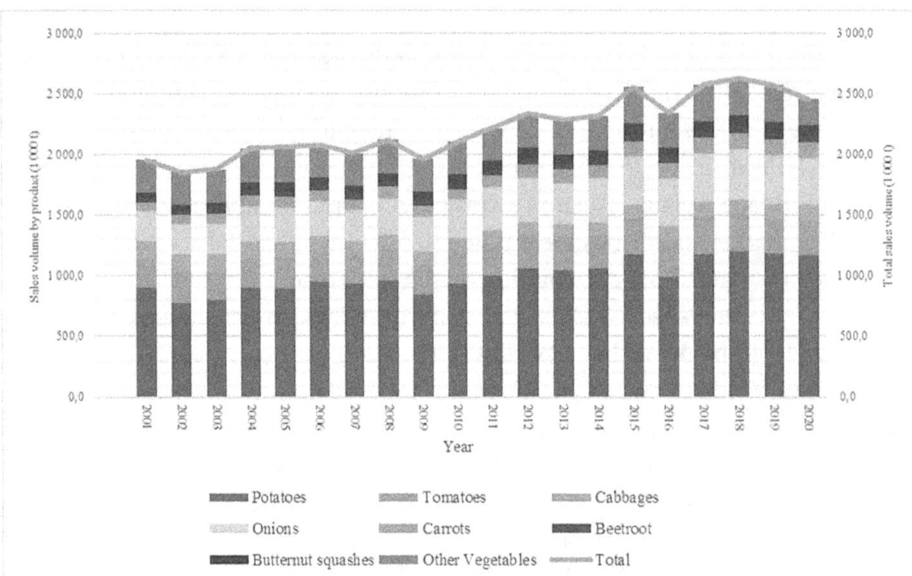

Figure 2 illustrates the total production of fruits and vegetables in South Africa over the past 20 years against the total sales of fruits and vegetables on the NFPMs during the same period. During the period under review (2001 to 2020), the total production of fruits and vegetables increased from 8.97 million tons in 2001 to 15.04 million tons in 2020 (an increase of 68%). During the same period, the sales of fruits and vegetables at the NFPMs increased from 2.72 million tons in 2001 to 3.18 million tons in 2020, representing an increase of 17%. This is an indication that although total sales of fruits and vegetables on the NFPMs have increased during the past 20 years, such an increase was outpaced by the increase in total production volumes for fruits and vegetables during the corresponding period. Specifically, there was a 5% decline in the volume of fruits sold on the NFPMs between 2001 and 2020, compared with an 86% increase in production in the country. The sales of vegetables on the NFPMs during the same period increased by 25%, but this was below the 45% increase in total production.

Figure 2. Total South African production of fruits and vegetables vs. total fruits and vegetable sales on the major fresh produce markets, 2001 to 2020

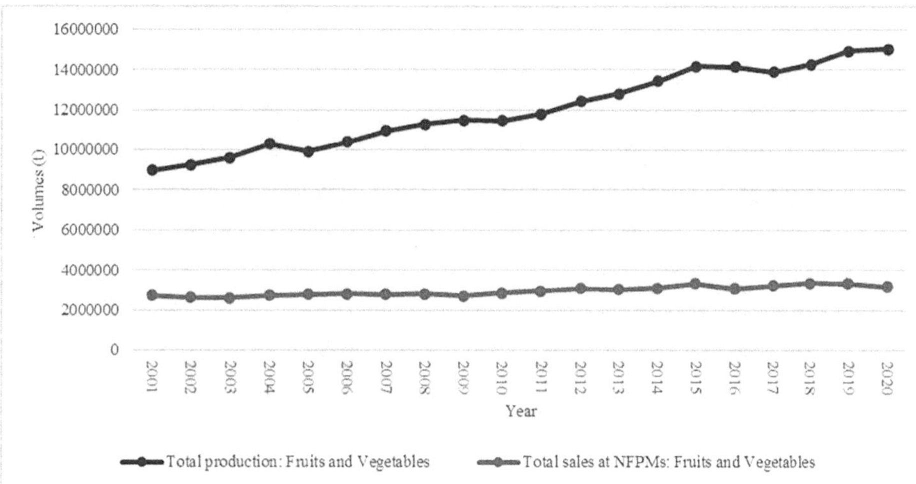

Data from the Department of Agriculture Land Reform and Rural Development (2020) and South African Union of Food Markets (2022) indicate that the JFPM was by far the largest fresh produce market in South Africa during 2020, accounting for 45% of the total turnover of all FPMs during the same year. It was followed by the Tshwane, Cape Town, and Durban FPMs with 19%, 10%, and 9% market shares, respectively. The section below refers to JFPMs to flag some operational challenges facing smallholder farmers. JFPMs are also used to demonstrate the role of informal traders around the market. This group is neglected, yet their economic activities contribute to the economy and support their livelihoods.

THE CASE OF THE JFPM

Overview of the JFPM

The JFPM can be traced to the discovery of gold in 1886 and the emergence of camps to accommodate miners in what became Johannesburg (Beavon, 2001). The miners and others involved in the rapidly growing industry needed food. Farmers and traders came to the camps to take advantage of that market opportunity. The burgeoning miners' camps and related food trade required physical space, and by December 1986 the camps comprised what was, unsurprisingly, named Market Square, which became the Johannesburg central business district. Perhaps seeing an opportunity for revenue collection, as well as a need for certain orders in the food market, the Kruger government issued a concession for the management of the market in 1887. Market buildings constructed in 1888 could not accommodate all the market activity, much of which continued in the open square as depicted in Figure 3.

Figure 3. Market Square, 1899

In response to rapid population growth and commensurate demand for food, the market moved to a new market hall on 11 hectares of land called Newtown Market in 1913 (Davie, 2018). Growth and changes continued until the market was moved again in 1974 to its current location on 63 hectares of land in City Deep. There is currently a substantial new extension to the Mandela market, which is under construction. But there are regular complaints that the main infrastructure and key facilities such as cold storage are not being maintained, which is affecting the competitiveness of the market, especially compared to the tightly managed supply chains and cold chain management of supermarket groups.

The JFPM is a wholesale and retail fresh produce market situated 5 km from Johannesburg's central business district in City Deep. It operates 24 hours per day and sells from 4 a.m. to 11 a.m. on weekdays and from 4 a.m. to 10 a.m. on Saturdays. The JFPM is entirely owned by the City of Johannesburg Metropolitan Municipality, although it operates with a semi-independent Board of Directors. It directly employs 313 people and facilitates the sale of produce from around 5,000 farmers through 19 registered market agents to an estimated 10,000 buyers per day (Johannesburg Fresh Produce Market, 2021). The farmers include the largest commercial farms in the country as well as small-scale farmers. The buyers at the JFPM include wholesalers, supermarkets, independent grocery stores, hospitality and catering businesses, households, school suppliers, exporters, and informal traders of various kinds. In January 2022, the JFPM had a turnover of R652 million and sold 108,987 metric tons of fresh produce (SAUFM, 2022).

The JFPM is organized into three main hubs, which are housed in large buildings, for each of the following produce types: potatoes and onions, fruit, and vegetables. Only registered market agents can sell in these hubs from trading areas designated for each market agency. There are also two other market spaces within the JFPM: the Mandela and Unity markets. In these markets, those who have bought in larger quantities in the main hubs, sell in smaller quantities to cash buyers who do not have cards to trade in the main hubs.

The farmers supplying the market are primarily from the Limpopo, Mpumalanga, and Gauteng provinces (Freidberg & Goldstein, 2011). With a delivery note, the produce is transported in lorries, trucks, bakkies, and private vehicles from farms to the JFPM. The delivery note is inspected at the entrance and marked with a stamp with the date and a tracking number, which allows the driver to enter the market. The driver delivers the produce to a market agent, an intermediary for farmers, or sellers. All JFPM market agents operate under a fresh produce agency and are registered with the Agriculture Produce Agents Council (APAC). The registration process involves completing an application form; achieving a minimum 75% pass rate in an online training course; and paying a once-off fee of R52,441.12, which includes the issuance of a certificate (R1,761.92), vetting and crediting fee (R679.20), and APAC fidelity fund contribution (R50,000), and an annual fee of R12,128.03. In addition, market agencies are also governed by the JFPM bylaws.

Most market agencies are owned by families of English, Portuguese, and Afrikaner descent and have been operational in the market for decades. Licensed market agents are primarily white males, and some are second and third-generation market agents. For example, Rodrigo immigrated as a child to South Africa from Portugal, moving with his father who helped his uncle operate his market agency at the JFPM in 1977. Over time, Rodrigo's father started his own fresh produce selling business operating under his uncle's market agency. After high school, Rodrigo joined the family business and helped his father operate the business. Subsequently, after learning the ropes, Rodrigo opened his own fresh produce business under the same market agency. In 2021, Rodrigo operated from the same sales floor as his two sons who partnered and opened their own business after learning the trade, and they share the use of equipment such as forklifts, cold storage, and jack lifts. This type of arrangement within a market agency is typical of the JFPM. Market agents that operate under the same market agency either share a related ethnicity or come from the same area. These relations extend to the farmers they represent at the Market. Furthermore, ethnic-based arrangements expand to buyers. Portuguese family-owned fresh produce retailers, familiar in Gauteng suburbs, tend to buy from Portuguese market agents. They have long-standing relationships and extend to social interactions outside their activities at the market such as attending parties, braais, and funerals.

The majority of black people selling on the JFPM are salespersons not registered with APAC. However, a handful of black licensed market agencies and agents can be found at the JFPM, like Dineo, a black female market agent. She came to the market in 2010 as an intern in the management offices of the JFPM. She was promoted within a year and worked for the JFPM for six years before becoming a market agent under one of the black-owned market agencies in 2016. Dineo currently employs five people and sources produce from farmers in the Western Cape, Gauteng, and Limpopo provinces. She is among just a few female market agents operating at the JFPM.

To transact on the sales floors, market agents and buyers are registered on an electronic payment system called Freshmark Systems and issued smart cards by the JFPM. Buyers need to provide identification before registering on the system such as a South African Identity Document (ID) or a passport. There are buyers who cannot provide proof of identification to register for a smart card, so they bypass this requirement by sharing smart cards or asking a buyer with a card to make a purchase on their behalf for a fee. Sales between market agents and buyers are concluded through a private treaty. The commission system is the same as that for other FPMs, with the agent getting 5-7.5%, the JFPM 5% and the balance (87.5-90%) going to the farmer.

Contribution of Informal Traders at the JFPM

Informal traders have become integral to the functioning of the JFPM, especially as the large supermarket groups have shifted their approach to acquiring more through direct supply chains that they control. The concept of the "informal sector" arose in the 1970s after it was coined by Keith Hart based on his research in Accra, Ghana. There is uncertainty about its definition, but the informal sector can be understood as being made up of economic activities which take place outside the framework of official institutions and are not part of "what we imagine constitutes the orthodox core of our own economy" (Hart, 1985, p. 57). The informal traders that buy at the JFPM range from energetic young men pushing a trolley full of fresh bananas shouting, "*amaBanana! R1!*" through the bustling streets of Johannesburg CBD to cross-border traders from as far north Zambia. They include street traders, bakkie traders, and spaza shop owners. Due to limited access to storage facilities, street traders visit the JFPM daily to buy their stock. They arrive as early as 4 a.m. on the market floor, having traveled from as far as Tembisa, about 40 km away. Market agents who were interviewed attribute 50-60% of their sales to informal traders and the report in 2006 estimated that the informal sector accounted for 63% of turnover at the JFPM. Taking the lower bound of 50%, informal traders accounted for R326 million of the sales in January 2022 and around R4 billion in 2021 (SAUFM, 2022). This makes the informal traders essential for the JFPM and for the farmers that supply it.

While there are many small-scale informal sector buyers, there are also large-scale buyers performing a wholesale function who are considered informal because they have no registered businesses, and they sell to others in the informal sector. These include the Maputo Mamas—Mozambican women who buy fresh produce in bulk, sometimes spending hundreds of thousands of Rands in a day to take across the border to supply markets such as Zimpeta in Maputo. There are also domestic informal wholesalers like Ntsetselelo, who sells in bulk, mostly to street traders, in a large township. Ntsetselelo peruses the produce herself and spends a minimum of R50,000 per day across the three formal hubs and employs five people in her operation. The JFPM, by being a consistent and cost-effective source of fresh produce, plays a key role in the creation of jobs in the informal sector.

The JFPM, together with the informal sector traders it supplies, makes an important contribution to food security in Gauteng by making fresh produce available and accessible, especially in low-income areas. A number of studies have clearly shown this through selling at lower prices; in flexible and small quantities that fit buyers' needs and ability to pay; at locations close to where people live, work, and travel; and at times, on interest-free credit (Battersby & Watson, 2018; Rudolph et al., 2021; M. C. A. Wegerif, 2020). Despite the importance of the informal sector to JFPM, which is owned by the municipality, there is almost no attention paid to it in the municipal integrated and spatial development plans (see, e.g., City of Johannesburg, 2016; COJ, 2020). Instead, our interviews with street traders and others in the informal sector revealed that many experience ongoing harassment by municipal officials, primarily the Johannesburg Metro Police Department. The lack of understanding and even negative stance toward the informal food sector was also strongly demonstrated at a national level during the first Covid-19 lockdown when, despite food being an essential service, street traders were banned from operating (Battersby, 2020; M. C. A. Wegerif, 2020).

GOVERNMENT POLICY AND REGULATORY ENVIRONMENT

The marketing of agricultural products in South Africa is regulated through the Marketing of Agricultural Products (MAP) Act, 1996 (Act No. 47 of 1996). To facilitate implementation, the MAP Act provides, among other things, for the establishment of the National Agricultural Marketing Council (NAMC). The MAP Act has four main objectives as described previously: 1) increasing market access for all market participants, 2) promotion of the efficiency of the marketing of agricultural products, 3) optimization of export earnings from agricultural products, and 4) enhancement of the viability of the agricultural sector. The MAP Act recognizes the importance of increasing market access for all market participants and places this as the first objective of the act.

In realizing the role that the government must play in the marketing of agricultural products within the country, the DALRRD developed and published the Agricultural Marketing Policy with the vision of creating a globally competitive and profitable agricultural sector with equitable access and participation in the opportunities offered by agricultural marketing in the various value chains (DALLRD, 2010a). According to the Agricultural Marketing Policy (DALRRD, 2010a), government intervention in agricultural marketing is premised on the following three key guiding principles:

Government intervention in agricultural marketing should be limited to the correction of market imperfections and socially unacceptable effects.

Government intervention in agricultural marketing should prioritize targeted non-market mechanisms over the correction of socially unacceptable conditions and

Government intervention in agricultural marketing should allow for export marketing arrangements which enhance the welfare of the nation as a whole. (p. 5)

The Agricultural Marketing Policy makes it clear that government intervention in agricultural marketing must be limited or confined to those actions that are essential for the correction of market imperfections. The passing of the MAP Act heralded in a significant shift in the regulation of agricultural marketing in South Africa in that the involvement of the state shifted from a highly regulatory role to a limited one, whereas the involvement of private enterprises in agricultural marketing was encouraged and supported.

Regulation of Fresh Produce Agents and Markets

Fresh produce agents are regulated through the Agricultural Produce Agents (APA) Act, 1992 (Act No. 12 of 1992). The APA Act, as amended by the Agricultural Produce Agents Amendment Act (No. 47 of 2003), provides for the establishment of the Agricultural Produce Agents Council (APAC) to regulate and oversee the market agents and manage a fidelity fund. The NFPMs that fall under the control of municipalities are also regulated through bylaws and municipal procurement and financial procedures. In effect, the regulatory responsibilities at the NFPMs are split between the APAC and market authorities (municipalities) in the 16 NFPMs still owned by municipalities. The privately owned markets still fall under the APA Act, but not the municipalities. The market authorities are responsible for ensuring that the agents observe the rules of the markets, while the APAC has a statutory responsibility to regulate the conduct of fresh produce agents (NAMC, 1998).

Many of those interviewed raised concerns about the constraints that market authorities face in having to work as part of and in compliance with municipal regulations and procedures. Some expressed the opinion that municipalities also use the NFPMs as cash cows, diverting income generated in markets for other activities. This situation has become more acute with the fiscal crisis brought on by the Covid-19 pandemic, which is also affecting municipalities. The semi-independence of the JFPM, which has its own board but remains fully owned by the municipality, does not seem to have freed it from these constraints, and allegations are made of political interference in board appointments and management. Certainly, the rapid turnover of CEOs in the last few years has indicated management challenges and has not helped to build the confidence of stakeholders in market management.

Pending and Draft Amendments to the Agricultural Produce Agents (APA) Act

At the time of writing this chapter, there are ongoing efforts by the government and other stakeholders to amend the APA Act. A draft Agricultural Produce Agents Amendment Bill (the APA Amendment Bill) is currently being consulted on and processed by Parliament's Portfolio Committee on Agriculture, Land Reform, and Rural Development (Agricultural Produce Agents Amendment Bill, 2020). The critical shortcomings identified in the current APA Act, which the amendments aim to address, are that it does not provide for the APAC to undertake appropriate investigations, inspections, and disciplinary actions against export and livestock agents who mistreat farmers. Furthermore, the current act does not provide for fresh produce and export agents to take out fidelity insurance to protect farmers against losses while the produce is in the care of agents. Export agents are also not compelled to separate their monies from those belonging to their farmers, as no producer trust accounts are kept by export agents.

Recent Developments in National Fresh Produce Markets Regulation: Project Rebirth

As early as 1997, there were concerns raised regarding the operations of the NFPMs arising from the significant deterioration in infrastructure and service standards together with the collapse in management, which seriously affected the rendering of efficient and commercially competitive services. This also undermined the ability of the NFPMs to transform and improve market access to smallholder producers (DALRRD, 2009, p. 8). These concerns triggered the then Minister of Agriculture and Land Affairs, Minister Thoko Didiza, through the National Agricultural Marketing Council to launch investigations termed Section 7 Committee Investigations (in terms of section 7 of the Marketing of Agricultural Products Act, 1996).

In 2013, the continuing deterioration in service standards in the NFPMs triggered key stakeholders in the fresh produce industry, fresh produce market agents through the Institute of Market Agents of South Africa (IMASA) and producers through Potatoes South Africa (PSA), to approach the DALRRD for immediate intervention to address the persistent challenges faced by the markets. This led to the evolution of a new initiative called "Project Rebirth" under the leadership of the DALRRD, which was aimed at improving the operations and service standards of the NFPMs and transforming the markets. Through Project Rebirth, the DALRRD concluded and published Codes of Best Practice (CoBPs) for NFPMs (DALRRD, 2015) and facilitated the implementation of these codes in a collaborative effort with municipalities that own the markets. A mentorship program under the leadership of the Joburg Market and Tshwane FPM was established to assist ailing markets in implementing the standard operating procedures (SOPs) and some provisions of the CoBPs. The mentorship program focuses on building the capacity of market management and its market staff on six key areas: 1) consignment (stock) control, 2) risk and fiscal management, 3) information management, 4) stakeholder management, 5) hygiene, and 6) cold rooms and ripening rooms, as well as compliance to the regulatory environment. During the past few years, a team from Tshwane FPM assisted the Vereeniging FPM, Pietermaritzburg FPM, East London, and Port Elizabeth FPM while a team from Joburg Market assisted the Sol Plaatjie FPM in remedying the current challenging situations in these markets with development and application of Standard Operating Procedures (SOPs), bylaws, and Service Level Agreements (SLAs).

To address the transformation pillar of the CoBPs and owing to the snail pace of the transformation agenda, particularly the transformation of market agents' landscape, the Project Rebirth Steering Committee (PRSC) established a transformation subcommittee to develop strategies and programs aimed at ensuring sector participants reflect the country's demographics. The subcommittee drafted a document titled "Empowering and Capacitating Black Market Agents," which details the challenges faced by black market agents; proposes strategies and programs aimed to promote the development, growth, and competitiveness of black market agents as part of efforts to ensure inclusivity; and drives the transformation agenda in NFPMs.

Given the gravity of the issues faced by several municipalities in addressing the challenges faced by various NFPMs, the DALRRD developed a draft bill to authorize the establishment of a National Fresh Produce Market Council (NFPMC). The main objective of the bill was to provide for the regulation of NFPMs through the creation of a centrally located and legally mandated entity that would coordinate efforts to develop, manage, and transform NFPMs. The Office of the Chief State Law Advisor (OCSLA), however, provided its legal argument and concluded that markets are a local government function and putting markets under the control of a national statutory council (the NFPMC) is unconstitutional. This has, to date, stymied that particular initiative.

MARKET CHALLENGES FOR SMALLHOLDER FARMERS

It is without any doubt that thriving smallholder farmers can enhance food security and contribute to higher agricultural productivity and income growth, which in turn can contribute to job creation, poverty reduction, and economic growth (Barrett, 2008; Bernard & Spielman, 2009; Markelova & Mwangi, 2010). In South Africa, the emergence and success of more black smallholder farmers have a particular urgency not just because of poverty and food insecurity, but also in the light of the unsustainable levels of racial inequality in land holding and the agricultural sector (Binswanger-Mkhize, 2014; Cousins, 2015; Hall & Aliber, 2010). The National Development Plan (NDP) has identified the development of smallholder farmers, especially in fresh produce/horticultural production, as critical to job creation and transformation in the agricultural sector (National Planning Commission [NPC], 2011). Increasing levels of urban food insecurity, as well as the large market opportunity in urban areas, highlights the need to effectively link productive black smallholder farmers with urban populations, especially those in poverty (Statistics South Africa, 2019; Wegerif, 2022).

It has become a widely accepted view that for smallholder farmers to thrive in the global dynamic economic environment, they need to be part of the conventional, corporate-controlled, value chains (see, e.g., DALRRD, 2021b; Gereffi et al., 2001). Such an argument implies, often without evidence, that smallholder farmers are striving to be commercialized through incorporation into value chains and that this will be beneficial. Given the widespread focus among governments and development agencies on incorporating farmers into value chains, there is surprisingly "a clear lack of high-quality impact assessments that would substantiate claims that VC interventions are capable of achieving the broader goals" (Humphrey & Navas-Alemán, 2010, p. 61). Even those claiming success and benefits for farmers realize that it tends to be the wealthier and better-educated farmers who are benefiting, while many others are excluded or adversely incorporated with negative outcomes, especially for the more vulnerable such as rural women (Coles & Mitchell, 2011; Du Toit, 2009; Riisgaard et al., 2010). The problem stems from the nature of value chains, which are about vertical integration under the control of lead firms, which aim to access high-value markets. Smallholder farmers are in a weak negotiating position in value chains because they face high expenses, such as the costs of third-party certification, and must meet demanding quality standards, while the logic for the lead firm is to increase profits by creating rent-seeking opportunities and exerting pressure on primary producers to deliver more for less (Kaplinsky & Morris, 2001; Markelova & Mwangi, 2010; Porter, 1998a, 1998b).

After apartheid, the South African government focused on building a unified sector to overcome the entrenched dualism created under apartheid (NPC, 2011). The problem is that the approach tended to always mean the incorporation of emerging black farmers into the established and white-dominated commercial agricultural sector "with the implication being that they are bona fide farmers only in so far as they begin to resemble large-scale white commercial farmers" (Aliber & Hall, 2012, p. 548). At the same time, it has been found that the high level of corporate concentration in the food system and the supermarket model in South Africa does not provide beneficial opportunities to anyone but a minority of farmers (Greenberg, 2015) and is in fact "inherently hostile towards smaller producers" (Heijden & Vink, 2013, p. 68). Despite these challenges, there is a tendency in South Africa to believe that the corporate food chains, with supermarkets in the lead, are the only option to feed the nation. Yet, in many parts of Africa and Asia, food systems based on small-scale producers are effectively supplying large cities through public markets, street vendors, and multitudes of small shops and traders (Van der Ploeg et al., 2016; Wegerif & Hebinck, 2016). For example, a study by Okello et al. (2012) found that despite the expansion of supermarkets in Kenya, more than 90% of fresh vegetable purchases are made through outlets such as open-air markets, roadside markets, kiosk vendors, and on-farm purchases.

The NFPMs provide an important market opportunity for all scales of farmers, even the small and distant (Wegerif, 2022), and they link farmers to the growing informal sector, as previously explained. There are challenges however. Although the rule of the APAC stipulates that agents must treat their principals equally and without discrimination based on the amount of produce that principals bring to the market, interviews with farmers for this study, and other studies (see, e.g., NAMC, 1998, 2006), have found that many small-scale black farmers feel they are not prioritized by agents, and they are dissatisfied with the prices they receive. The challenge of high relative transport costs when delivering small quantities to the market and waiting for an uncertain payment adds to the dissatisfaction. The complaints of farmers have some basis as, during 2017, a number of market agencies were investigated and charged by the Competition Commission for price-fixing and fixing trading conditions (Competition Commission South Africa, 2017). The investigation followed a complaint alleging that certain market agencies may have been behaving in an anticompetitive manner by, among other things, practicing product reservation, engaging in late sales, and colluding to keep emerging and previously disadvantaged agencies out of the market.

Where the NFPMs have worked well for small-scale farmers, it has been due to good relations established with particular agents (Wegerif, 2022), showing both the importance and possibility of these relationships. NFPMs are an important part of the price-setting mechanism, distribution, and marketing of fresh produce in South Africa. Price determination, in simple terms, is the outcome of the interaction of economic forces of supply and demand—the price being the point of equilibrium found between sellers and buyers seeking to maximize utility—in a situation of perfect competition (Breimyer, 1957). According to Stigler (1957), "Perfect market competition will prevail when there are an indefinite number of traders (no one of which controls an appreciable share of demand or supply) acting independently in a perfect market. A perfect market is one in which traders have full knowledge of all offers and bid prices" (p. 7). Further, he states that for perfect market competition to exist the following axioms ought to be met: 1) rational decision-making; 2) economic mobility (i.e., no transaction costs to enter the market); 3) perfect, continuous, costless intercommunication among participants; and 4) atomization—every participant acts in their own interest (Stigler, 1957, p. 12). Some economists have long argued, particularly in relation to agricultural markets, that perfect competition rarely if ever exists (Breimyer, 1957), while sociologists have argued that social structures, morals, and established norms also influence the functioning of markets and economies (Granovetter, 1985; Ostrom, 2010; Zak, 2011).

It is the functioning of the market system that poses a number of challenges that could be seen as making far less than perfect market conditions for the participants in the NFPMs. Among other challenges, there are a limited number of market agencies that operate on the market floor, especially in some of the smaller markets, with some having only a few agencies and the Mthatha market having only one in 2021. Both the number of NFPMs and the number of agencies operating in them have been declining over the last few years. In 2021, the Mpumalanga market stopped trading and the George market was sold to a private company (SAUFM, 2022). This can negatively affect competition in the market even if there are many buyers and sellers. Buyers and agents also do not have access to the same information at the same time, as agents have access to the computer trading systems used in the markets. The farmer is also highly dependent on the agent. Information such as the previous days' prices do not inform the farmer of things such as the comparable quality of their produce to that of another farmer or the fast-changing circumstances on the market floor where agents have make decisions to cut prices to move stock when appropriate. One of the options identified in response to such challenges with market agents and the commission system is to allow wholesalers onto the market floor (NAMC, 1998). The wholesaler operating model involved the purchase of produce with upfront payments to the farmer, with the wholesaler, instead of the farmer, carrying the risk of finding the final buyer. The research findings (NAMC, 1998) also indicate wider factors beyond the economic rationale that shape decision-making and transactions in the markets. Actors who participated in the research at the JFPM were also influenced by social and cultural factors that coexisted with the economic forces in an ongoing contest. Other factors identified from the interviews with farmers, agents, and buyers include relationships and trust as the main factors that shaped market transactions. Trust is even more important between the farmer and the agent selling their produce. While the buyers meet and interact with the agent face to face, the often sits at a distance. They load produce on bakkies or trucks and send it to the market hoping it will be sold for a profit with no guarantee of the price, realizing the payment after the sale is made.

Other researchers have also identified the importance of relationships and trust between farmers and agents for the functioning of the NFPMs in South Africa (see, e.g., Range, 2015; Range & Leonard, 2015). The same phenomena has been found in other countries; for example, a study of a fresh produce market in Australia that also operates with the agent commissions system found that "market agents must endeavor to build long-term relationships with growers" and "personal relationships are important in building trust in fresh produce markets" (Batt, 2003, p. 67). A study in Tanzania found that interdependent relationships, among farmers and traders who were at least familiar with each other and have common cultural identities, were essential to the functioning of the food system and markets (Wegerif, 2018).

The problem in South Africa, with its extraordinarily divided past and massive inequality, is that those who do not have a social network will find it extremely hard to succeed. It is difficult for black farmers and new black market agents to get customers and succeed in the markets when they often have to operate with a lack of capital and smaller scale operations while dealing with entrenched social networks of which they are not a part.

DISCUSSIONS AND CONCLUSION

The NFPMs perform a crucial role in the distribution of fresh produce, providing a market outlet for farmers, access to produce for retailers, which all contribute to increased economic activity, betterment of livelihoods, and substantial food security. So, keeping markets functioning effectively as public institutions with the mandate of contributing and investing in them is vital. These markets, however, face fundamental challenges evidenced in their overall loss of market share and the lack of transformation. If the loss of market share continues, the competitiveness and viability of these markets will be threatened. This will have extremely negative effects on overall competitiveness regarding the fresh produce sector in South Africa, and it would be particularly devastating for the small-scale farmers who cannot hope to beneficially access corporate supply chains. The degradation of NFPMs would also be detrimental to the informal sector, which depends on these farmers their supplies. The lack of transformation in the NFPMs is not sustainable in South Africa today and undermines their ability to support needed transformation in the agricultural sector. The challenge is to bring change without undermining what is working in these markets. We suggest that the solutions to these challenges lie in three main areas: 1) regulatory reforms, 2) recognizing and factoring the importance of social relations to the functioning of the markets, and 3) valuing and building upon the importance of the informal food sector for the markets. It is however important to note that the use of a case of JFPM may be regarded as a weakness and should not be generalized over the entire functioning of the NFPMs in the country.

Key regulatory reforms need to include creating greater autonomy for the markets from municipalities that are constraining their operations. The markets must be able to function in a nimble and competitive way. Funds generated in the markets also need to remain dedicated to the operation and reinvestment in the market. There should be caution, however, not to let municipalities ignore markets that perform a key function in local economies and food systems, for in some cases, they need more investment from the municipalities. At the national level, the position of the Chief State Law Adviser on the bill that DALRRD drafted, suggests that the DALRRD needs to enter into discussions with the Department of Cooperative Governance and Traditional Affairs (DCOGTA). These discussions should address issues such as the possibility of standardizing and rationalizing FPM market bylaws, publishing the CoBPs by COGTA, ring-fencing and ensuring reinvestment of a portion of the revenues generated by markets for market operations and infrastructure improvements, using the Municipal Infrastructure Grant (MIG) for market infrastructure upgrades and expansion of fresh produce markets, and providing FPMs with more operational independence for greater efficiency.

The implication of social factors on markets is often overlooked in the formulation of policy and regulation due to a narrow view of a market, such as the JFPM, as only being a platform where buyers and sellers meet to trade fresh produce. The research shows that the JFPM is more than the economic definition of a market. It is a market which facilitates myriad dynamic relationships between actors based on both economic and social interactions. Understanding and intervening in the markets requires an appreciation of the social relations involved, which need more nuanced programs and interventions. The entrenched social relations, particularly the relations of trust between farmers and agents, are essential to the functioning of the markets, and so, should not be tampered with lightly. At the same time, these relationships that facilitate smooth access to the markets for some, end up becoming a barrier to new entrants that come from different racial and social groups that do not have financial and production muscle. The collaborative initiatives among many of the informal traders show how social networks can play a role in helping smaller-scale and new entrants.

There will never be a perfect market in the economic sense, so the challenge is to enhance the positive characteristics of social relations and overcome the problematic ones. There are no easy solutions, but a greater awareness of these functions, and more research and discussion, will help inform interventions that have a better chance of success. Certainly, the approach taken in the Agricultural Marketing Policy and the MAP Act, by limiting government interventions and leaving things to market forces and private sector self-regulation, does not reflect the reality of how markets are functioning, and the entrenched interests and relations involved. As there are currently different models of market management in operation, from full privatization to full municipal management, there would be great value in conducting in-depth comparative studies to assess the different operations and their outcomes, especially for farmers of all scales, retailers in informal and small-scale sectors, and regarding food accessibility. We can also learn more from the international experiences of markets and their regulation.

It would also be valuable to obtain more research; to document and learn from the experiments conducted in different markets such as the Mandela and Unity markets at the JFPM, the farmers market at Durban, and the soon-to-be-implemented introduction of a wholesaler onto the main trading floor in Durban. More testing of different options and learning from the experiences is one way to bring about incremental but effective change with less risk of substantial disruptions to the markets. Valuing the informal sector requires doing more to understand it and foregrounding the needs and nature of the informal sector in all the market and municipal development planning. Creating a more enabling environment for the informal food sector activities requires incorporating their needs into integrated and spatial development plans and amending regulations and bylaws that are unnecessarily used to criminalize and undermine their activities. This could include planning interventions similar to the crawling-peg approach in China that reserves more public space for food and certain fresh produce trading in cities (Zhong et al., 2019). These challenges need to be met as a matter of urgency if the decline in the service standards and significance of the NFPMs is to be halted and remedied effectively. We offer ideas on solutions as a contribution to what should facilitate constructive and inclusive discussions on the organization of food markets in South Africa and beyond.

REFERENCES

Agricultural Produce Agents Amendment Bill. B22-2020. (2020). https://www.parliament.gov.za/storage/app/media/Bills/2020/B33_2020_Agricultural_Produce_Agents_Amendment_Bill/B33_2020_Agricultural_Produce_Agents_Amendment_Bill.pdf

Agricultural Produce Agents (APA) Act. 1992 (Act No. 12 of 1992). https://www.gov.za/sites/default/files/gcis_document/201409/a121992.pdf

Aliber, M., & Hall, R. (2012). Support for smallholder farmers in South Africa: Challenges of scale and strategy. *Development Southern Africa*, *29*(4), 548–562. doi:10.1080/0376835X.2012.715441

Baeumler, A., D'Aoust, O., Das, M. B., Gapihan, A., Goga, S., Lakovits, C., Restrepo Cavadid, P., Singh, G., & Terraza, H. (2021). *Demographic trends and urbanization*. W. B. Group. https://documents1.worldbank.org/curated/en/260581617988607640/pdf/Demographic-Trends-and-Urbanization.pdf

Balineau, G., Bauer, A., Kessler, M., & Madariaga, N. (2021). *Food systems in Africa: Rethinking the role of markets*. World Bank Group. doi:10.1596/978-1-4648-1588-1

Barrett, C. B. (2008). Food Security in Africa: market and trade policy for staple foods in Eastern and Southern Africa. In *Smallholder market participation: Concepts and evidence from Eastern and Southern Africa*. Edward Elgar Publishing.

Batt, P. J. (2003). Building trust between growers and market agents. *Supply Chain Management*, *8*(1), 65–78. doi:10.1108/13598540310463378

Battersby, J. (2020). South Africa's lockdown regulations and the reinforcement of anti-informality bias. *Agriculture and Human Values*, *37*(3), 543–544. doi:10.100710460-020-10078-w PMID:32395011

Battersby, J., & Watson, V. (2018). *Urban food systems governance and poverty in African Cities*. Routledge. doi:10.4324/9781315191195

Beavon, K. (2001). *The role of transport in the rise and decline of the Johannesburg CBD, 1886-2001* [Paper presentation]. 20th South African Transport Conference (SATC 2001): Meeting the Transport Challenges in Southern Africa, South Africa. https://repository.up.ac.za/bitstream/handle/2263/8191/5b4.pdf?sequence=1

Benton, T. G., & Bailey, R. (2019). The paradox of productivity: Agricultural productivity promotes food system inefficiency. *Global Sustainability*, *2*(e6), 1–8. doi:10.1017us.2019.3

Benton, T. G., Bieg, C., Harwatt, H., Pudasaini, R., & Wellesley, L. (2021). *Food system impacts on biodiversity loss: Three levers for food system transformation in support of nature*. Chatham House. https://www.ciwf.com/media/7443948/food-system-impacts-on-biodiversity-loss-feb-2021.pdf

Bernard, T., & Spielman, D. J. (2009). Reaching the rural poor through rural producer organizations? A study of agricultural marketing cooperatives in Ethiopia. *Food Policy*, *34*(1), 60–69. doi:10.1016/j.foodpol.2008.08.001

Binswanger-Mkhize, H. P. (2014). From failure to success in South African land reform. *African Journal of Agricultural and Resource Economics*, *9*(4), 253–269.

Breimyer, H. F. (1957). On price determination and aggregate price theory. *Journal of Farm Economics*, *39*(3), 676–694. doi:10.2307/1234424

Calatayud, A., Mangan, J., & Christopher, M. (2019). The self-thinking supply chain. *Supply Chain Management*, *24*(1), 22–38. doi:10.1108/SCM-03-2018-0136

Chikazunga, D., Deall, S., Louw, A., & Van Deventer, J. (2008). *The role of fresh produce markets in South Africa*. Regoverning Markets. https://www.up.ac.za/media/shared/Legacy/sitefiles/file/48/2052/5_roleoffreshproducemarkets.pdf

Christopher, M. (2016). *Logistics & supply chain management*. Academic Press.

City of Johannesburg. (2016). *Spatial development framework 2040*. https://www.joburg.org.za/documents_/Pages/Key%20Documents/policies/Development%20Planning%20%EF%BC%86%20Urban%20Management/Citywide%20Spatial%20Policies/Spatial-Development-Framework-2040.aspx

City of Johannesburg. (2020). *The city of Johannesburg final integrated development plan: 2020/21*. https://www.joburg.org.za/documents_/Documents/Intergrated%20Development%20Plan/IDP%202021-23/ITEM%2004%20ANNEXURE%20IDP.pdf

Civil Society Mechanism. (2016). *Connecting smallholders to markets: An analytical guide*. https://www.csm4cfs.org/connecting-smallholders-markets-analytical-guide/

Coles, C., & Mitchell, J. (2011). *Gender and agricultural value chains: A review of current knowledge*. ESA Working Paper No. 11-05.

Competition Commission South Africa. (2017). *14 fresh produce market agents & the institute charged for price fixing*. https://www.compcom.co.za/wp-content/uploads/2017/01/fruit-and-veg-referral-draft-003.pdf

Cousins, B. (2015). 'Through a glass darkly': Towards agrarian reform in South Africa. In B. Cousins & C. Walker (Eds.), *Land divided, land restored. Land reform in South Africa for the 21st century* (pp. 250–269). Jacana.

Cripps, E. A. (2012). *Provisioning Johannesburg, 1886-1906*. University of South Africa. https://earthwormexpress.com/wp-content/uploads/2019/08/provisioning-johannesburg-1886-e28093-1906-.pdf

Davie, L. (2018). *Joburg's market square defined the town*. The Heritage Portal. https://www.theheritageportal.co.za/article/joburgs-market-square-defined-town

Department of Agriculture Land Reform and Rural Development. (2009). *Report of the ministerial interim committee on the restructuring of fresh produce markets in South Africa*. http://webapps.daff.gov.za/AmisAdmin/upload/Report%20of%20the%20Ministerial%20Interim%20Committee%20on%20restructuring%20of%20fresh%20produce%20markets%20in%20SA.pdf

Department of Agriculture Land Reform and Rural Development. (2020). *Statistics on Fresh Produce Markets: Report 58*. https://www.dalrrd.gov.za/Portals/0/Statistics%20and%20Economic%20Analysis/Statistical%20Information/Statistics%20on%20Fresh%20Produce%20Markets%202020.xlsx

Department of Agriculture Land Reform and Rural Development. (2021a). *Abstract of agricultural statistics 2021*. https://www.dalrrd.gov.za/Portals/0/Statistics%20and%20Economic%20Analysis/Statistical%20Information/Abstract%202021.pdf

Department of Agriculture Land Reform and Rural Development. (2021b). *Agriculture and agro-processing master plan (AAMP)*. Author.

Du Toit, A. (2009). *Adverse incorporation and agrarian policy in South Africa, or, how not to connect the rural poor to growth*. Institute of Poverty, Land and Agrarian Studies.

Food and Agriculture Organization, IFAD, UNICEF, WFP, & WHO. (2021). The state of food security and nutrition in the world 2021: Transforming food systems for food security, improved nutrition and affordable healthy diets for all. Food and Agriculture Organization of the United Nations.

Freidberg, S., & Goldstein, L. (2011). Alternative food in the global south: Reflections on a direct marketing initiative in Kenya. *Journal of Rural Studies*, *27*(1), 24–34. doi:10.1016/j.jrurstud.2010.07.003

Gereffi, G., Humphrey, J., Kaplinsky, R., & Sturgeon, T. J. (2001). Introduction: Globalization, value chains and development. *IDS Bulletin*, *32*(3), 1–8. doi:10.1111/j.1759-5436.2001.mp32003001.x

Goodson, I. F. (2012). *Developing narrative theory: Life histories and personal representation*. Routledge. doi:10.4324/9780203817704

Granovetter, M. (1985). Economic action and social structure: The problem of embeddedness. *American Journal of Sociology*, *91*(3), 481–510. doi:10.1086/228311

Greenberg, S. (2015). *Corporate concentration and food security in south Africa: Is the commercial agro-food system delivering?* Academic Press.

Guarín, A. (2013). The value of domestic supply chains: Producers, wholesalers, and urban consumers in Colombia. *Development Policy Review*, *31*(5), 511–530. doi:10.1111/dpr.12023

Hall, R., & Aliber, M. (2010). *The case for re-strategizing spending priorities to support small-scale farmers in South Africa*. Working Paper, 17.

Hart, K. (1985). The informal economy. *Cambridge Anthropology*, 54–58.

Harvard School of Public Health. (2011). *Healthy eating plate*. https://www.hsph.harvard.edu/nutritionsource/healthy-eating-plate/

Heijden, T., & Vink, N. (2013). Good for whom? Supermarkets and small farmers in South Africa: A critical review of current approaches to increasing access to modern markets. *Agrekon*, *52*(1), 68–86. doi:10.1080/03031853.2013.778466

Humphrey, J., & Navas-Alemán, L. (2010). *Value chains, donor interventions and poverty reduction: A review of donor practice*. IDS Research Reports, 2040-0217.

JFPM. (2021). *Johannesburg Fresh Produce Market website*. Accessed 30 April 2022. Available from: http://www.joburgmarket.co.za/index.php

Kaplinsky, R., & Morris, M. (2001). *A handbook for value chain research* (Vol. 113). IDRC.

Lang, T., & Heasman, M. (2004). *Food Wars: The global battle for mouths, minds and markets: Earthscan*. James & James.

Markelova, H., & Mwangi, E. (2010). Collective action for smallholder market access: Evidence and implications for Africa. *The Review of Policy Research*, *27*(5), 621–640. doi:10.1111/j.1541-1338.2010.00462.x

Marketing of Agricultural Products Act 47 of 1996. https://www.gov.za/sites/default/files/gcis_document/201409/act47of1996.pdf

National Agricultural Marketing Council. (1998). *Report on the investigation into fresh produce marketing: Report 1 Johannesburg fresh produce market*. https://www.namc.co.za/wp-content/uploads/2017/09/FreshProduceStudy1.pdf

National Agricultural Marketing Council. (2000). *Report on the Investigation into Fresh Produce Marketing*. Report 2: Fresh Produce Marketing in South Africa. http://www.namc.co.za/ASSETS/PDF/published/ FreshProduceStudy2.pdf

National Agricultural Marketing Council. (2005). *National Fresh Produce Markets. Part of the NAMC Section 7 Section Committee Investigation on Fresh Produce Marketing in South Africa*. https://www.namc.co.za/the-investigation-into-fresh-produce-markerting-report

National Agricultural Marketing Council. (2006). *Section 7 investigation: National fresh produce markets*. https://www.namc.co.za/wp-content/uploads/2017/09/NationalFreshProduceMarkets.Section7InvestigationFinalReport.pdf

National Agricultural Marketing Council. (2021). *Smallholder Market Access Research Report*. https://www.namc.co.za/wp-content/uploads/2022/05/SHMA-Research-Report_2021-22.pdf

National Planning Commission. (2011). *National development plan 2030: Our future - make it work*. The Presidency, Republic Of South Africa.

Ojermark, A. (2007). *Presenting life histories: A literature review and annotated bibliography*. Chronic Poverty Research Centre.

Okello, J. J., Lagerkvist, C.-J., Hess, S., Ngigi, M., & Karanja, N. (2012). Choice of fresh vegetable retail outlets by developing-country urban consumers: The case of kale consumers in Nairobi, Kenya. *European Journal of Development Research*, *24*(3), 434–449. doi:10.1057/ejdr.2011.58

Ostrom, E. (2010). Beyond markets and states: Polycentric governance of complex economic systems. *The American Economic Review*, *100*(3), 408–444. doi:10.1257/aer.100.3.641

Patel, R. (2007). *Stuffed & starved*. Black Inc.

Pieterse, M. (2021). Anatomy of a crisis: Structural factors contributing to the collapse of urban municipal governance in Emfuleni, South Africa. Urban Forum, 32, 1-15. doi:10.100712132-020-09406-4

Popke, E. J., & Ballard, R. (2004). Dislocating modernity: Identity, space and representations of street trade in Durban, South Africa. *Geoforum*, *35*(1), 99–110. doi:10.1016/S0016-7185(03)00002-2

Porter, M. E. (1998a). *Competitive advantage: Creating and sustaining superior performance*. The Free Press. doi:10.1007/978-1-349-14865-3

Porter, M. E. (1998b). *Competitive strategy: Techniques for analyzing industries and competitors*. The Free Press.

Range, J. (2015). *A conceptual framework for the enhancement of trust in South Africa's fresh produce markets through the use of self-service technologies*. University of Pretoria. https://repository.up.ac.za/bitstream/handle/2263/52999/Range_Conceptual_2015.pdf?sequence=1

Range, J. J., & Leonard, A. (2015). Progressive trust formation within the South African fresh produce industry through the use of self-service technologies. *The African Journal of Information Systems*, *7*(3), 1–26. https://digitalcommons.kennesaw.edu/ajis/vol7/iss3/1/

Riisgaard, L., Fibla, A. M. E., & Ponte, S. (2010). *Gender and value chain development: Evaluation study*. Academic Press.

Rudolph, M., Kroll, F., Muchesa, E., Paiker, M., & Fatti, P. (2021). Food security in urban cities: A case study conducted in Johannesburg, South Africa. *Journal of Food Security*, *9*(2), 46–55. doi:10.12691/jfs-9-2-2

Skinner, C., & Watson, V. (2020). The informal economy in urban Africa: Challenging planning theory and praxis. In M. Chen & F. Carré (Eds.), *The Informal economy revisited: Examining the past, envisioning the future* (pp. 123–131). Routledge. doi:10.4324/9780429200724-21

Smit, S., & Madubela, A. (2021, October 25). Terribly scary: Dysfunctional municipalities are a threat to South Africa's economy. *Mail & Guardian*. https://mg.co.za/politics/2021-10-25-terribly-scary-dysfunctional-municipalities-are-a-threat-to-south-africas-economy/

South African Union of Food Markets. (2022). *Comparative figure of turnover and mass of the 17 fresh produce markets*. Author.

Staff, W. (2021). *8 reasons why South Africa's municipalities are completely broken*. Business Tech. https://businesstech.co.za/news/government/528846/8-reasons-why-south-africas-municipalities-are-completely-broken/

Statistics South Africa. (2019). *Towards measuring the extent of food security in South Africa: An examination of hunger and food adequacy*. http://www.statssa.gov.za/publications/03-00-14/03-00-142017.pdf

Stigler, G. J. (1957). Perfect Competition, Historically Contemplated. *Journal of Political Economy*, *65*(1), 1–17. doi:10.1086/257878

Van der Ploeg, J. D., Huifang, W., & Congzhi, H. (2016). Markets. In J. D. Van der Ploeg & Y. Jingzhong (Eds.), *China's peasant agriculture and rural society: Changing paradigms of farming*. doi:10.4324/9781315643250 van der Ploeg, J. D., Jingzhong, Y., & Schneider, S. (2012). Rural development through the construction of new, nested, markets: Comparative perspectives from China, Brazil and the European Union. *The Journal of Peasant Studies*, *39*(1), 133–173. doi:10.1080/03066150.2011.652619

Van der Ploeg, J. D., Ye, J., & Schneider, S. (2022). Reading markets politically: On the transformability and relevance of peasant markets. *The Journal of Peasant Studies*, 1–26. doi:10.1080/03066150.2021.2020258

Wegerif, M. (2020). The symbiotic food system. In J. Duncan, J. S. Wiskerke, & M. S. Carolan (Eds.), *Routledge handbook of sustainable and regenerative food systems* (pp. 188–203). Routledge. doi:10.4324/9780429466823-14

Wegerif, M. (2022). The impact of Covid-19 on black farmers in South Africa. *Agrekon*, *61*(1), 52–66. doi:10.1080/03031853.2021.1971097

Wegerif, M. C. A. (2018). An ethnographic exploration of food and the city. *Anthropology Today*, *34*(5), 16–19. doi:10.1111/1467-8322.12460

Wegerif, M. C. A. (2020). "Informal" food traders and food security: Experiences from the Covid-19 response in South Africa. *Food Security*, *12*(4), 797–800. doi:10.100712571-020-01078-z PMID:32837657

Wegerif, M. C. A., & Hebinck, P. (2016). The symbiotic food system: An 'alternative' agri-food system already working at scale. *Agriculture*, *6*(3), 40. doi:10.3390/agriculture6030040

Willett, W., Rockström, J., Loken, B., Springmann, M., Lang, T., Vermeulen, S., Garnett, T., Tilman, D., DeClerck, F., & Wood, A. (2019). Food in the anthropocene: the EAT–Lancet Commission on healthy diets from sustainable food systems. *Lancet*. https://openaccess.city.ac.uk/id/eprint/21633/

Zak, P. J. (2011). Moral markets. *Journal of Economic Behavior & Organization*, *77*(2), 212–233. doi:10.1016/j.jebo.2010.09.004

Zhong, T., Si, Z., Crush, J., Scott, S., & Huang, X. (2019). Achieving urban food security through a hybrid public-private food provisioning system: The case of Nanjing, China. *Food Security*, *11*(5), 1071–1086. doi:10.100712571-019-00961-8

Chapter 8
Market Access Enablers for Small-Scale Crop Farmers:
Evidence From Limpopo Agricultural Cooperatives

Aluwani Maiwashe-Tagwi
https://orcid.org/0000-0001-9701-3171
University of South Africa, South Africa

ABSTRACT

Cooperatives are deemed as vehicles for rural development, especially in developing countries. Smallholder farmers generally experience huge marketing transactional costs, and to reduce this they normally resort to becoming members of cooperatives. In South Africa, most small-scale farmers do not have access to markets. With no markets in place, the transformation would be almost impossible. The study's aim was to assess the determinants of market access in agricultural cooperatives in Limpopo Province. A face-to-face survey was conducted with 146 agricultural cooperatives members. The results showed that market access was associated with age, training, workers in a household, knowledge of the cooperatives principles and act, annual income, and access to funding. The study recommended younger farmers be targeted when designing market access interventions as the impact of job creation is huge amongst this group. Tailor-made training on market access and cooperative principles and acts should be designed. Funding should be made more accessible to small-scale farmers in cooperatives.

INTRODUCTION

Small-scale **farmers** (SCF) play a significant role in Africa, in Sub-Saharan Africa (SSA). About 80% of the food production comes from SCF and accounts for 90% of food production in the region (Unsal et al. 2020). However, SSA's storage capacity is constrained, despite recent crop yield advancements, resulting in large food losses (Okou et al., 2022). The role of SCF in food production cannot be underestimated in the SSA (Hlophe-Ginindza & Mpandeli, 2021). Moreover, the role can be amplified if there

DOI: 10.4018/978-1-6684-4780-2.ch008

are opportunities for markets. Small-scale farmers typically face a wide range of complex challenges in developing countries, including high transaction costs for gaining access to markets, and weak bargaining power in the markets (Liu et al., 2021; Zheng et al., 2021). Trends show that the major share of food production will continue to come from smallholder farmers and will remain critical to global population food security immensely (Giller et al., 2021).

The above is an indication that small-scale farmers have the potential to produce food for the local and international markets if proper support is put in place. The potential is not harnessed in financial terms, as these farmers lose money as they are faced with several market access-related challenges. Security of formal markets through contracts has been confirmed to enhance the welfare of farmers (Ma & Abdulai, 2016a). Unfortunately, such security is not commonly available to farmers and as a result, small-scale farmers resort to selling their produce at the farmgate for convenience (Hao et al., 2018), further excluding themselves from the lucrative opportunities in established markets (Omiti, et al. 2009; Maiwashe, 2012).

Participation in established markets is hampered by various factors, among others, low pricing has been flagged (Gyau, Mbugua & Oduol, 2016). In Uganda, market access was associated with holding a market contract, land size, production diversity and value addition in a potato study (Kyomugisha et al., 2018). However, securing **a market** contract for a farmer demands a set of skills, which is usually scarce. Farmers' inability to bargain on the price of the produce deprives them of the opportunity to grow. In the long run, farmers produce quality that is traded mostly by agents and less profit is received by the producer. In Africa, there is a rising concern that smallholder farmers could be excluded from new market opportunities as a result of increased globalization and market liberalization (Alemu et al., 2021). The concern is based on the current low participation of SCF in local formal markets which might suggest that participation in global markets will be even tougher.

On the demand side, the population projections indicate that producers will have more opportunities to trade as more demand for food is expected in the future. It has been observed **that more** people move from villages to urban areas in **search for** better opportunities. The argument is that this influx is influenced by the welfare systems and urban biased-public policy which usually re-direct more resources to urban areas leading to over-urbanization (Glaeser, 2014; De Jong & De Valk, 2020; Black et al., 2022). Trends also suggest that more people could move from urban areas to villages in the future if the same opportunities or policy changes are presented. The majority of people in Sub-Saharan Africa (SSA) live in rural areas, where poverty rates are high. For most farmers, the scale of production is usually constrained by a lack of market access.

The access is capital-intensive and rural farmers lack access to the credit market (Shuaibu & Nchake, 2021). Utility maximization theory postulates that farmers will maximize their expected utility of net returns through formal market channel use (Ito et al., 2012; Fischer & Qaim, 2012; Ma & Abdulai, 2016), suggesting that farmers will fully benefit when exposed to formal markets. To allay the challenges of market access, studies show that small-scale farmers resort to cooperative membership for solutions (Chagwiza et al., 2016; Maiwashe, 2017; Manda et al., 2020; Sebhatu et al., 2021; Ma et al., 2022). Research shows that **membership** in agricultural cooperatives may have an impact on smallholder farmers' financial performance and market participation (Zhang et al., 2020).

Generally, cooperatives face many problems such as free-riding, group dynamics, farmer heterogeneity and strategic planning (Elliott et al., 2018; Omar et al., 2022) amongst others. For small-scale farmers with limited resources the benefits outweigh the challenges. There is a cost to access market information and cooperatives have proven to reduce such cost, especially for smallholder farmers (Serra & Davidson,

2021). Despite the fact that the number of established farmer cooperatives is increasing every year in developing countries, their performance, particularly in terms of value addition and access to better markets, overall remains lacking yet it has a promising future for smallholder farmers (Alemu et al., 2021). The participation of farmers in the markets creates more opportunities in the commodity value chains. In South Africa, small-scale farmers' major struggle is market access (Hlatshwayo et al., 2021). With high youth unemployment, the participation of farmers in the value chain can open up job opportunities (processing, grading, value addition and transportation) along with the value chains (Chipfupa & Tagwi, 2021) attractive enough to the younger generation. Research demonstrates that market activity promotes rural development through the building of rural roads, electrification, and industrialization over time (Mmbando et al., 2015; Tarekegn et al., 2017).

Historically, cooperatives were used as development vehicles in South Africa during the apartheid administration regime. The history of cooperatives in South Africa, however, has two schools of thought, i.e. two different angles of cooperative growth that Okem and Lawrence (2013) identified as being pre- and post-apartheid. The pre-and post-apartheid cooperative reforms are comparable since racial barriers were used to construct both of them (Satgar and Williams, 2011). While black farming declined and became part-time and unprofitable, pre-apartheid events gave the white minority access to production and market protection (Satgar, 2011). Although the current discourse might be that market protection was removed after 1994, the fact remains that the structural economic market protection has not been removed, primarily because the assets accumulated during the years of support even matured over the years, providing further leverage to white cooperatives in South Africa while the cooperatives established post-1994 lacked the same capital injection.

Although post-1994, the South African government prioritized cooperatives as one of the rural development vehicles, studies have shown that cooperatives in South Africa are inefficient (Yobe et al., 2020). The common problems associated with the failure of cooperatives previously reported point to obstacles related to history, ideology, organizations, operations, stemming mainly **from lack** of backing from the government (Kanyane & Ilorah, 2015). Other problems identified earlier on by the government relate to lack of access to land, poor produce quality and quantity, lack of production knowledge, poor infrastructure, the absence of markets, information access issues, high transaction **costs, lack** of assets, financial services and selling skills, as well as illiteracy (DAFF, 2012). The present-day challenges are similar, with capital injection being the most pressing. Budget cuts made by the government in response to austerity measures do not improve the situation. Although farmers are resilient and can overcome some of these obstacles; the main one is market instability, which makes things worse for farmers. The dynamic nature of markets makes it more difficult for farmers to improvise if planning was not done as local markets are linked with international markets (Elleby et al., 2020). The effect of globalization is double-edged and as such, opportunities can easily be created for farmers but at the same time opportunities can be wiped away (Zahonogo, 2018). This is because market gaps do not last as they can easily be closed by international trade. This explains how countries could have a shortage of a given commodity but still have access to it locally without experiencing price increases. This is because the nature of international trade agreements was designed not to discriminate among trading countries under the World Trade Organization (WTO) agreement but to be inclusive (Trebilcock et al., 2013; Van den Bossche et al., 2021). This benefit also applies to South African small-scale farmers, **thus, if** they have the capacity, the market is open for them internationally.

Recently, some positive impacts have also been reported in achieving nine Sustainable Development Goals (SDG) targets connected to the environment with the help of international trade (Xu et al., 2020).

Further building a business case for market integration in the 21st century. At the local production level, the downside is that, if farmers do not supply some of the crops, the crops can be supplied by other international trade partners in the WTO. The onus is therefore on the farmers themselves to close the supply gaps in the local market. Notwithstanding that, such opportunities require economies of scale, something that small-scale farmers can achieve collectively through farming cooperatives but with government involvement. This is because governmental processes have undergone substantial modifications as a result of globalization (Glushkova et al., 2019).

Due to the development of the global integrated industrial systems, globalization assumes that nations become interconnected and therefore government involvement in market development becomes key. Market globalization not only presents opportunities but also threats which demand government involvement. Studies also show that the advantage of the emerging markets which drive production will come at a cost for developing countries, especially Sub-Saharan Africa and others in the near future. The lower world market prices are estimated to **favor** Europe and North America mostly in the future while not jeopardizing their own biodiversity due to the concept of carbon emission importation. This is when rich countries resort to less **industrialized** activities that **produce more** emissions and rather import such products. The major concern is that biodiversity will be endangered in developing countries as a result (Zabel et al., 2019). This threat requires a calculated environmental sustainability approach from the farmer cooperatives and the government side. Cooperatives are not only suitable vehicles for consolidating supply and securing the market but also can be used as vehicles for climate-friendly farming approaches and educating platforms.

Cooperatives have been hailed as market solutions in developed countries. Studies have also reported that cooperatives suffer from poor economic performance (Hirsch et al., 2020). This problem is usually prevalent in developing countries. The global importance of agricultural cooperatives cannot be overemphasized in developed countries. There is no doubt that collective bargaining derives more benefits for individual farming members. A clear case is in the USA, where 70% of the nation's dairy produce was marketed by a cooperative in 2017 which is an indication of the huge potential in agricultural cooperatives elsewhere (Wadsworth, 2019). In western nations, agricultural cooperatives hold large market shares in the supply chains for agri-food systems.

Some of the problems experienced by cooperatives in developed countries slightly differ from problems experienced by developing countries. Oversupply of produce which runs the risk of perishability is one of the potential problems highlighted in Candemir et al. (2021). While this is possible anywhere, it is less likely in developing countries as cooperatives usually struggle to create sustainable economies of scale primarily because of land tenure issues and factors of production costs that are high, to begin with. However, in contrast with developed countries, there is a history of agricultural subsidization which has over the years given farmers a competitive advantage in developing countries.

Although farmers who are members of cooperatives usually operate smaller land sizes in Africa, they devote a bigger portion of their production to food and account for a greater variety of crops globally. This affirms that government and stakeholders should continue to support small-scale farmers as their efforts meaningfully contribute to hunger mitigation (Ricciardi et al., 2018). The study's background provided insight into the role that SCF plays in food production in SSA. It was evident that market access is one of SCF's major challenges impeding better farming returns. Due to future population growth projections, SCF stands to benefit from increased demand provided they have market access. The lack of market access has made the use of cooperatives a desirable alternative tool for SCF for various beneficial reasons. In South Africa, cooperatives have experienced many challenges leading to poor performance.

The challenges have **however not** reduced the economic importance of the smallholder farming sector. In addition, the SCF sector is faced with the greater challenge of a dynamic environment where local markets are interlinked with global markets. This environment requires SCF to be competitive in order to benefit from open markets.

The use of cooperatives as vehicles for market access has been endorsed by the South African government for the agricultural sector due to its ability to create economies of scale and resource pooling benefits. There is no doubt that smallholder farmer's participation in markets is low in most sub-Saharan African countries due to various constraints (Otekunrin et al., 2019) and in South Africa, market access is one of the major challenges facing the smallholder farming sector (Ruwanza et al., 2022). Smallholder's role in Limpopo province is crucial in the creation of job opportunities, food security and poverty alleviation (Maponya et al., 2015). Cooperatives can bridge the gap of market access; however, this is threatened by the documented poor performance of agricultural cooperatives. It is crucial to assess if SCF belonging to agricultural cooperatives are able to access markets for their cash crops (spinach, cabbage, tomatoes, peppers, butternuts and beetroot). This will provide relevant information for policymakers in relation to agricultural cooperatives as vehicles for market access. To the best of the author's knowledge, little to no studies have assessed the role of cooperatives in enabling SCF to access markets in the Limpopo province.

RESEARCH METHODOLOGY

Description of the Study Area

The Limpopo Province is situated in the **Northeastern** corner of South Africa and shares borders with Botswana, Zimbabwe, and Mozambique. It forms a link between South Africa and countries in sub-Saharan Africa. Limpopo offers a mixture of exceptional scenic landscapes, an attractive cultural heritage, an abundance of wildlife species and many nature-based tourism opportunities. The Limpopo Province is the bread and fruit basket of South Africa **and** one of South Africa's richest agricultural areas. The province is endowed with abundant agricultural resources. It is one of the country's prime agricultural regions noted for the production of livestock, fruits, vegetables, cereals and tea. The province produces about 75% of South Africa's mangoes, 65% of its papayas, 36% of its tea, 35% of its oranges, **bananas,** and litchis, 60% of its avocados, and two-thirds of its tomatoes (The Limpopo Provincial Government, 2022). The population of Limpopo Province consists of various ethnic groups distinguished by culture, **language,** and race. The population size of Limpopo Province was 5 852 553 in 2019 of which women constitute about 52.6% of the provincial. The province is categorized by high out-migration of young people to Gauteng province in search of **jobs, creating** population pyramids with a bigger proportion for older age groups in the province.

People migrate out of Limpopo to Gauteng because economic activity and job opportunities are highest in that province. Limpopo is ranked the third highest province with out-migration compared to the other provinces in the country. The trend suggests low levels of economic activity in the province due to vast rural expanses. Though farming only constitutes a small component of the province's economic activities, it forms an important part of livelihoods in the province. Mixed farming is the most common form of farming **practiced** by farmers in the province. Limpopo Province is basically a rural area without large cities except for the capital city, Polokwane.

Approximately 12% of the population is situated in urban areas while 88% is situated in rural towns and villages in former homelands areas. About 4860 farms in excess of 4 million hectares have been redistributed to black people and communities. There are four main land tenure types in Limpopo Province that could be divided into commercial land (owned by banks), privately owned land, state land and communal land (under the leadership of traditional authorities). After COVID-19, the mining and government services industry along with Agriculture were identified as key sectors that needed government support (2020-2025 Limpopo Development Plan, 2022).

Figure 1. A map of Limpopo Province

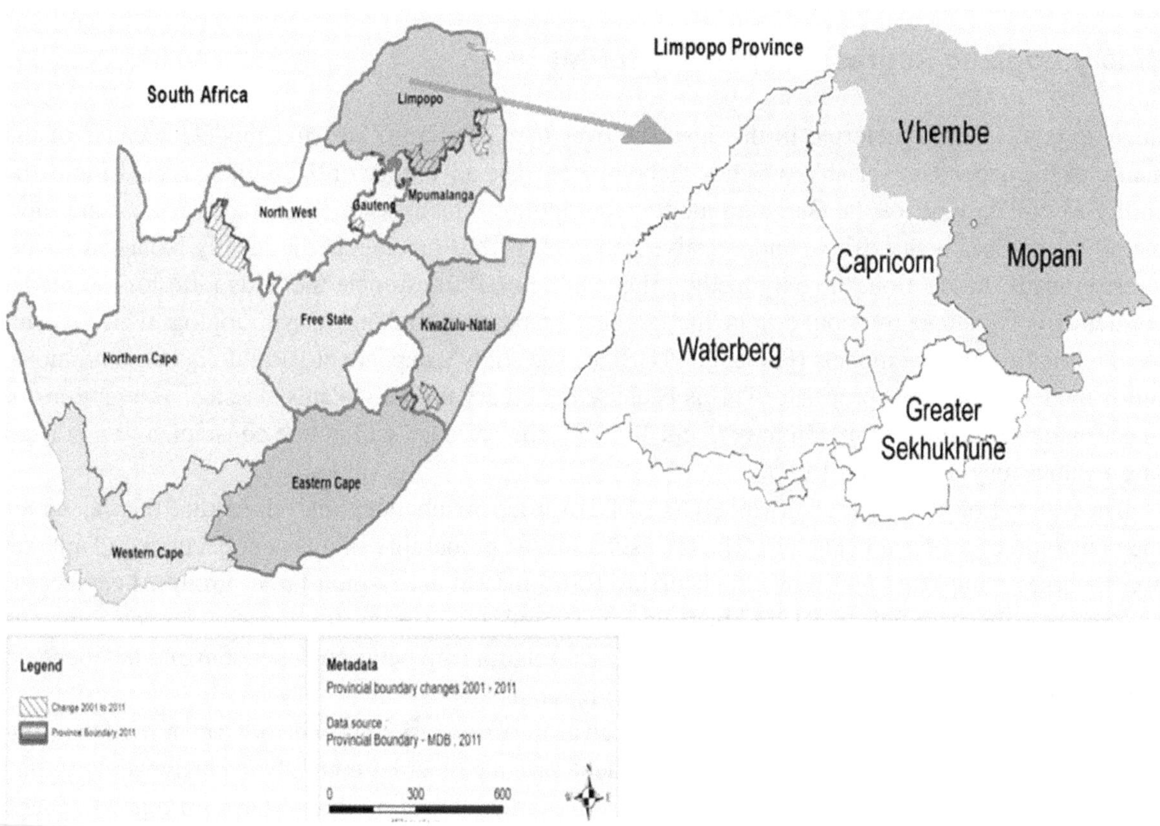

Sampling and Data Collection

The study focused on cash crops (spinach, cabbage, tomatoes, peppers, butternuts, and beetroot) small-scale farmers who were members **of agricultural** cooperatives in Limpopo Province. There were 268 registered small-scale agricultural cooperatives in CODAS in Limpopo (Cooperative Data Analysis System). The intended sample size was a population size of 268. However, it was discovered at the district level that most cooperatives from the national database were not known at the district and municipal level and therefore were not yet operational but simply registered on the national database. From the 268 there were only 146 functional cash crops cooperatives from 21 municipalities known by the agricultural extension officials which were used as a sample. Thus, a cooperative leader from each cooperative was used as a respondent and semi-structured questionnaires were used to collect data. Data was also collected through one focus group discussion from the Mopani district. The focus group's aim was to solicit farmers' views on success, and failure factors and solutions for cooperatives. Limpopo Department of Agriculture facilitated contact with the members of the cooperatives. Extension officers in the municipalities were used to introduce the research team to the cooperatives. The data were **analyzed** using SPSS version 28.

ANALYTICAL FRAMEWORK

Binary Logistic Regression Model (BLRM)

Logistic regression is preferred in this instance over the Gaussian regression models because of the nature of the dependent variable which is dichotomous. The advantage of the logistic regression is the ability to violate most of the Gaussian model assumptions which assume constant variance and normality (comprises both positive and negative numbers, with 0 serving as the ideally balanced mean) in error terms. In addition, the only method that can be used to estimate the odds ratio for the model predictors is logistic regression, despite the fact that probit, is complementary to loglog, which can all handle binary response models (Hilbe, 2009). Since the study's dependent variable is not continuous, linear modelling is not appropriate. This is because the binary model violates classical assumptions of linear models, the error term will have 0 and 1 values, the variance will not be constant, and predicted mean values may differ from 1 to 0.

The Binary Logistic Regression Model (BLRM) is a group membership analysis method that calculates the odds ratio of an event occurring. Thus, it estimates the probability of an event occurring (1-market access-target group) and an event not occurring (0-no market access-non-target group). **i.e.,** success over the probability of failure. **BLRM method** is also preferred over linear regression models due to its relaxed assumptions. BLRM does not require a linear relationship between dependent and independent variables due to the dichotomous nature of the regressand.

Assumptions of the linear regression model such as linearity, residuals or error term normal distribution, and equal variance are relaxed. The dependent variable must be mutually exclusive and exhaustive, thus, in this case, the farmer must either have market access or should have no market access. The exponential value of B serves as a gauge for the probability ratio (odds), Exp(B), which should be greater than 1, showing that the likelihood of an event occurring increases as a variable increase. **It is** the likelihood of being included in the target population of farmers having access to markets. The converse is true if Exp(B) is smaller than 1, it **will show** that the likelihood of farmers belonging to the market access group is decreasing as the variable increases. The chance of belonging to the group of farmers with market access is the same as the likelihood of belonging to the group of farmers without market access when the Exp(B) is equal to 1.

The BLRM is useful in determining variables or drivers that increase the odds of belonging to the group of farmers with market access within the agricultural cooperative. Given the poor performance that cooperatives have had, it is important to know the drivers of market access especially to reflect areas of intervention that policymakers can target to improve market access for cooperative members.

To address the research question, we estimated a binary logistic model of cooperatives' market access. The BLRM was used to determine whether socioeconomic factors had an association with the cooperatives' market access. The main advantage of BLRM is that it allows analysis across two categories. In BLRM, a single outcome variable Yi (i =1,..., n) takes a Bernoulli probability function that takes on the value of 1 with probability Pi and 0 with probability $1\text{-}P_i$. $P_i/1\text{-}P_1$ is referred to as the odds of an event occurring (Greene, 2003).

In the regression context, it is assumed that there is a set of predictor variables, $X_1,\ldots, X_k$, that are related to Y and, therefore, provide additional information for predicting Y (Greene, 2003).

$$\text{Logit}\,(P_i) = \ln\,(P_i\,/\,1\text{-}P_i) = \alpha + \beta_1 X_1 + \ldots + \beta_n X_n + U_t \quad (1)$$

$$\ln\,(P_i\,/\,1\text{-}P_i) = \alpha + \beta_1 X_1 + \beta_2 X_2 + \beta_3 X_3 + \beta_4 X_4 \ldots\ldots + U_t \quad (2)$$

When the variables are fitted into the model, the model is expressed as:

$$\ln\,(\text{market access/ no market access}) = \alpha + \beta_1\,AGE + \beta_2\,MAS + \beta_3\,TRA + \beta_4\,NHS + \beta_5\,KCO + \beta_6\,EXT + \beta_7\,AGM + \beta_8\,ASF + \beta_9\,ANI + \beta_{10}\,MFC + \beta_{11}\,ACF + \ldots U_t \quad (3),$$

with symbols as expressed in Table 1.

Dependent Variable

In this study, the outcome variable was market access. An ordinal variable with two outcomes was constructed namely, market access and no market access with a value of 1 and 0 respectively. Market access refers to the participation in lucrative **formalized** markets. In this study market access referred to the active participation of farmers in retail supply, contract marketing and agency marketing in Fresh Produce Markets (FPM) only. Farmers who sold their produce through any of the three channels were deemed as having market access to **formalized** markets. While farmers who sold their produce outside formal arrangements were deemed not to have access to markets. Table 1 results show that the assumption of no multicollinearity was met as tolerance levels were above 1.0 and VIF levels of less than 10 were observed. The inspection of **standardized** residual values in the **case wise** list indicated that there were five outliers in the data, due to the insignificant number, the outliers were kept in the data. As indicated by the Omnibus test, the model was statistically significant suggesting that it could distinguish between those who had market access and those who did not. The model explained between 27% (Cox & Snell R Square) and 54% (Nagelkerke R Square) of the variance in the dependent variable and correctly classified 90% of the cases as shown in Table 1.

Independent Variables Choice

The various variable symbols that appear below in Table 1 are explained. **The selected** variables were proven to be key to market access by a number of studies, also as further explained below. The demographic characteristics of the farmers were represented by the gender of the farmer (GEN) (Sebatta et al., 2014). The variable was important primarily because in the developmental discourse gender has been found to play a pivotal role in determining farmer success. Traditionally access to resources is usually skewed toward male farmers. The study wanted to ascertain if gender could also predict market access. Age (AGE) (Nwafor, 2020) has a double bearing which could be negative or positive, that is, considering that farming is dominated by ageing farmers across the African continent, this could be a disadvantage or an advantage. The implication is that older farmers get limited mobility, especially physically over time. The limitation also extends to exposure to information, especially in modern media streams.

The positive side of the older farmers could also be a wealth of experience and long-term established networks that could work as goodwill and information capital. On the other hand, younger farmers could possess rich current relevant information, especially in the digital space which can supplement the lack of experience. Marital status (MAS) (Egbetokun & Omonona, 2012; Zondi et al., 2022) was included because in rural communities there is a general belief that married individuals are more stable and can easily access resources. The study wanted to assess if marriage status could have a bearing on market access. TRA (Jari & Fraser, 2012) represents training in bookkeeping and production. Human capital development is crucial in farming as it determines how well a farmer can interpret available information (both documented and verbally) shared. In the study, the variable training representing human capital was preferred as it measures the direct skills needed for farming that a farmer should have. The study wanted to check if training in farming-related aspects could predict market access. NHS (Hlatshwayo et al., 2021) measures the number of workers in the household of the cooperative member. The number of people working in a household represents potential financial, physical and human capital for a farmer. If the farmer has more workers in the household this would then increase the potential resources that a farmer could access within the household. Considering that market access on its own is an activity that requires resources, the expectation would be that farmers with on-farm working household members would be at an advantage when it comes to accessing markets.

KCO represents whether a farmer had knowledge of cooperatives principles as contained in the South African and International Cooperative **legislations**. The International Cooperative Alliance (1995) ha within globally. The principles are *Open and voluntary membership; democratic member control; member economic participation; autonomy and independence; education and training and information; cooperation among co-operatives; and concern for the community.* A cooperative as an organization should be founded on these seven international principles which give guidance on the values and principles that should be followed when running a cooperative across the world. The lack of this knowledge has been seen as one of the problems contributing to the failure of cooperatives' governance leading to dysfunctional cooperatives. The variable then was crucial in potentially predicting market access. Accessibility to information from government-based structures was measured by the variable EXTENSION (Kyaw et al., 2018; Hlatshwayo et al., 2021). The role of extension services amongst small-scale farmers who generally have limited access to information and cannot afford the cost of information is vital and as a result, the variable was of interest in the study.

AS represented the affiliation of the cooperatives to other external secondary organizations as a form of non-government support. This type of support would normally attract some financial obligations from the farmers as opposed to government information support services. However, the services from these external organizations are expected to expose farmers to more opportunities and therefore such affiliation was expected to predict market access for farmers. ANI (Abay & Hirvonen, 2017) represented the annual income of the farmer from the cooperative. The importance of the variable was that economic activities of the farmer such as income generation during the financial year could provide an indication of the farmer's **activity**. This is because generally cooperatives have been reported to be dysfunctional owing to group dynamics problems. The variable then could act as a pseudo-functional cooperative, the expectation being that such activeness should arise to market access. MFC represented member contribution financially to the cooperative (Jacoby & Minten, 2009; Usman & Callo-Concha, 2021) a proxy for farmers' livelihood position. This variable also measured the commitment of farmers to their respective cooperatives. Cooperatives are self-sustaining organizations and therefore farmers are expected to sustain the cooperatives financially in order to run day-to-day activities such as travelling to events, market contract negotiations, information sessions, training and attendance to other informal gatherings and administrative activities.

The financial commitment from a farmer could predict the cooperative abilities and consequently market access for the farmer. ACF represented access to funding by the cooperative (Aliber & Hall, 2012). The ability of a farmer to produce will depend on the available resources and therefore, access to the market for such products will also depend on the latter. Usually, small-scale farmers do not have access to such resources and access to funding enables them to take care of the short-run farming financial obligations, similar to that for commercial farmers. The study therefore deemed it fit to include this **variable.**

RESULTS AND DISCUSSION

The study results are presented in Table 1 that **shows key** variables associated with market access within cooperatives in the study area. The table indicates six independent variables out of eleven such as age, training in farming received, number of workers in the household, knowledge of the cooperative's principles and Act, annual income and access to funding were statistically significant.

Age significantly contributed to the model at a 5% significance level (P=0.039, B=-1.550, OR=0.212), indicating an association with the market accessibility of cooperatives. However, the odds ratio of 0.212 suggests that an increase in the age of the farmer was 0.212 times less likely to have market access. The negative coefficient (-1.550) and the odds ratio (Exp (β)) of less than 1 indicates that the likelihood of a shift from no market access to market access is less likely. This was consistent with findings from Hlatshwayo (2021) who found that age was less likely to predict market access. This shows that in the study area, older farmers were less likely to have access to the market. This could be attributed to less mobility that comes with ageing, i.e. as farmers grow older, their mobility decreases as well. Older farmers also have limited exposure to social media platforms which have also become a source of market information. While younger farmers might have mobility as an advantage, they also have access to modern-day media with more exposure to information and market information.

Training in farming received was statistically significant at a 1% level (P=0.008, B=-1.388, OR=0.250). The results indicate that farmer training could significantly predict whether the farmer could have access to the market or not. These are the farmers who indicated they had received training in bookkeeping and production from the government. Contrary to the expectations in the study, the negative coefficient (B) and odds ratio of 0.250 suggests that it is less likely by 0.250 times for farmers to have access to the market as a result of having received the training. It was expected that training should have a positive influence on market access, precisely because training is a source of information, a critical factor in market access. The results support Kangile et al. (2020) findings in a farmer's market choice study conducted in Tanzania where training was found to be a determinant in marketing. The contrary findings in the study area could be indicating that the training received did not afford farmers the necessary market acumen.

Table 1. Regression results of factors influencing market access

Code	Variable	B	Sig.	Tolerance (>0.1)	VIF (<10)	Exp(B)/ODDS RATIO
GEN	Gender	0.430	0.548	0.980	1.020	1.537
AGE	Age	-1.550	0.039	0.842	1.188	0.212**
MAS	Marital status	0.370	0.311	0.783	1.278	1.448
TRA	Farming training received	-1.388	0.008	0.871	1.148	0.250***
NHS	No. of workers in household	0.715	0.010	0.849	1.178	2.044***
KCO	knowledge of coop principles & act	0.624	0.029	0.795	1.258	1.867**
EXT	Extension services	-0.457	0.576	0.812	1.232	0.633
ASF	Association Affiliation	-0.593	0.287	0.932	1.073	0.553
ANI	Annual income	-2.570	0.013	0.757	1.321	0.077**
MFC	Member Financial contribution	0.952	0.430	0.799	1.251	2.591
ACF	Access to funding	0.989	0.050	0.829	1.206	2.688**
	Constant	0.524	0.857			1.688
Omnibus test			Model Summary			
Chi-square		46.838	Cox & Snell R Square			0.274
P-value		<,001	Nagelkerke R Square			0.535
Classification						
Overall Percentage		90.4				

*Denotes 10%, **5%, ***1% level of significance.

This suggests that the general training currently offered to farmers does not address marketing-specific issues that help farmers access markets in the study area.

The number of workers in the household indicated a significant contribution in the model at a 1% level (P=0.010, B=0.715, OR=2.044). This parameter had an association with the market accessibility of the farmer. As expected, the odds ratio of 2.044 suggests that for every increase in the number of people working in a household of the farmer it was 2.044 times more likely to have market access. The household size variable was important as it not only indicates household size but the number of members who can contribute financially to the household. In the literature, it was also indicated that generally, farmers with larger household sizes are more likely to join cooperatives (Mojo et al. 2017). The variable could also work as a pseudo-free labor availability meaning that the farmer could spend less on labor affording more resources for other farming activities by using financial resources from members of the family to hire more workers. As research has shown, marketing activities are costly. Another possible explanation could be when household members of a farmer are able to contribute financially to the household, maybe through buying food, paying utility bills and other household costs, the farmer can commit more financial resources to farming activities.

Knowledge of the cooperatives' international principles and Act was associated with the market accessibility of the farmer at a 5% level (P=0.029, B=0.624. OR=1.867). It was expected that farmers who had knowledge of these principles and values would predict market access. The odds ratio of 1.867 suggests that knowledge of cooperative principles could significantly predict the market accessibility of the farmer. The odds ratio indicates that it was more likely by 1.867 times for a farmer to have market access. This variable was key in the study as research has shown a relationship between cooperative practices and the non-financial performance of cooperatives (Kyazze et al. 2020). Farmers' knowledge of these principles would be an indication of an understanding of cooperatives' governance and the expectations thereof.

Annual income was significantly associated with market access at a 5% level of significance (P=0.013, B=-2.570, OR=0.077). Contrary to the expectations in the study, the odds ratio of 0.077 suggested that it was less likely by 0.077 times for a farmer to have market access as their annual income increased. The implication could be that fewer financial resources were ploughed back into marketing activities due to other household commitments. The possible explanation could also come from the focus group discussions held. During the focus group discussions, some of the pertinent issues raised were the high cost of electricity in the area. Farmers indicated that they were literally working for Eskom (the local national energy-producing enterprise) meaning that their profits were wiped out by high electricity costs related to irrigation. This was largely due to the three-phase transformers which were too big for their farming operations and also attracted hefty running costs. The implication is that when farmers do not have financial resources, marketing activities could be affected negatively as markets also determine farmers' revenue (Stifel & Minten, 2017).

Access to funding significantly and positively predicted market access of farmers at a 5% level (P=0.050, B=0.989, OR=2.688). As expected in the study, the odds ratio of 2.688 suggests that it is more likely by 2.688 times for farmers to have market access when they have access to funding. The implication is that when farmers have financial resources to move around to negotiate and also to identify market opportunities, markets can be secured. This is something that is difficult to do in the absence of income considering that funding has been found to be associated with an improvement in the income of cooperative members (Ma et al., 2016b).

CONCLUSIONS AND POLICY IMPLICATIONS

The goal of the study was to assess variables that have an impact on small-scale cash crops (spinach, cabbage, tomatoes, peppers, butternuts, and beetroot) farmers' access to markets. Results showed that although age had little impact on market access, possibly because as farmers age, they become less exposed to farming imperatives, including market information. The results suggested that farmers who were older were not better off than younger farmers in the study. On the other hand, farmers who are older could enjoy business networks established over time which could afford them market opportunities. It is therefore crucial for stakeholders to target younger farmers when dealing with market access interventions as market access is a financial incentive for a farmer. Younger farmers need secure markets to be encouraged. A visible growth in market access could attract more younger farmers to participate in farming and consequently reduce the current high youth unemployment. The government and stakeholders should consider tailor-made programs targeting youth-owned agricultural cooperatives' market development.

Farmers who received training in bookkeeping and production reflected reduced market access. Although the odds of guaranteed market access were less likely, the findings suggested that training in farm management aspects played a lessor in accessing the market. However, it could be important for stakeholders to modify training programs as part of market access intervention efforts especially designing tailor-made market access training programs for optimum impact. The study findings indicated a positive association between the number of workers in the farmers' households and market access. This indicates an indirect existence of increased resources for a farmer. During the olden days, farming families greatly benefited from free labor, a trend that has since faded after industrialization as more people migrate from rural to urban areas looking for better opportunities. Market access brings the same opportunities for farmers through the development of the value chain and could increase the participation of farmers in these ventures. More farmer participation in the crop value chains consequently creates more jobs. The government should focus on building market access through resilient and current functional cooperatives for farmers to fully unpack this job creation opportunity. When a farmer has financial backup from families, it is easier to commit resources to the farm or market activities.

Numerous cooperatives around the world have failed due to problems related to free-riding and a lack of understanding of the purpose for cooperatives. The international cooperatives principles were created to provide a common understanding of what cooperatives should and not do. The study findings suggest that knowledge about these principles could significantly predict whether the farmer had access to the market or not. This indicates that market access interventions from stakeholders should enforce workshops on the international principles of cooperatives. This intervention could play a major role in reviving the current dysfunctional cooperatives in the Limpopo Province. This is because most dysfunctional cooperatives were subjected to misunderstandings related to cooperative governance. Some cooperatives believe that the finances of their organization are entirely up to the state. While others join cooperatives to access financial support from the state. This narrative could be corrected through training intervention on the principles and values of cooperatives.

The prediction that annual income was less likely for farmers to access markets is worrying as it reflects that farmers who can sell their products are unlikely to stay in business. The latter could have been impacted mostly by lack of knowledge of bookkeeping as reflected in this study, calling for a concerted effort at increased exposure to record keeping. Funding or credit access could also significantly predict whether the farmer had market access or not. This further confirms that the growth of farmers relied on access to credit and in turn, the growth affords farmers more credit for bigger expansions. Farmers with market access had this opportunity to expand suggesting that market access interventions could create positive spin-offs for farmers in many ways. Although the government and other stakeholders should focus on the creation of markets for cooperatives, funding should be made more accessible to small-scale farmers in cooperatives. The focus group discussions have shown that government should take an active role in facilitating contractual market agreements between cooperatives and retailers to minimize contractual misinterpretations and manipulations. In the future, this study could be expanded to incorporate market players such as retailers and market agencies to gain a broader understanding of the dynamics of market access. The activities between farmer cooperatives and the retail sector should be assessed as to the extent to which these affect farmers' market access and profit.

REFERENCES

Abay, K., & Hirvonen, K. (2017). Does market access mitigate the impact of seasonality on child growth? Panel data evidence from northern Ethiopia. *The Journal of Development Studies*, *53*(9), 1414–1429. doi:10.1080/00220388.2016.1251586

Alemu, D., Guinan, A., & Hermanson, J. (2021). Contract farming, cooperatives and challenges of side selling malt barley value-chain development in Ethiopia. *Development in Practice*, *31*(4), 496–510. doi:10.1080/09614524.2020.1860194

Aliber, M., & Hall, R. (2012). Support for smallholder farmers in South Africa: Challenges of scale and strategy. *Development Southern Africa*, *29*(4), 548–562. doi:10.1080/0376835X.2012.715441

Black, R., Bellagamba, A., Botta, E., Ceesay, E., Cissokho, D., Engeler, M., Lenoël, A., Oelgemöller, C., Riccio, B., Sakho, P., Somparé, A. W., Vitturini, E., & Zingari, G. N. (2022). Migration drivers and migration choice: Interrogating responses to migration and development interventions in West Africa. *Comparative Migration Studies*, *10*(1), 1–18. doi:10.118640878-022-00283-3 PMID:35284235

Candemir, A., Duvaleix, S., & Latruffe, L. (2021). Agricultural cooperatives and farm sustainability–A literature review. *Journal of Economic Surveys*, *35*(4), 1118–1144. doi:10.1111/joes.12417

Chagwiza, C., Muradian, R., & Ruben, R. (2016). Cooperative membership and dairy performance among smallholders in Ethiopia. *Food Policy*, *59*, 165–173. doi:10.1016/j.foodpol.2016.01.008

Chipfupa, U., & Tagwi, A. (2021). Youth's participation in agriculture: A fallacy or achievable possibility? Evidence from rural South Africa. *Suid-Afrikaanse Tydskrif vir Ekonomiese en Bestuurswetenskappe*, *24*(1), 1–12. doi:10.4102ajems.v24i1.4004

De Jong, P. W., & de Valk, H. A. (2020). Intra-European migration decisions and welfare systems: The missing life course link. *Journal of Ethnic and Migration Studies*, *46*(9), 1773–1791. doi:10.1080/1369183X.2019.1611421

Department of Agriculture, Forestry, and fisheries (DAFF), (2012). National Agricultural Cooperatives Indaba, Directorate Co-operative and Enterprise Development. Pretoria.

Egbetokun, O. A., & Omonona, B. T. (2012). Determinants of farmers' participation in food market in Ogun State. *Global Journal of Science Frontier Research Agriculture and Veterinary Sciences*, *12*(9), 24–30.

Elleby, C., Domínguez, I. P., Adenauer, M., & Genovese, G. (2020). Impacts of the COVID-19 pandemic on the global agricultural markets. *Environmental and Resource Economics*, *76*(4), 1067–1079. doi:10.100710640-020-00473-6 PMID:32836856

Elliott, M., Elliott, L., & Sluis, E. V. D. (2018). A predictive analytics understanding of cooperative membership heterogeneity and sustainability. *Sustainability*, *10*(6), 2048. doi:10.3390u10062048

Fischer, E., & Qaim, M. (2012). Linking smallholders to markets: Determinants and impacts of farmer collective action in Kenya. *World Development*, *40*(6), 1255–1268. doi:10.1016/j.worlddev.2011.11.018

Giller, K. E., Delaune, T., Silva, J. V., Descheemaeker, K., van de Ven, G., Schut, A. G., van Wijk, M., Hammond, J., Hochman, Z., Taulya, G., Chikowo, R., Narayanan, S., Kishore, A., Bresciani, F., Teixeira, H. M., Andersson, J. A., & van Ittersum, M. K. (2021). The future of farming: Who will produce our food? *Food Security*, *13*(5), 1073–1099. doi:10.100712571-021-01184-6

Glaeser, E. L. (2014). A world of cities: The causes and consequences of urbanization in poorer countries. *Journal of the European Economic Association*, *12*(5), 1154–1199. doi:10.1111/jeea.12100

Glushkova, S., Lomakina, O., & Sakulyeva, T. (2019). The economy of developing countries in the context of globalization: Global supply chain management. *International Journal of Supply Chain Management, 8*(1), 876–884.

Gyau, A., Mbugua, M., & Oduol, J. (2016). Determinants of participation and intensity of participation in collective action: Evidence from smallholder avocado farmers in Kenya. *Journal on Chain and Network Science, 16*(2), 147–156. doi:10.3920/JCNS2015.0011

Hao, J., Bijman, J., Gardebroek, C., Heerink, N., Heijman, W., & Huo, X. (2018). Cooperative membership and farmers' choice of marketing channels–Evidence from apple farmers in Shaanxi and Shandong Provinces, China. *Food Policy, 74*, 53–64. doi:10.1016/j.foodpol.2017.11.004

Hilbe, J. M. (2009). *Logistic regression models*. Chapman and hall/CRC. doi:10.1201/9781420075779

Hirsch, S., Mishra, A., Möhring, N., & Finger, R. (2020). Revisiting firm flexibility and efficiency: Evidence from the EU dairy processing industry. *European Review of Agriculture Economics, 47*(3), 971–1008. doi:10.1093/erae/jbz003

Hlatshwayo, S. I., Ngidi, M., Ojo, T., Modi, A. T., Mabhaudhi, T., & Slotow, R. (2021). A typology of the level of market participation among smallholder farmers in South Africa: Limpopo and Mpumalanga Provinces. *Sustainability, 13*(14), 7699. doi:10.3390u13147699

Hlophe-Ginindza, S. N., & Mpandeli, N. S. (2021). *The Role of Small-Scale Farmers in Ensuring Food Security in Africa*. Food Secur. Afr. doi:10.5772/intechopen.91694

Ito, J., Bao, Z., & Su, Q. (2012). Distributional effects of agricultural cooperatives in China: Exclusion of smallholders and potential gains on participation. *Food Policy, 37*(6), 700–709. doi:10.1016/j.foodpol.2012.07.009

Jacoby, H. G., & Minten, B. (2009). On measuring the benefits of lower transport costs. *Journal of Development Economics, 89*(1), 28–38. doi:10.1016/j.jdeveco.2008.06.004

Jari, B., & Fraser, G. (2012). Influence of institutional and technical factors on market choices of smallholder farmers in the Kat River Valley. In *Mansholt publication series-Volume 10 edited by* (p. 59). Herman D. van Schalkwyk Jan A. Groenewald. doi:10.3920/978-90-8686-168-2_3

Kangile, R. J., Mgeni, C. P., Mpenda, Z. T., & Sieber, S. (2020). The determinants of farmers' choice of markets for staple food commodities in Dodoma and Morogoro, Tanzania. *Agriculture, 10*(5), 142. doi:10.3390/agriculture10050142

Kanyane, M. H., & Ilorah, R. (2015). The cooperative landscape in South Africa: Analyzing critical issues. *Africa Today, 61*(3), 3–14. doi:10.2979/africatoday.61.3.3

Kyaw, N. N., Ahn, S., & Lee, S. H. (2018). Analysis of the factors influencing market participation among smallholder rice farmers in magway region, central dry zone of Myanmar. *Sustainability, 10*(12), 4441. doi:10.3390u10124441

Kyazze, L.M., Nsereko, I. and Nkote, I., (2020). Cooperative practices and non-financial performance of savings and credit cooperative societies. *International Journal of Ethics and Systems*.

Kyomugisha, H., Sebatta, C., & Mugisha, J. (2018). Potato market access, marketing efficiency and on-farm value addition in Uganda. *Scientific American, 1*, e00013.

Limpopo Development Plan 2020-2025 (LDP). (2022). LPG. http://www.limpopo.gov.za

Liu, M., Min, S., Ma, W., & Liu, T. (2021). The adoption and impact of E-commerce in rural China: Application of an endogenous switching regression model. *Journal of Rural Studies, 83*, 106–116. doi:10.1016/j.jrurstud.2021.02.021

Ma, W., & Abdulai, A. (2016a). Linking apple farmers to markets: Determinants and impacts of marketing contracts in China. *China Agricultural Economic Review, 8*(1), 2–21. doi:10.1108/CAER-04-2015-0035

Ma, W., & Abdulai, A. (2016b). Does cooperative membership improve household welfare? Evidence from apple farmers in China. *Food Policy, 58*, 94–102. doi:10.1016/j.foodpol.2015.12.002

Ma, W., Zheng, H., Zhu, Y., & Qi, J. (2022). Effects of cooperative membership on financial performance of banana farmers in China: A heterogeneous analysis. *Annals of Public and Cooperative Economics, 93*(1), 5–27. doi:10.1111/apce.12326

Maiwashe, A. (2012). *Assessment of Determinants of Agricultural Commodity Marketing Channels Choice for Smallholder Farmers in Thohoyandou Service Centre, Limpopo Province, South Africa* [Master's dissertation, University of Venda].

Maiwashe, A. (2017). Towards a framework for effective performance of smallholder agricultural cooperatives in Limpopo Province, South Africa [Doctoral dissertation, University of Venda].

Manda, J., Khonje, M. G., Alene, A. D., Tufa, A. H., Abdoulaye, T., Mutenje, M., Setimela, P., & Manyong, V. (2020). Does cooperative membership increase and accelerate agricultural technology adoption? Empirical evidence from Zambia. *Technological Forecasting and Social Change, 158*, 120160. doi:10.1016/j.techfore.2020.120160

Maponya, P., Venter, S. L., Modise, D., Heever, E. V. D., Kekana, V., Ngqandu, A., Ntanjana, N., & Pefile, A. (2015). Determinants of Agricultural Market Participation in the Sarah Baartman District, Eastern Cape of South Africa. *Journal of Human Ecology (Delhi, India), 50*(1), 1–9. doi:10.1080/09709274.2015.11906854

Mmbando, F. E., Wale, E. Z., & Baiyegunhi, L. J. (2015). Determinants of smallholder farmers' participation in maize and pigeonpea markets in Tanzania. *Agrekon, 54*(1), 96–119. doi:10.1080/03031853.2014.974630

Mojo, D., Fischer, C., & Degefa, T. (2017). The determinants and economic impacts of membership in coffee farmer cooperatives: Recent evidence from rural Ethiopia. *Journal of Rural Studies, 50*, 84–94. doi:10.1016/j.jrurstud.2016.12.010

Nwafor, C. (2020). Market Participation Variables in Commercializing South African Smallholder Farmers. (Preprints). doi:10.20944/preprints202003.0171.v1

Okem, A. E., & Lawrence, R. (2013). Exploring the opportunities and challenges of network formation for cooperatives in South Africa. *KCA Journal of Business Management, 5*(1), 16–33.

Okou, C., Spray, J., & Unsal, D. F. (2022). Staple Food Prices in Sub-Saharan Africa: An Empirical Assessment. *IMF Working Papers, WP/22/135*

Omar, A. R. C., Ishak, S., Othman, A. S., & Sum, S. M. (2022). Exploring the Catalyst and Constraint Factors of Agriculture Cooperative Performance. *International Journal of Business and Society*, *23*(1), 476–495. doi:10.33736/ijbs.4626.2022

Omiti, J.M., Otieno, D.J., Nyanamba, T.O. and McCullough, E.B., (2009). Factors influencing the intensity of market participation by smallholder farmers: A case study of rural and peri-urban areas of Kenya. *African Journal of Agricultural and Resource Economics, 3*(311-2016-5509), 57-82.

Otekunrin, O. A., Momoh, S., & Ayinde, I. A. (2019). Smallholder farmers' market participation: Concepts and methodological approach from Sub-Saharan Africa. *Current Agriculture Research Journal*, *7*(2), 139–157. doi:10.12944/CARJ.7.2.02

Ricciardi, V., Ramankutty, N., Mehrabi, Z., Jarvis, L., & Chookolingo, B. (2018). How much of the world's food do smallholders produce? *Global Food Security*, *17*, 64–72. doi:10.1016/j.gfs.2018.05.002

Ruwanza, S., Thondhlana, G., & Falayi, M. (2022). Research progress and conceptual insights on drought impacts and responses among smallholder farmers in South Africa: A review. *Land (Basel)*, *11*(2), 159. doi:10.3390/land11020159

Satgar, V. (2011). Challenging the globalized agro-food complex: Farming cooperatives and the emerging solidarity economy alternative in South Africa. *The Journal of Labour and Society*, *14*(2), 177–190. doi:10.1111/j.1743-4580.2011.00331.x

Satgar, V., & Williams, M. (2008). The passion of the people: Successful cooperative experience in Africa. Johannesburg: Cooperative and Policy Alternative Centre (COPAC).

Sebatta, C., Mugisha, J., Katungi, E., Kashaaru, A. and Kyomugisha, H., (2014). Smallholder farmers' decision and level of participation in the potato market in Uganda.

Sebhatu, K. T., Gezahegn, T. W., Berhanu, T., Maertens, M., Van Passel, S., & D'Haese, M. (2021). Exploring variability across cooperatives: Economic performance of agricultural cooperatives in northern Ethiopia. *The International Food and Agribusiness Management Review*, *24*(3), 397–419. doi:10.22434/IFAMR2019.0215

Serra, R., & Davidson, K. A. (2021). Selling together: The benefits of cooperatives to women honey producers in Ethiopia. *Journal of Agricultural Economics*, *72*(1), 202–223. doi:10.1111/1477-9552.12399

Shuaibu, M., & Nchake, M. (2021). Impact of credit market conditions on agriculture productivity in Sub-Saharan Africa. *Agricultural Finance Review*, *81*(4), 520–534. doi:10.1108/AFR-05-2020-0063

Stifel, D., & Minten, B. (2017). Market access, well-being, and nutrition: Evidence from Ethiopia. *World Development*, *90*, 229–241. doi:10.1016/j.worlddev.2016.09.009

Tarekegn, K., Haji, J., & Tegegne, B. (2017). Determinants of honey producer market outlet choice in Chena District, southern Ethiopia: A multivariate probit regression analysis. *Agricultural and Food Economics*, *5*(1), 1–14. doi:10.118640100-017-0090-0

The Limpopo Provincial Government. (2022). *Homepage.* LPG. http://www.limpopo.gov.za

Trebilcock, M., Howse, R. and Eliason, A., (2013). The Regulation of International Trade.

Unsal, F., Garbers, H., Bogmans, C., & Prifiti, E. (2020). *Food (in)security in SSA under COVID-19: a macro perspective.* http://dmedms.imf.org/cyberdocs/viewdocument.asp?doc=6981213&lib=DMSDR1S

Usman, M. A., & Callo-Concha, D. (2021). Does market access improve dietary diversity and food security? Evidence from Southwestern Ethiopian smallholder coffee producers. *Agricultural and Food Economics*, *9*(1), 1–21. doi:10.118640100-021-00190-8

Van den Bossche, P., & Zdouc, W. (2021). *The Law and Policy of the World Trade Organization*. Cambridge Books. doi:10.1017/9781108784542

Wadsworth, J.J. (2019*). Marketing Operations of Dairy Cooperatives, (No. 1502-2021-2279).*

Xu, Z., Li, Y., Chau, S. N., Dietz, T., Li, C., Wan, L., Zhang, J., Zhang, L., Li, Y., Chung, M. G., & Liu, J. (2020). Impacts of international trade on global sustainable development. *Nature Sustainability*, *3*(11), 964–971. doi:10.103841893-020-0572-z

Yobe, C. L., Ferrer, S. R. D., & Mudhara, M. (2020). Measuring the financial efficiency of agricultural cooperatives in South Africa: An application of the Simar–Wilson methodology. *Agrekon*, *59*(3), 269–286. doi:10.1080/03031853.2020.1761845

Zabel, F., Delzeit, R., Schneider, J. M., Seppelt, R., Mauser, W., & Václavík, T. (2019). Global impacts of future cropland expansion and intensification on agricultural markets and biodiversity. *Nature Communications*, *10*(1), 1–10. doi:10.103841467-019-10775-z PMID:31253787

Zahonogo, P. (2018). Globalization and economic growth in developing countries: Evidence from Sub-Saharan Africa. *The International Trade Journal*, *32*(2), 189–208. doi:10.1080/08853908.2017.1333933

Zhang, S., Sun, Z., Ma, W., & Valentinov, V. (2020). The effect of cooperative membership on agricultural technology adoption in Sichuan, China. *China Economic Review*, *62*, 101334. doi:10.1016/j.chieco.2019.101334

Zheng, H., Ma, W., Wang, F., & Li, G. (2021). Does internet use improve technical efficiency of banana production in China? Evidence from a selectivity-corrected analysis. *Food Policy*, *102*, 102044. doi:10.1016/j.foodpol.2021.102044

Zondi, N. T. B., Ngidi, M. S. C., Ojo, T. O., & Hlatshwayo, S. I. (2022). Impact of Market Participation of Indigenous Crops on Household Food Security of Smallholder Farmers of South Africa. *Sustainability*, *14*(22), 15194. doi:10.3390u142215194

Chapter 9
Commercialization of Smallholder Farmers in Mutale Municipality of Vhembe District of Limpopo

Elliot Mahlengule Zwane
University of Limpopo, South Africa

Elekanyani Nekhavhambe
Milk South Africa, South Africa

ABSTRACT

This chapter seeks to discuss and assess the concept of commercialization of smallholder farmers, and whether it is a reality or a myth. The focus of the study area was the former Mutale Local Municipality, in Vhembe district of Limpopo Province in South Africa, drawn from 4 areas namely Tshipise, Tshixwadza, Tshishivhe, and Masisi agricultural zones by using a sample of 153 small holders drawn randomly. The problem of the study was that there are more than 4 million households engaged in agriculture in South Africa who in are engaged in subsistence farming. If these farmers can be assisted, they can feed the nation. Data were collected through a structured qualitative and quantitative questionnaire that was administered face-to-face to respondents and the data was analysed by using a computer programme of the SPSS Version 24. The findings suggest that farmers need support in different aspects such as training, extension, and comprehensive support by the government

INTRODUCTION

Smallholder farmers play a very important role in feeding millions of people not only in South Africa but around the globe, yet they are not always given the recognition they deserve; the question is why? This chapter, therefore, discusses some of the factors that need to be considered that can bring smallholders on the pedestal of development such as being commercialized to play a much bigger role in development.

DOI: 10.4018/978-1-6684-4780-2.ch009

The concept of smallholder commercialization is explored in detail, and whether commercialization is a reality or a myth in South Africa. Smallholder farming can be commercialized, but doing so sustainably and keeping production at that level is a challenge (Kirsten et al., 2013). There are more than 4 million households engaged in subsistence agriculture in South Africa (Stats South Africa, 2016). If these farmers were to be commercialized, they would contribute beyond food security, grow the economy, and create more job opportunities.

The small-scale farming sector is characterised by a lack of capital, human, physical, financial and technological resources, which makes it difficult for farmers to fully participate. In addition to resource support, there is also a need to first unpack conceptualization as a concept for farmers to understand what is at stake. Several authors have identified constraints and barriers facing smallholder farmers in South Africa which include poor market access, unavailability of credit, lack of institutional support, high transaction costs, lack of training, and inadequate property rights (Abdulai & Birachi 2009; Thamaga-Chitja & Morojele, 2014). A similar study conducted in Europe reported similar findings, although the emphasis differs with context (Laurent, 2005).

What does commercialization and smallholder mean? Is the idea of commercialization feasible to begin with considering the socio-economic context of smallholder farmers? Is there a country which has successfully commercialized its smallholder farmers? This chapter provides evidence based on literature and the empirical findings from Mutale Local Municipality in the Vhembe district, Limpopo Province. The chapter contextualizes the meaning of commercialization and covers problems faced by smallholder farmers when they commercialize and make recommendations. The next section provides the theoretical background of the chapter, methodology, and results followed by a conclusion and recommendations.

THEORETICAL BACKGROUND

The literature revealed that the subject of smallholder commercialization has received attention and studies were conducted in different countries such as Europe; (Labarthe & Laurent, 2013; Eicher & Staatz, 1985); Sub-Saharan countries and South Africa (Kirstin, et al., 2013; Ghana: Martey, et al., 2012; Kenya: Wanjohi & Karani, 2017); Ethiopia (Gebre-Ab. 2006; Abera, 2009), Swaziland (Kiribege, 2016) and Ghana, Tanzania, Nigeria and Zimbabwe (Saha et al., 2021). Several studies were also conducted in South Africa on the concept of smallholder commercialization (Zantsi et al., 2019; Carelsen et al., 2021 and Zondi et al., 2022). Lessons drawn from these studies are discussed in this chapter. Agriculture is considered very important in all the countries where the studies were carried out. The South African government has also identified it as a key mover in rural development and in creating jobs (National Development Plan, 2012). Apart from job creation, it is noted that agriculture produces faster growth, reduces poverty and sustains the environment if it is made to work in harmony with other sectors of the economy (World Bank, 2007). As noted by various studies, agriculture contributes to the development of a country. Abera (2009) and Zondi et al. (2022) categorised three levels of agriculture importance. On the first level, it serves as an economic activity, at the second level, it provides a source of livelihood and on the third level is a provider of environmental services. According to Galal (2022), over six million households were engaged in smallholder agriculture in South Africa from the South African Labour Force Survey. Provinces with larger portions of rural areas, such as Limpopo (37.5 per cent) and Eastern Cape (35.9 per cent), were more likely to share large numbers of households involved in some sort of agricultural production.

In order to realize transformation in the smallholder's agricultural sector, farmers have to increase income and welfare outcomes as this will reduce poverty and promote food security. Prior to 1994, there were 55 000 commercial farmers in South Africa (Stats South Africa, 2000). However, the size has been reduced to 32 000 producing 80% of the country's food. These commercial farmers are vital in ensuring the food security of almost 60 million South Africans (Cronje, 2015; International Trade Administration, 2021). The shrinking of the commercial sector further indicates a need to capacitate the small-scale sector more. Fenji Materechera and Mary Scholes (2022) argued that both smallholder and commercial farmers are important for food security. Therefore support should be geared towards capacitating both sectors. However, the focus should be on the under-resourced small-scale sector.

One of the approaches that can achieve agricultural productivity growth is commercialization. The concept of commercialization of smallholder farmers is broad, according to Mahaliyanaarachchi et al. (2006), it is a complex phenomenon. However, it can best be explained from the input perspective side of production. The input side considers commercial farming as being profit oriented; food is produced by advanced technological means for sale in the market, often employing very few workers. The concept is also complicated by various perceptions about what it is, and how it can be carried out in practice. Also, smallholder farming involves millions of households who are scattered in rural areas where they need to be organized and supported. These small-scale farmers focus on production for home consumption, only marketing surplus in order to generate income to fulfil other households' needs, especially in South Africa.

Stevens (2017) asserted that it is not sufficient to focus on production agriculture but to link it to agribusiness and agripreneurship. The authors have no doubt that this can be interpreted as commercialization. The above attests to a different understanding of the subject of smallholder farming and commercialization. Different authors have made contributions to render the concept more understandable. According to Kirsten et al., (2013) small holder farming is divided into three levels, namely, subsistence, semi-commercial and commercial. This was asserted by Pingali and Rosegrant (1995) who call the semi-commercialized stage a "transition stage "to commercial farming.

Some evidence asserts that smallholder farmers who became ptimizeizati have suffered repeated sustainability failure, especially in Sub-Saharan African countries (Kirsten et al., 2013). The World Bank has identified some causes such as insufficient or lacking comprehensive support to assist smallholder farmers to overcome pre-existing bottlenecks along the agricultural value chain (World Bank, 2008). However, in the South African context, a programme called the Comprehensive Agricultural Support Programme (CASP) was introduced to assist farmers. As outlined in the country's National Treasury (2020), a huge sum of money amounting to R4,9 billion, against the total budget of R5.4 billion was spent towards the commercialization of smallholder farmers between 2017/18 and 2019/20. However, the same report alluded to the poor performance of CASP on its key indicators. And that several smallholder farmers across the country still required infrastructure capacity for them to participate meaningfully in the sector. Furthermore, the ptimizeization of underperforming land reform projects was also introduced to increase spending by supporting the plight of smallholder and new-entrant farmers. The point that has been argued by researchers revolved around whether smallholder farmers require land reform programmes and initiatives that are largely responsive to and enhance the assets and capabilities of the local people (Kirsten et al., 2013 and Chitja, et al., 2014).

Commercialization of smallholder agriculture is a long journey that requires the involvement of all directly affected groups throughout the entire value chain. Noteworthy, there has been a good partnership between government and private sector commodity group associations. A study was undertaken by the

National Agricultural Marketing Council in an attempt to assess the impact of statutory levies and trusts expenditure on transformation in the South African agricultural sector. There is evidence that both statutory levies and trust funds have been contributing positively and significantly to the transformation of the South African agricultural sector. The results of the study show that over the past 16 years, industries collected approximately R5.6 billion through statutory levies and spent approximately R1.12 billion on transformation activities. On the other side, the same study shows information regarding agricultural trusts' expenditure on the transformation that approximately R405 million was spent on transformation by the relevant trusts. Initially, the main activities funded around transformation were mainly training, mentorship, social projects, and the development of training material. However, in recent years, industries have focused more on enterprise and skills development. Enterprise development has been related to strengthening and accelerating the development of the operational and financial capacity of black entrepreneurs; providing mentoring and access to inputs, credit, infrastructure, markets, and technology, as well as providing extension services where applicable to smallholder farmers of amongst the respective commodity sectors which statutory and trusts funds are applicable (NAMC, 2019; Mazibuko et al., 2016).

RESEARCH METHODOLOGY

The study was conducted in the Mutale Local Municipality, Vhembe District of South Africa on a proportionally randomly selected sample of 153 smallholder farmers after clustering them into agricultural zones (Tshipise, Tshixwadza, Tshishivhe and Masisi) and commodity groupings (vegetables under irrigation, dryland maize and citrus fruit farming). Data were collected through a structured qualitative and quantitative questionnaire that was administered face-to-face to respondents (smallholders and extension officers) and captured into the SPSS Version 24 computer program. Secondary data sources were also used to explore some concepts related to this study.

RESULTS AND DISCUSSION

Demography

Socio-demographic characteristics relating to gender, age, and educational level of farmers, household, and their main income sources are shown in Table 1.

According to Table 4. The study revealed that most respondents (47.1%) were between the age of 36 and 59 years with relatively low youth participation (9.8%). Females were found to be the dominant (56.2%) gender in the study area. The dominance of female farmers is appreciated as it is in line with the government policy of women's empowerment and emancipation (Alunga and William, 2013). The results also show that the majority (88.2%) of farmers were characterized by low monthly income ranging between R100-R5000 while a few (3.3%) were within the high-income group (above R10000 per month). Most (45.1%) farmers generated their income from social grants rather than from farming (17%). Informal- and formal employment activities were the main income sources of households (8.5% and 11.1% respectively).

Table 1. Socio-economic characteristics of respondents

Farmers attributes		Percentage
Gender	Male	43.8
	Female	56.2
Age	Youth: £ 35	9.8
	Aged: 36-59	47.1
	Elderly: > 59	43.1
Educational level	Never attended	23.5
	Primary	32.7
	Secondary	32.0
	Tertiary	11.8
Household income/month	Low income:<5000	88.2
	Middle income:5000-10000	8.5
	High income:>10000	3.3
Main sources of income	Agriculture	17.0
	Formal employment	8.5
	Informal employment	11.1
	Social grant	45.1
	Mixed	5.2
	Other	13.1

The Importance of Commercialization

The importance of agriculture and agricultural extension in development has been identified by a number of researchers both in developed as well as in developing countries (Van den Ban and Hawkins, 1990). However, the need to commercialize agriculture has been seen as one of the solutions to grow the economy and alleviate poverty. The need to link smallholder farmers to markets was also stressed (Stevens, 2017). In fully commercialized agriculture, it is noted that inputs are predominantly obtained from markets and profit maximization becomes the farm household's driving objective (Pingali and Rosegrant, 1995).

The Meaning of the Smallholder Farming Concept

Smallholder farmers are those marginal and sub-marginal farm households that own or/and cultivate less than 2.0 hectares of land and constitute about 78 per cent of Sub-Saharan countr''s farmers (FAO, 2012). Subsistence farming is a form of agriculture where almost all production is consumed by the household, often characterized by low-input use, with labour generally provided by the family. Farming is generally rain-fed and production is mainly based on staple crops with low yields. Their main target is self-consumption (FAO, 2010). The meaning differs from country to country, for example, Laurent (2005) indicated that in Europe the smallholder farmers differ due to their structures because they are heterogeneous. This factor should be kept in mind when discussing the meaning of the smallholder farming concept.

What is Meant by Commercialization?

Different authors see the definition differently for example Martey et al (2012) and Pradhan et al (2010) noted that agricultural commercialization refers to the process of increasing the proportion of agricultural production that is sold by farmers. Commercialization is assumed to take place if it is producing a significant amount of cash commodities, allocating a proportion of its resources to marketable commodities, or selling a considerable proportion of its agricultural outputs (Jaleta, et al., 2009). The question to be asked could be to check whether there is any success story elsewhere to learn from.

Literature reported pockets of success stories in the following countries: Malawi, Zambia, Tanzania, Uganda, and Mozambique, within the International Centre for Tropical Agriculture (CIAT) under a program entitled the Enabling Rural Innovation (ERI) whichptimizd on building small-scale farmers to have the capacity to produce for the market (Kirsten, et al., 2013). One recorded success story is that from South-Western Uganda where smallholder potato farmers were successfully linked to high-value markets. Looking at the classification of small and commercial enterprises in South Africa, one could argue that the majority of households involved in agriculture could be far from being commercialized if commercialization is measured by their annual total revenue. The Department of Trade and Industry (2019) describes a smallholding enterprise as an enterprise with annual total revenue of between R10 million and R50 million whereas a large enterprise was defined as any enterprise with annual total revenue of R50 Million and above. Therefore, market access is one of the crucial determinants of commercialization where smallholder farmers could be able to trade their farm produce for higher revenue. Jaleta et al. (2009) argued that smallholder farmers need to sell a considerable volume of their products which could align them to achieve the highest annual total revenue as classified by the Department of Trade and Industry report undertaken in 2019.

FACTORS PROMOTING SMALLHOLDER COMMERCIALIZATION

Several studies were conducted to assess their level of commercialization in countries like Kenya, Ghana Matey, and South Africa (Kirstin et al., 2013), Ethiopia (Gebre-Ab, 2006) and Mutale Local Municipality in Limpopo, South Africa (Nekavhambe, et al., 2017). It is noted that factors that promote smallholder commercialization include improved physical, infrastructure such as roads, railways and ICT facilities (Sibale, 2010). Improved access to natural resources, increased adoption of new technologies, level of specialization in fewer staple food and cash crops coupled with availability of assured markets through contracts. Legal agreements are also vital in promoting the commercialization of smallholder farmers (Sibale, 2010). In the study, respondents were asked to indicate determinants of commercialization and the findings are reflected in Table 2.

Table 2 show that on many occasions support was found either inadequate or lacking in the farmer's environments. What might worsen the situation could be a lack of agro–industries or well-functioning markets locally. In the case of South Africa, there are effective markets in big cities, including an efficient transport system for hauling products to markets, although at a high cost. Under these circumstances, a smallholder farmer may compete with established commercial farmers, however, if the quality of the product is questionable the farmer suffers a loss. Apart from these factors, a smallholder farmer who aspires to commercialize needs to have an entrepreneurial spirit, coupled with managerial skills (Eicher and Staatz, 1985).

Table 2. Determinants of commercialization

Determinants for commercialization		Percentage (%)
Credit access	Access to credit	13.1
	No access	86.9
Training access	Access to training	46.4
	No access to training	53.6
Membership to agricultural organisations	Access to agricultural organisations	22.9
	No access to agricultural organisations	77.1
Market support service access	With access	54.6
	With no access	43.6
Farming inputs access	With access	28.1
	With no access	71.9

Criteria to Commercialize Farmers

It should be noted that commercialization is a long process. There must be some characteristics which need to be observed over a period. Some of the things to be checked are whether there is a marketable surplus available. According to Mahaliyanaarachchi (et al., 2006), the marketable surplus in the context of agricultural produce refers to the quantities of products available for consumption by the non-farming population and also as raw materials for manufacturing and processing industries. If one considers this, one can determine whether the farm can be commercialised or not. A production level that is less than 25% of the potential total production is regarded as subsistence. However, a marketable surplus ranging between 25-50% of total production, or more than 50% of the total production qualifies for commercial farming (Mahaliyanaarachchi et al., 2006). Some literature sources have noted that the socio-economic characteristics of smallholder producers are critical determinants of farm commercialization (Pingali and Rosegrant, 1995). It was demonstrated by Kirsten et al., (2013) that crops mainly produced for the market (tobacco, paprika and cotton) are controlled by male members of the households, while the female-headed household focuses on food security-related crops that usually do not attract the market. The availability of labour from the household is another important determinant.

Stages in the Commercialization

Experience suggested that the process of agricultural commercialization usually takes a long transformation process from subsistence to semi-commercial and then to fully commercialized agriculture (Pingali and Rosegrant, 1995). Three stages of farming have been identified namely; the subsistence, the transition stage, and the commercialization. Farmers have to be assisted if they wish to move out of subsistence to semi-commercialization. Many smallholder farmers were found to be operating under subsistence levels with a dependence on rain-fed farming.

CHALLENGES OF COMMERCIALIZATION

Inadequate Availability of Produce

Sibale (2012) observed that one critical challenge of commercialization is that agricultural commodities have been inadequately produced to meet industrial demand. One of the causes could be their scattered nature, and thus the need for proper organization. Specifically, farmers need to be organized so that they work as a team to satisfy the needs of the market that they serve. If they work individually, they may not be able to produce the required quantity for the market.

Low Productivity

Smallholder farmers have tendencies to produce less despite favourable ecological conditions. Kirsten et al. (2013) found that some of the challenges of low-productive farming include declining soil fertility, pest and disease outbreaks and land fragmentation. Smallholder farmers are typically poor and practice low-input agriculture, not by choice but due to poverty. This results in low productivity and production. Big buyers find it problematic to deal with this category of farmers due to the inconsistency and unpredictability of supply (Sibale, 2010). Other authors have reported that the average maize yield per hectare in wealthy countries like Canada is three times higher than the average maize yield in Sub-Saharan African low-income countries (Verdier-Chouchane and Charlotte- Karagueuzian, 2016).

Poor Access to Resources

Kibirige (2016), found that the household commercialization index (HCI) was low, and this was attributed to several factors including limited access to natural, physical, labour, and financial resources, and import for increased production of marketable surplus. Therefore, policies and rural development programmes geared towards improved access to these resources by resourced -poor smallholders should be initiated or catalysed by the government and other development agencies to boost productivity and market access (Kibirige, 2016). It was further noted that many smallholder farmers practice rain-fed agriculture, with poor formal education, and this fact plus the inherent low resource base relegates smallholder farmers to lower productivity (Dorward, et al., 2003). Without resources, it would be challenging to commercialize smallholder farmers. For the present study farmers were asked to indicate the challenges that they experience. Results are indicated in Table 3.

The following challenges emerged: pests and diseases, lack of access to inputs and distance to markets. For example, mechanization (technology) was regarded as the main problem facing many respondents (92.8%). Many respondents (80.4%) ranked shortage of water as the second main constraint in their daily farming activities. More than half of the farmers also pointed out that infrastructure and access to the market and long distance from the point of farming is still a challenge in the Mutale area, 69.3% and 60.7% respectively.

Table 3. Major challenges faced by farmers in the mutale local municipality

Challenges faced by farmers in their farming activities	Percentages (%)
Low mechanization	92.8
Shortage of water	80.4
Poor infrastructure	69.3
Poor safety and security on farm property	17.0
Pest and disease	17.6
Distance to markets	60.7

STRATEGIES TO SUPPORT COMMERCIALIZATION

Commercialization of smallholders has been found possible in sub-Saharan countries and in other countries of Africa such as Tanzania, Malawi, and Kenya. However, it was found that not all successful smallholder commercialization efforts are sustainable over the long term (Kirsten et al., 2013). Sustainable ones were those that received private or government assistance especially in enhancing their capacity to manage their enterprise from a business perspective. One strategy asserted by Kirsten et al. (2013) which could be used to promote sustainable commercialization is when authorities such as governments and the private sector agree to provide consistent and comprehensive support to smallholder farmers, especially to overcome pre-existing bottlenecks along the agricultural value chain. Further strategies are provided below:

Provision of Knowledge and Training

Some well-known knowledge systems needed by smallholder farmers include technical and general knowledge (which sometimes is known as common sense) and most importantly indigenous knowledge transferred to smallholder farmers from previous generations although this type of knowledge is usually undermined by some technocrats. Access to reliable knowledge for smallholder farmers is crucial (Mahaliyanaarachchi, et al., 2006). Exposing smallholder farmers to reliable knowledge will ultimately be equated to knowledge empowerment.

Farmers will treasure information when availed as this will enable them to make effective decisions. Karani and Wanjoh (2017) identified information needs such as gross margins for a particular farm produce, possible markets, stability of the product in the market, availability and price of inputs and projected transportation costs for inputs as critical components of a knowledge system. In addition to price information, it was found that farmers also require information such as weather forecasts, advice on crop production, marketing and the use of appropriate seeds and fertilizers (Awasthi, 2007).

Leadership and Directing

Leadership can make or break the attitude of farmers. It is the observation of the authors that general behaviour patterns which may not mean much to an ordinary person could be critical in ensuring that the right relationship develops for an aspiring farmer to become an agripreneur, a critical component of effective leadership. An effective agripreneur should be able to reflect and avoid those common pitfalls that come with the willingness to change and become better. In most cases, it is hard to unlearn bad credit practices such as failure to honour credit payments or using credit for something else other than for the purpose it was acquired for. In situations where challenges occur, a good agripreneur should be able to offer an apology and promise to correct such anomalies in future. The ability to exercise some level of credit discipline shows effective leadership traits.

It is not proper to assume that the other person will understand, hiding under the fact that "I am a small-scale farmer" when in a real business transaction, good behaviour counts. It is the authors' observation that this behaviour needs to be improved. Smallholder farmers should understand issues of quality and management of their products if they are to maintain a commercialized business enterprise. Experience has shown that some farmers are not worried about ensuring that they produce quality products which could fetch high market prices. Membership in farmer organizations with proven records could be vital for smallholder farmers, especially for compliance with standards set out by the Global Good Agricultural Practice (Global GAP).

Providing Cost-Effective Production Advisory Services

Farmers need regular information on production advisory services. Information Communication Technology (ICTs) offers a cost-effective and efficient means of providing this information (Kibirige, 2009). Agriculture is becoming increasingly knowledge-intensive and high-technology, requiring that an advisory service be able to guide farmers with such knowledge. It is reported that both public and private institutions have traditionally targeted smallholder services with SMS as empowerment tools that possess appropriate production tips (World Bank, 2011). Commercialization will not be an overnight issue but a process which needs to be guided through several activities such as those reflected below.

Political Will or State Intervention

It is noted that for the smallholder to be commercialized it is important that the government should intervene to provide investment that would enable the process to take place (Kirsten, et al, 2013).

Use of Weather Information

Smallholder agriculture in Sub Sahara Africa is largely dependent on the rainwater supply. This, therefore, implies that for farmers toptimizee their productivity there is a need to know in advance the rainfall season forecast with a constant update of changing weather patterns (Verdier-Chouchane and Karagueuzian, 2016). Greater access to information seems to help farmers make better decisions around transportation and logistics, price, location, supply and demand, diversification of their product base, and access to inputs (Dixie and Jayaraman. 2011).

CONCLUSION AND RECOMMENDATION

The chapter has demonstrated that commercialization of the smallholder farming sector is possible. However, it is not something which can be achieved overnight, it is a long process. Critical elements outlined in this study were the concept of commercialization, the criteria to be considered before it is promoted, the challenges of farmers towards commercialization, and general challenges encountered by farmers, especially in the Mutale local municipality. It can be concluded that commercialization can be done but its sustainability depends on the availability of comprehensive farmer support on an ongoing basis.

The following can be recommended:

- Policy to support the initiatives towards commercialization- the government should demonstrate its willingness to support the process through state intervention which will provide investment.
- Empowering farmers- this can be done through different means and tools to access critical information needed to take decisions in farming. It can further be used in transiting towards commercial farming, proven information generation which is risk proof and easily accessible.
- Training farmers- strengthening extension services which will train smallholders in all aspects that will promote and sustain their managerial capacity to handle high-value crops, and developing a programme which uplifts women farmers to pursue the commercial route. This is so because female-headed households tend to focus on food security while men pursue commercialized cash crops.

REFERENCES

Abdulai, A., & Birachi, E. A. (2009). Choice of coordination mechanism in the Kenyan fresh milk supply chain. *Applied Economic Perspectives and Policy*, *31*(1), 103–121.

Abera, G. (2009). Commercialization of smallholder farming: Determinants and welfare outcomes. *A cross-sectional study in Enderata District, Tigari, Ethiopia.* [Unpublished master thesis, University of Agder, Kristiansand, Norway].

Alunga, J. U., & Jaluo, W. M. (2013). Gender roles and agribusiness in the Kenyan communities: The case of Likuyani District. *Journal of Emerging Trends in Educational Research and Policy Studies*, *4*(5), 733–738.

Awasthi, B. D. (2007). Relevance of market information system to environment protection. *Journal of Agriculture and Environment*, *8*, 46–54. doi:10.3126/aej.v8i0.726

Carelsen, C. P. R., Ncube, B., & Fanadzo, M. (2021). Classification and characterisation of smallholder farmers in South Africa: A brief review. *South African Journal of Agricultural Extension*, *49*(2), 97–106. doi:10.17159/2413-3221/2021/v49n2a12821

Cronje, F. (2015). Threats to commercial Farms. *Farmers Weekly*. https: //www.farmersweekly.co.za.

Dixie & Jaraman. (2011). Promotion of information. World Bank. http://siteresources.worldbank.org/EXTINFORMATIONANDCOMMUNIC ATIONANDTECHNOLOGIES/Resources/IC4D-2012-Chapter-2.pdf

Dorward, A., Poole, N., Morrison, J., Kydd, J., & Urey, I. (2003). Markets, institutions and technology: Missing links in livelihoods analysis. *Development Policy Review*, *21*(3), 319–332. doi:10.1111/1467-7679.00213

DTI. (2019). Codes of good practice on Broad Based Black Economic Empowerment. DTI. https://www.gov.za/sites/default/files/gcis_document/201905/42496gen306.pdf

Eicher, C. K., & Staatz, J. (1985). *Agricultural development in the third world.* The Johns Hopkins University Press.

Gebre-ab, N. (2006). Commercialization of smallholder agriculture in Ethiopia. Ethiopian Development Research Institute Notes and Papers Series No. 3. Addis Ababa, Ethiopia.

International Trade Administration. (2021). *South Africa - Country Commercial Guide.* ITA. https://www.trade.gov/country-commercial-guides/south-africa-agricultural sector#:~:text=There%20are%20approximately%2032%2C000%20commercial,80%20percent%20of%20agricultural%20output

Jaleta, M., Gebremedhin, B., & Hoekstra, D. (2009). Smallholder commercialization: Processes, determinants and impact. *ILRI Discussion Paper.*

Karani, K. D., & Wanjoh, J. (2017). Factors Influencing Marketing Of Agricultural Produce Among Small-Scale Farmers, A Case Of Sorhgum In Giaki Location, Meru County Kenya. *International Journal of Economics, Commerce and Management, 8*, 664-678. https://ijecm.co.uk/wp-content/uploads/2017/08/5838.pdf

Kibirige, D. (2016). Smallholder commercialization of maize and social capital in the Eastern Cape Province of South Africa. *International Journal of Economics. Commerce and Management, 4*(9), 236–252.

Kirsten, J., Mapila, M., Okello, J. J., & De, S. (2013). *Managing agricultural commercialization for inclusive growth in Sub-Saharan Africa* (No. 206518). University of Pretoria, Department of Agricultural Economics, Extension and Rural Development.http://ageconsearch.umn.edu/record/206518/files/2013.pdf

Labarthe, P., & Laurent, C. (2013). Privatization of agricultural extension services in the EU: Towards a lack of adequate knowledge for small-scale farms? *Food Policy, 38*, 240–252. doi:10.1016/j.foodpol.2012.10.005

Laurent, C. (2005). Les agricultures de l'Union européenne et leur territoire. In C. Laurent & P. Thinon (Eds.), *Agricultures et territoires* (pp. 31–55). Hermès Science.

Magesa, M. M. (2015). *Linking rural farmers to markets using ICTs.* CTA. Accessed on 12 April 2018.https://publications.cta.int/media/publications/downloads/1875_PDF.pdf

Mahaliyanaarachchi, R. P., & Bandara, R. M. A. S. (2006). Commercialization of agriculture and role of agricultural extension. https://suslj.sljol.info/article/10.4038/suslj.v6i1.1686/galley/1419/download/

Martey, E., Al-Hassan, R. M., & Kuwornu, J. K. (2012). Commercialization of smallholder agriculture in Ghana: A Tobit regression analysis. *African Journal of Agricultural Research, 7*(14), 2131–2141.

Martey, E., Al-Hassan, R. M., & Kuwornu, J. K. (2012). Commercialization of smallholder agriculture in Ghana: A Tobit regression analysis. *African Journal of Agricultural Research*, *7*(14), 2131–2141.

Materechera, F., & Scholes, M. (2022). There's a place for big and small farms in securing South Africa's food supply. *Academic rigour, journalistic flair*. https://theconversation.com/theres-a-place-for-big-and-small-farms-in-securing-south-africas-food-supply-180826

Mazibuko, N., Balarane, A., & Matsobane, M. (2016). *South Africa's statutory levies in context of same levies elsewhere- are we funding the correct area?* Research Gate. https://www.researchgate.net/publication/320164621_SOUTH_AFRICA'S_STATUTORY_LEVIES_IN_CONTEXT_OF_SAME_LEVIES_ELSEWHERE_ARE_WE_FUNDING_THE_CORRECT_AREA

NAMC. (2019). *Status Report On Statutory Measures Implemented In Terms Of The Marketing Of Agricultural Products Act, Act No. 47 Of 199.* NAMC. https://www.namc.co.za/wp-content/uploads/2020/03/2019-Stat-measures-REPORT.pdf

National Treasury. (2020). *Comprehensive Agriculture Support Programme*. National Treasury. https://www.gtac.gov.za/pepa/wp-content/uploads/2021/11/1.-Agriculture-Support-Programme-Spending-Review-Report.pdf

Pingali, P. L., & Rosegrant, M. W. (1995). Agricultural commercialization and diversification: Processes and policies. *Food Policy*, *20*(3), 171–185. doi:10.1016/0306-9192(95)00012-4

Pradhan, K. D. R. & MINSTEN, B. (2010). Agricultural Commercialization and Diversification in Bhutan. IFPRI (International Food Policy Research Institute), Washington, DC, USA.

Saha, A., Sabates-Wheeler, R., & Thompson, J. (2022). Insights into smallholder capacity for agricultural commercialisation: Evidence from four African contexts. *European Journal of Development Research*, *34*(4), 1757–1802. doi:10.105741287-021-00414-z

Sibale, E. (2010). Strategies for commercialising smallholder agriculture. *How We Made it in Africa*. .https://www.howwemadeitinafrica.com/strategies-for-commercialising-smallholder-agr

STATS SA. (2016). *Community survey Agricultural households – statistics South Africa*. SA Stats. www.statssa.gov.za>publications>prese

Stevens, J. B. (2017). Is agricultural extension positioned to promote agripreneurship in South Africa? *South African Journal of Agricultural Extension*, *45*(2), 86–94. doi:10.17159/2413-3221/2017/v45n1a437

Thamaga-Chitja, J. M., & Morojele, P. (2014). The context of smallholder farming in South Africa: Towards a livelihood asset building framework. *Journal of Human Ecology (Delhi, India)*, *45*(2), 147–155. doi:10.1080/09709274.2014.11906688

Van Den Ban and Hawkins. (1990). *Agricultural Extension. Longman Scientific & Technical*. John Wiley & Sons, Inc.

Verdier-Chouchane, A. & Karagueuzian, C. (2016). Economic development in the Middle East and North Africa. Challenges and Prospects. 8TS.

World Bank. (2008). World Development Report 2008: Agriculture for Development. Washington DC. World Bank. http://econ.worldbank.org/WBSITE/EXTERNAL/EXTDEC/EXTRESEARCH/EXTWDRS/EXTWDR2008/0,contentMDK:21410054~menuPK:3149676~pagePK:64167689~piPK:64167673~theSitePK:2795143,00.html

Zantsi, S., Greyling, J. C., & Vink, N. (2019). Towards a common understanding of 'emerging farmer' in a South African context using data from a survey of three district municipalities in the Eastern Cape Province. *South African Journal of Agricultural Extension*, *47*(2), 81–93. doi:10.17159/2413-3221/2019/v47n2a505

Zondi, N. T. B., Ngidi, M. S. C., Ojo, T. O., & Hlatshwayo, S. I. (2022). Factors Influencing the Extent of the Commercialization of Indigenous Crops Among Smallholder Farmers in the Limpopo and Mpumalanga Provinces of South Africa. *Frontiers in Sustainable Food Systems*, *5*, 542. doi:10.3389/fsufs.2021.777790

Chapter 10
Agricultural Marketing Dynamics in the Face of Climate Change:
Insights of Rural-Based Cash-Crop Emerging Farmers

Mpho Tshikororo
University of Venda, South Africa

ABSTRACT

Climate change is instrumental for agricultural performance, with its production and marketing being the leading cores. Although numerous climate change adaptive strategies have been identified, there is a need to explore how climate change impacts agricultural marketing. The focus of the study was to investigate the agricultural marketing dynamics brought by the impact of climate change among rural-based emerging farmers. The study was conducted in the Vhembe district of the Limpopo province. A simple random sampling technique was used to select the participants, while a structured questionnaire was used to collect data. Discriminant analysis was used to draw a distinct difference between farmers who had observed the impact of climate change on their marketing dynamics from those who didn't. The study results show that unstable selling prices have been a negative impact of climate change on the market. The study recommended that collaborative efforts be directed toward assisting farmers in enhancing their marketing practices.

Inroduction

Agriculture relies on climate, with its performance significantly influenced by climatic conditions. A study conducted by Hristov et al. (2021), revealed that climate change has substantial negative impacts on agricultural performance, including yields and sales volumes. Also, agricultural development has been significantly influenced by the performance of agricultural marketing. It has also been noted that

DOI: 10.4018/978-1-6684-4780-2.ch010

the marketing function should be an integral part of accelerating agricultural growth. A study by Tshikororo et al. (2021) ascertained climate change as a global phenomenon that has been of great concern as it continuously contributes to food insecurity and increased malnutrition, particularly in developing countries.

Agriculture is highly affected by climate change; hence unfavourable climatic changes bring undesired shocks to the sector. It has been noted that climate change will continuously impact agricultural production negatively (Kogo et al., 2018). It has also been noted that agricultural supply chain systems are disrupted by climate change, with farm production, marketability, processing, retailing, and human consumption being among other vital areas that suffer the most (Godde et al.,2021). Agricultural markets have been hampered by the impacts of climate change, with most farmers severely affected at the farm level. Consequently, farmer productivity has been significantly challenged, leading to poor market performance (Zabel et al., 2019).

A study conducted by Zhang et al. (2020) discovered that fluctuations in agricultural commodity prices have negatively obstructed society in its bid for food security in recent years. Despite climate change negatively impacting agriculture, population growth has mounted severe pressure on the agricultural sector. Whereas there is a need to ensure food and nutritional security for humankind, these have been negatively impacted by climate change over time (Malhi et al., 2021). The future of rural livelihoods hangs in the balance through agricultural uncertainties brought by the impacts of climate change (Yalew et al., 2018). Forecasting agricultural commodity prices is essential for market performance and the sustainability of both farmers and agribusiness firms (Manogna and Mishra, 2021). Farming has become a considerable risk, mainly due to the impacts of climate change, which contributes to higher interest rates by financial institutions. Ultimately, farmers are mostly limited to effectively adapting to climate change for remaining competitive in the market (Nyoro, 2019). The author further outlines the frequent change in market information with farmers failing to access such information timely being disadvantaged. The marketability of agricultural products harvested under severe climatic changes has proven costly; with fewer resources leading to farmers losing their marketing share (Rasul et al., 2021).

In a study conducted by Seppelt et al. (2014), it was revealed through projections of climate change that agricultural productivity growth is unlikely to continue; on the other hand, it has been noted that prices of agricultural produce have been dropping gradually, mainly due to products quality concerns (Kastner et al., 2012). In comparison, a study by Tilman et al. (2011) has emphasized the common trend of climate change is a threat to food security, especially for the poor. It has been noted that the change in drought and flood occurrence threaten food security (Gomez-Zavaglia et al., 2020). Emerging farmers were the focal point of the present study in that they are the epitome of the rural economy and vital players in food security, yet the most affected group of farmers by climate change. Therefore, this chapter precisely unpacked agricultural marketing dynamics resulting from the impacts of climate change.

CONTEXT OF THE PROBLEM

As noted above climate change has severely impacted the agricultural chain through its critical determinants, such as unfavourable weather patterns, which influence numerous agricultural operations. Marketing of agricultural produce has been left in a dire situation because of climate change, with diminishing world prices of agricultural commodities and increasing prices of agricultural inputs being other negative impacts of climate change felt in agricultural markets around the globe (Nyoro, 2019).

Literature has shown that the pace of climate change determines its impact and the cost of adjustments, which significantly influence the marketing of agricultural produce (Zilberman et al., 2004). Studies have shown the reduction of agricultural productivity and yield is the common trend of the negative impact of climate change towards agriculture (Long et al., 2006; Aydinalp and Cresser, 2008). The study sought to pinpoint the agricultural marketing dynamics due to climate change. It provides insight into how agricultural marketing dynamics such as selling price, quantity sold, access to market and access to storage facilities are impacted by climate change. The study further sought to draw a distinct difference relating to marketing dynamics among farmers who had observed the impacts of climate change as against those who did not observe the phenomenon, especially as regards their marketing activities. The latter part of the chapter provided insights into the type of markets and transactional arrangements that farmers within the study sample were involved in. Challenges facing farmers are also outlined, to showcase the dynamics which these farmers were facing.

LITERATURE REVIEW

Agricultural commodities are essential for food security (Darekar and Reddy, 2017). The fluctuations in agricultural input prices negatively impact farmers due to climate change (Sabu & Kumar, 2020). However, efforts to mitigate climate change through comprehensive, economy-wide greenhouse gas (GHG) emissions reductions may also negatively affect food security due to indirect impacts on prices and supplies of essential agricultural commodities (Havlik et al., 2014). The current emission rate of greenhouse gases may lead to stringent climate mitigation policies that could severely influence the operations of various sectors, amounting to a great deal of global hunger and food consumption than the direct impacts of climate change on agricultural productivity (Hasegawa et al., 2018).

Study findings show that years with high occurrences of heat lead to a reduction in agricultural production. This subsequently impacts agriculture, mortality rate, labour productivity and economic growth (Van Meijl et al., 2018). A study on agricultural contribution towards economic development noted that a sharp decline in economic growth causes a decrease in international meat prices. Furthermore, the study pointed out that the income losses and local supply chain disruptions associated with the pandemic led to an increase in food insecurity in many developing countries (Elleby et al., 2020). In another study, it was noted that farm net revenues are very sensitive to seasonal changes in temperature and precipitation as livestock and crop farms have different responses to climate variability (Bozzola et al., 2018). Agricultural production measured by precipitation and temperature has been found to have a closer association with climate change which significantly increase the export of agricultural commodities and decreases of imports (Yu et al., 2020). It has been forecasted that climate change would undoubtedly change the natural conditions of crop growth through its critical determinants such as temperature and precipitation (Parry et al., 2004). In a related study, it was discovered that climate-induced yield changes generate large price movements that incentivize adjustments in production and trade. Both production and trade adjustments reduce welfare losses globally, with production adjustments making an enormous contribution (Gouel & Laborde, 2021).

Agricultural markets are severely impacted by adverse productivity shocks and increased global food prices (Hristov et al., 2021). Input market distance, cooperative farmer meetings, off-farm income, education, and the number of male family members are other vital determinants that significantly influence agricultural market performance (Mahmood et al., 2021). It has been proposed that farmers need to receive institutional support to sharpen their access to farm inputs, agricultural information, and extension services for smooth climate change adaptation, which could enhance their market participation (Fahad & Wang, 2018). The importance of institutional support has also been noted in a study that identified the provision of formal education on the national languages and assistance to access agricultural markets to have positively impacted the adaptive capacity of emerging farmers towards climate change (Ifejika, 2010).

The input market provides a platform for farmers to access the latest information about various inputs and even adaptation practices from input dealers and fellow farmers (Mahmood et al., 2021). Climate change also has implications for the processing, storage, transport, retailing, and consumption of livestock products. This chapter is of paramount importance in that there needs to be more information on how climate change impacts the agribusiness supply chain, particularly post-harvest activities leading to market performance (Godde et al., 2021).

METHODOLOGY

Study Area

The study was carried out in Vhembe District. It is one of the five districts of Limpopo province of South Africa. Vhembe district consists of four local municipalities: Thulamela, Makhado, Musina, and Collins Chabane. It covers about 25,597 km2, with a population estimate of about 1 402 779 people (Cooperative Governance Traditional Affairs, 2020). Being part of Limpopo province, the Vhembe district is one of its prime agricultural regions, well-known for its production of livestock, fruits, vegetables, cereals and tea (Limpopo Travel Agency, 2014). The study area is shown in figures 1 and 2 below.

Sampling Technique And Data Collection

The population of emerging cash crop farmers in the Vhembe District was estimated to be 476 during the data collection. The study used a simple random sampling technique to select 229 emerging farmers. The sample size was calculated using Raosoft's (2014) sample size calculator. Smallholder farmers from all four local municipalities of the Vhembe district had an equal opportunity to participate in the study. Face-to-face interviews were carried out with the participants, whereby structured questionnaires were administered.

Empirical Model

The study employed a Discriminant function analysis procedure. This analysis was chosen to establish differences between two groups concerning several variables simultaneously, determining whether meaningful differences exist between the groups and identifying the discriminating power of each variable (Klecka, 1980). The discriminant analysis includes the determination of a linear equation like a regression that will predict which group the case belongs to. The discriminant model employed in this study can be specified as follows:

$$D = v_1X_1 + v_2X_2 + v_3X_3 + v_i X_i + a$$

Figure 1. Map showing Limpopo Province of South Africa

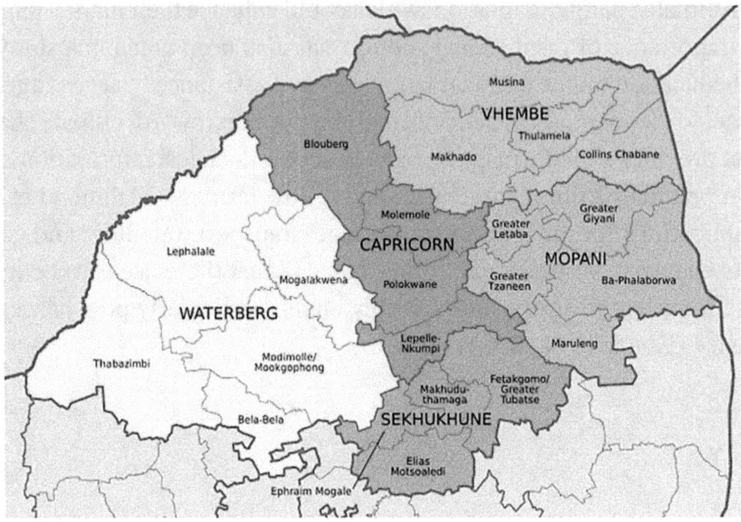

Where:

D = discriminant function.
v_i = discriminant coefficient of each explanatory variable (X_1, X_2, X_3, X_4, etc.);
X_i = explanatory variables; and
a = constant

Figure 2. Map showing Vhembe District

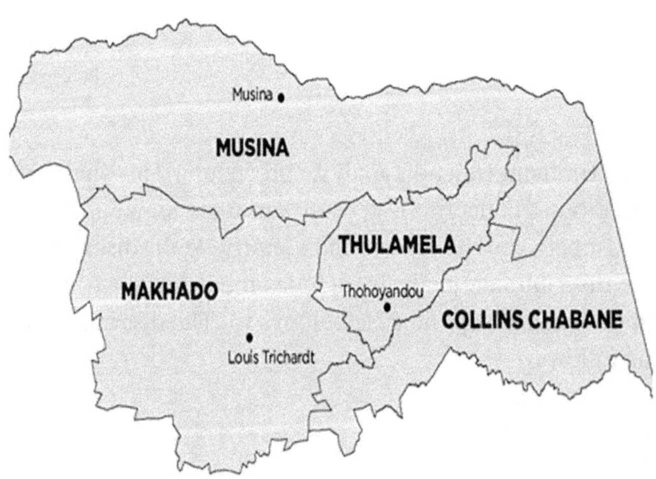

Table 1. Description of the explanatory variables

Variable	Description	Expected sign
X1	Assembling	+/-
X2	Access to proper storage facilities	+/-
X3	Transportation	+/-
X4	Perishability	-
X5	Market access	+
X6	Marketing costs	-
X7	Market distance	-
X8	Target market	+/-
X9	Selling price	+/-
X10	Product quality	+
X11	Increased quantity sold	+
X12	Decline in quantity sold	+/-
X13	Grading	+

RESULTS AND DISCUSSION

Results on Farmers' Awareness of Climate Change and Socio-Economic Characteristics

Table 2 shows the results of farmers' awareness of climate change in the study sample. The study results show that most farmers accounting for 82.1%, were aware of climate change. Farmers who indicated that they are not aware of climate change account for 12.7%, while those who were unsure were about 5.2%. However, most farmers were found to be aware of climate change; the combination of those who were not aware and those who were not sure suggest that there is still a huge role to play to ensure that all farmers are aware of climatic changes taking place from time to time. Climatic changes significantly impact agricultural performance, and critical role players like farmers should be aware of such changes. A recent study conducted by Akano et al. (2022) showed that farmers around various parts of the globe are growing in their knowledge of climate change and its critical determinants, such as prolonged droughts and unexpected changes in temperature and rainfall patterns.

Table 3 shows that most farmers, accounting for 75.5%, were familiar with the impacts of climate change. This could be because most farmers within the study sample were aware of climate change. Socio-economic characteristics deemed predictor variables in the study were gender, age group, marital status, formal education, farming experience, and off-farm income. Regarding farmers' gender, the study results show that male farmers dominated the study sample while the age group was dominated by those aged 60 and above. The study results further show that most farmers within the study sample were married.

In contrast, when it comes to educational level, most farmers did not have any formal education (50, 7%) followed by those with tertiary qualifications accounting for 21.4%. Most farmers within the study sample had more than ten years of farming experience; this could be associated with the dominant age group, which was found to be 60 years and above. The second highest group were farmers who had one

Table 2. Distribution of awareness of climate change among farmers

Variable	Frequency	Percentage
Yes, I am aware	188	82.1
I am not aware	29	12.7
I'm not sure	12	5.2
Total	**229**	**100**

Source: author's survey results.

to five years of farming experience. Relating to off-farm income, the study uncovered that most farmers within the study sample relied on government pension grants. These findings can also be associated with the dominant elderly age group

For the study to explore how climate change has impacted the marketing of agricultural produce, it was deemed necessary to first investigate the association between farmers' knowledge about the impacts of climate change and the socio-economic characteristics of farmers within the study sample. The study used the Pearson correlation coefficient (as shown in Table 4) to quantify the strength of the association between variables, with the focus of the study being familiar with the impacts of climate change and its impact on the agricultural market performance of farmers. From the study, the results in Table 4 above show a significant and robust association between possessing a formal education and being aware of the impacts of climate change. Furthermore, the findings of the study also uncovered that there is also a strong relationship between possessing a farming experience and being aware of the impacts of climate change. This finding could be that as farmers carry out their operations, they can note the extent to which agriculture has been affected by climate change. The off-farm occupation was also found to have a significant and robust relationship with being aware of the impacts of climate change.

Results on Impact of Climate Change on Agricultural Marketing

The study investigated farmers' insight of how climate change impacts their agricultural marketing activities. Table 5 presents pooled sample means and the group means for all the explanatory variables used in the analysis. To achieve this objective, the study compared farmers who have observed the impact of climate change on their marketing practices with those who had not observed any impact. The study used the P-value to draw their differences in the level of significance, and their means were used to compare the two groups. The study results revealed that 75.5% of cash crop emerging farmers in the study sample were of the view that climate change significantly impacts their marketing practices. The study results also indicated that the following variables, accessing proper storage facilities and grading of agricultural produce, were found to be statistically significant at a 10% level of significance. Market access and product quality were statistically significant at a 5% significance level. At the same time, perishability, marketing costs, selling price, and decline in quantity sold were statistically significant at a 1% level of significance.

Table 3. Distribution of socio-economic characteristics among farmers

Variable	Frequency	Percentage
Familiar with the impacts of climate change		
Yes	173	75.5
No	56	24.5
Gender		
Male	137	59.8
Female	92	40.2
Age group		
21-39 years	67	29.3
40-59 years	69	30.1
60 and above years	93	40.6
Marital status		
Married	149	65.1
Unmarried	80	34.9
Formal education		
No formal education	116	50.7
Primary education	22	9.6
Secondary education	42	18.3
Tertiary education	49	21.4
Farming experience		
Less than a year	20	8.7
1-5 years	59	25.8
6-10 years	33	14.4
more than 10 years		51.1
Off-farm income		
Employed		8.3
Pension		41.1
Remittances		12.2
Other		38.0
Total		**100**

Access to Proper Storage Facilities

Access to proper storage facilities was statistically significant at 10%. Farmers in the study sample perceived access to storage facilities as one of their marketing elements being significantly impacted by climate change. The study revealed that in the postharvest scenarios, having access to proper storage facilities is impacted by climate change. Most farmers believe that climatic conditions impact their product quality; it also impacts their means to secure storage facilities timely. A study conducted by Shin et al. (2019) highlighted that the food industry faces storage challenges due to the perishable nature of

Table 4. Pair-wise correlation matrix of descriptor variables

Impacts of climate change	Impacts of climate change	Gender	Age group	Marital status	Formal education	Farming experience	Off-farm occupation
	1						
Gender	.052	1					
Age group	.041	.435**	1				
Marital status	.116	-.093	-.285**	1			
Formal education	.897**	.204**	-.398**	.594**	1		
Farming experience	.976**	.123	.078	-.013	-.532**	1	
Off-farm occupation	.687**	-.165*	.198**	-.681**	-.147*	.005	1

Note: **, correlation is significant at the 0.01 level (2-tailed). Source: Survey results.

agricultural produce. One of the most harmful elements that impacts accessing storage facilities timely is farmers' inability to accurately estimate their yield and the grade of their produce before harvest. This is in line with the study conducted by Liu et al. (2018), who stated that farmers' profit is significantly influenced by the amount of produce retrieved from the storage during each selling period. Agricultural produce possesses unique characteristics such as high perishability and bulkiness, making storage costly. Furthermore, the storage of agricultural produce also differs with time, with some products being stored for a short period while some require an extended period. It has also been noted that a considerable number of products are damaged and lost due to improper storage facilities, leading to farmers settling for a low remunerative price (Pani and Jena, 2020).

Perishability

The perishability of agricultural produce is one of the unique features that render the marketing of agricultural produce challenging and costly. Agricultural produce with a high perishability rate poses several risks to farmers, including a high possibility of incurring a loss. This finding is in line with the study of Kumar et al. (2021), who noted that agricultural commodities are under more significant wastage and perishability due to climatic changes. Farmers perceived climate-changing impacting the perishability of agricultural produce. Most farmers within the study sample believed that climate change significantly contributes to agricultural produce being more perishable. Constant increases and decreases in temperature and rainfall were critical determinants of climate change found to be the most contributing factors toward agricultural produce being more perishable.

Agricultural produce that is highly perishable gives farmers a short period for marketing. A longer time to market them leads to a complete loss as their quality will not be suitable for market standards. This is alluded to by the findings that discovered that there had been a rising in price instability among agricultural produce (Anoopkumar, 2021). Although informal markets can give a lifeline for most emerging farmers, their dynamics usually threaten such farmers. Although farmers can sell their produce, it is nearly impossible to determine the pace and duration at which a farmer can promptly have his or her produce sold out. High perishability makes agricultural marketing difficult as farmers sell their products out of fear of losing all the investment.

Table 5. Group means of variables employed in the analysis

Variables	Observed impact of climate change on agricultural marketing				
	No	Yes	All	F	P-value
Assembling	0.45(0.4892)	0.39(0.502)	0.40(0.491)	0.612	0.435
Access to proper storage facilities	0.32(0.467)	0.45(0.502)	0.35(0.478)	3.088	0.080*
Transportation	0.84(0.369)	0.84(0.371)	0.84(0.369)	0.000	0.984
Perishability	0.67(0.484)	0.75(0.437)	0.69(0.473)	0.733	0.005***
Market access	0.58(0.495)	0.64(0.483)	0.59(0.492)	14.706	0.050**
Marketing costs	0.12(0.328)	0.34(0.478)	0.17(0.381)	6.748	0.000***
Market distance	0.29(0.457)	0.48(0.504)	0.34(0.475)	0.191	0.151
Target market	0.70(0.473)	0.66(0.464)	0.67(0.470)	13.036	0.662
Selling price	0.08(0.264)	0.25(0.437)	0.12(0.323)	2.292	0.007***
Product quality	0.24(0.426)	0.34(0.478)	0.26(0.441)	0.005	0.045**
Increased quantity sold	0.75(0.938)	0.74(0.858)	0.74(0.917)	0.823	0.365
Decline in quantity sold	0.36(0.505)	0.43(0.499)	0.38(0.503)	2.911	0.000***
Grading	0.23(0.419)	0.34(0.478)	0.25(0.436)	0.441	0.089*
Number of cases (n)			173	56	229

***, **, *Signiðcant at 1%, 5%, and 10% probability level, respectively.

Market forces such as demand and supply also impact market performance for most farmers, with the perishability of different products being a burden for most farmers without market contracts. There has been a call for the aggregation of farm produce and integration of farmers with the market to address the rising food quality and safety concerns to avoid alarming food wastage and losses (Kumar et al., 2021).

Market Access

From the study, market access was statistically significant at a 5% level of significance. The implication is that most farmers believed that climate change had an impact on their access to the market. This finding is supported by the study by Fahad and Wang, (2018), which revealed that some farmers could not access desired markets due to their inability to afford suitable seeds and fertilizers to adapt to climate change. Most farmers without any marketing contract arrangements, particularly informal markets, solely depend on product quality for market penetration. Climate change negatively impacts agricultural production; as farmers are to cope with the impacts of climate change to ensure sustained production and maintain product quality to meet market standards. When climate change negatively impacts product quality, it significantly hinders farmers from participating in specific markets. Informal markets are the only platform left for most emerging farmers. However, when climate change has significantly impacted product quality, most farmers are left with no option but to exit the market. When product quality does not suit the market standards, market access is deemed limited if not close to impossible. The study results show that farmers view climate change as significantly impacting their market access. Climate change is the primary determinant of agricultural performance; in situations where farmers cannot cope with climate change, their performance in the market declines significantly. In the final analysis, it can be concluded that climate change has significantly contributed to low agricultural productivity, rendering most farmers incapable of participating in the market.

Marketing Costs

The results from the study show that farmers are of the view that their marketing practices are also impacted by climate change in that it significantly impacts their marketing costs. Climate change impacts marketing costs in several ways. Agricultural produce possesses unique features that make its marketing dynamics tricky; different costs are incurred to ensure that farmers participate in the market. When farmers harvest, it is ideal for them to know well in advance where they will sell their products, yet farmers in remote areas must go out in search of a market after harvesting. In a study regarding perceived barriers to farmers' market access, it has been noted that some farmers could not participate due to transportation and marketing costs (Ritter et al., 2019). These farmers end up incurring costs in the search for the market and financing activities that will make their products stand a better chance of being selected in the market. Agricultural produce possesses unique traits such as bulkiness; farmers spend a lot on packaging such products, which contributes to increased production costs. At the same time, it implies extra costs associated with marketing. This finding is supported by another study's findings, which noted that input price fluctuation, unregulated markets, middleman exploitation, transportation costs, and unauthorized deductions accumulate high marketing costs that most emerging farmers cannot match (Ahmad, 2019).

Selling Price

Regarding the selling price, the study uncovered that farmers who had observed the impact of climate change on their marketing practices believed that their selling prices were significantly affected. When unfavourable climatic changes temper the product quality, farmers are left with fewer chances to be competitive in the market, which leads to most of them settling for lower prices, such as a guaranteed minimum price. Climate change also impacts product grades to the extent that it becomes hard to have desired product variations through which consumers will have a wide range of products to choose from. Farmers who are adversely impacted by climate change charge lower selling prices so as to gain some market share. This finding is supported by Liu et al. (2018), who revealed that the demand for agricultural produce significantly relies on selling prices. Farmers who observed the impact of climate change on their marketing chain within the study sample noted that the trend of their selling price is significantly unstable as it mainly depends on market demand and product quality. A similar trend was noted in a study conducted by Zhang et al. (2020), who revealed that prices of agricultural commodities often fluctuate, which significantly impacts agricultural production outputs and, ultimately, their marketability. There has also been rising discontentment among emerging farmers due to declining profitability emanating from fluctuations in agricultural commodity prices (Reddy, 2018).

Product Quality

The study results reveal that farmers within the study sample believed that climate change significantly affects product quality. This finding is in line with the study findings conducted by Kogo et al. (2018), who noted that climate change has the most devastating impacts on agricultural production. Unfavourable climatic conditions significantly influence the production process, ultimately affecting product quality.

Farmers located in the Limpopo province of South Africa have been experiencing excessive heat and prolonged drought. The quality of their products has been significantly impacted by climate change. Although some farmers have some means to cope with adverse climatic changes, most emerging farmers, particularly in remote areas, are less resourced to cope with such changes. Subsequently, this impacts their marketing in that, at times, they fail to participate in specific markets. Under normal circumstances, they end up selling their products at a meagre price to cover the cost of production. Consumers in the market look out for specific qualities, and when products do not match such qualities, they stand slim chances of being picked up. Under unfavourable climatic conditions, it is unlikely for most farmers to achieve the most desired standard of quality, which will excel in the market, making marketing such agricultural produce challenging. The most impacted group of farmers with quality constraints are those with fewer resources, as most of them cannot afford resources that can improve their efficiency and standard.

Declined Quantity Sold

On sales of agricultural produce, most farmers within the study sample who have observed the impact of climate change on their marketing believed that there had been a constant decline in the quantity sold over the years. There are many contributing factors leading to a gradual decrease in the quantity of agricultural produce sold. With many emerging farmers dependent on rainfall, the gradual decrease in the annual rainfall patterns around the globe poses a considerable challenge for most farmers since their accessibility to water is minimal. Water scarcity has become a global issue; with farmers facing prolonged droughts and minimal agricultural production. Over the years, climate change has been noted to negatively affect agricultural production, making it difficult for farmers to meet their production targets. When less is produced, less is supplied; hence there has been a decline in the number of agricultural products sold. The study's findings are in line with the results of the study by Hasegawa et al. (2018), who revealed that food security had been hampered by prevailing warmer and drier climatic conditions that have significantly affected agricultural production. Furthermore, when farmers are not meeting market standards due to the undesirable quality of produce, most fail to regularly market and sell their produce. This can be attested by the study findings of Yakavenka et al. (2020), who noted that the supply of fresh products had been disrupted by the environmental phenomenon that tempers the mandatory consumption security standards.

Grading

From the study sample, farmers who had observed climate change being impactful on agricultural marketing further noted that agricultural produce grading was significantly impacted. Unfavourable climatic conditions significantly impact the grading of agricultural produce during the production phase, which ultimately impacts the market performance of the commodity. This was noted by Pani and Jena (2020), who discovered that farmers' failure to grade their produce during postharvest leads to them earning a low remunerative price. Another challenge imposed by climate change in grading agricultural produce is that production dynamics primarily influence product quality. In most cases, farmers produce products that have slight variations, which makes it difficult to set different categories while marketing such produce.

Furthermore, during the grading of agricultural produce, unique characteristics such as taste, size, color, and variation play a vital role. Little can be achieved within the stressful environment under which

Table 6. Standardized canonical discriminant function coefficients

Explanatory variable	Coefficient	Wilks' Lambda
Assembling	0.011	0.996
Access to proper storage facilities	0.236	0.978
Transportation	0.066	1.000
Perishability	0.557	0.992
Market access	0.281	0.978
Marketing costs	0.197	0.982
Market distance	0.152	0.944
Target market	0.652	0.930
Selling price	0.494	0.996
Product quality	-0.218	0.974
Increased quantity sold	0.205	0.996
Decline in quantity sold	-0.204	0.966
Grading	0.013	1.000
Statistics:		
Eigen value		0.180
Canonical correlation		0.391
Walks' Lambda		0.847
Chi-square		33.049
df		9
P-value		0.000
% grouped cases		75.5%

farmers tend to operate. Grading with little differentiation leaves farmers with little room to price-tag their produce and ultimately leads to them selling their produce under local conditions.

The results of Discriminant scores and levels of explanatory variables are presented in Table 6. The eigenvalue was used as one of the statistics to assess the magnitude of the discriminant analysis model used. The results' low eigenvalue (0.180) implies that the group differences were lower than within the group. Wilks` Lambda was used to determine the model's accuracy used for analysis. From the results, Wilks` Lambda scored a low value of 0.847, implying a high percentage of explained variance of the dependent variable, in this case, having observed the impacts of climate change on the marketing of agricultural produce as against non-observance of such an impact.

In this study, Wilks' Lambda in the case where all the functions are in the analysis, was estimated as 0.847, indicating 75.5 percent differences between two groups of farmers who either observed the impacts of climate change on their marketing practices or not. The null hypothesis was tested using the Chi-square statistics. The null hypothesis was not rejected as Chi-square scored 33.049 with degrees of freedom, df = 9. The percentage grouped cases show the model's prediction accuracy as 72.8%, evidence that the discriminant function has a high precision prediction. The results showed a 1 percent significance level ($P<0.000$), implying that the discriminant function used was appropriate for the analysis. The stan-

Table 7. Distribution of type of agricultural markets among farmers

Variable	Frequency	Percent
Local/Grower's market	71	31
Primary market	138	60.2
Secondary market	18	7.9
Terminal market	2	0.9
Total	**229**	**100.0**

Source: author's survey results.

dardized canonical discriminant coefficients of the explanatory variables showed that the variables that separated the two groups of farmers the most were: perishability nature, target market, and selling price.

Results on Current Marketing Practices and Challenges Among Farmers

Types of Agricultural Markets

The study investigated markets in which smallholder farmers are selling their produce. After summarizing the most used markets only primary, secondary, and terminal markets emerged as shown in Table 7. The study sample shows that most farmers accounting for 60.2%, were selling their produce in the primary market. This type of market is periodical, wherein it operates on certain days and times. Farmers in rural areas who do not have a secured market make primary markets their priority as it allows them to have a better market share. The primary market assists farmers as an assembling centre for the local produce and distributive centres. Farmers' preference for primary markets is usually based on their produce's perishability and the ability to sell their produce within the allocated period. Apart from the primary market, farmers also sold their products at local markets known as growers' markets. Farmers who participated in local markets believed that such markets minimize their transportation and storage costs. Furthermore, emerging farmers prefer local markets as they are deemed convenient and quick cash converters. From the results, some farmers had a market share at secondary markets. These markets are also considered convenient as they offer permanent platforms for daily transactions. Very few farmers, accounting for 2%, participated in terminal markets from the study sample. Terminal markets serve farmers with better incentives as they are mostly located in cities. The findings of this study are in line with those by Marwa et al. (2022) who noted that farmers participate in different types of markets. The study further noted that the high difference that exists between different types of markets is shaped by numerous factors including long distances from farm units to markets, middlemen costs and expensive storage facilities.

Farmers' Transactional Arrangements

The study explored different transactional arrangements, which their marketing practices usually conclude. Various transactional mechanisms among farmers where the actual transaction with the ultimate customer occurs are shown in Table 8 and Figure 4. The study uncovered that most farmers within the study sample conducted their transactions at farmers' markets. This finding of the study could be significantly influenced by the type of market, a primary market in which most farmers within the study

Figure 3. Types of agricultural markets

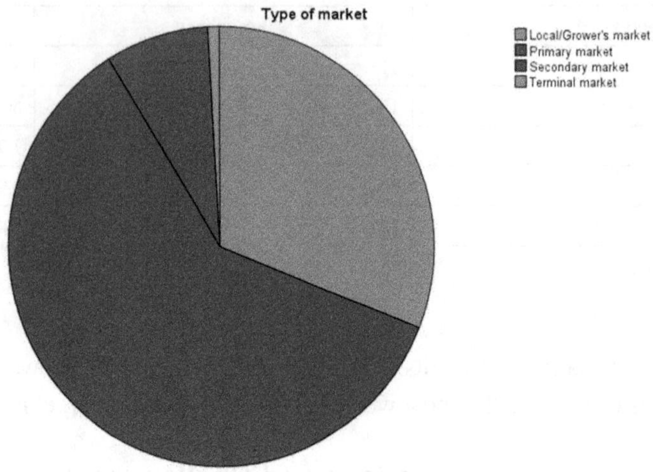

sample were found to be participating. Another dominating platform where the actual transactions took place was roadside outlets. Farmers who did not have access to desired markets mostly opted to sell their products on the roadside.

Table 8. Distribution of agricultural transactional arrangements among farmers

Variable	Frequency	Percent
Farmers' market	108	47.2
Farm gate	40	17.5
Distributors	28	12.2
Roadside outlet	45	19.7
Contractual	8	3.5
Total	**229**	**100**

Farmers who transacted at the farm gate accounted for 17.5% while those that sold their produce through contracts accounted for 3.5%. It was also noted that most farmers did not participate in marketing contracts, with most transactions taking place in informal markets. The findings of this study are in line with those of Ton et al. (2018) who noted that smallholder and less-resourced farmers are largely excluded from various marketing contracts.

Agricultural Marketing Challenges Among Farmers

The study investigated challenges that farmers faced when marketing their produce. The predictor variables were transportation challenges, access to storage facilities, lack of market access, selling price instability, undesired product quality, high level of competition, lack of market information, and declined quantity sold, as indicated in Table 9 and Figure 5 above. The study sample shows that transport-related chal-

Figure 4. Agricultural transactional arrangements

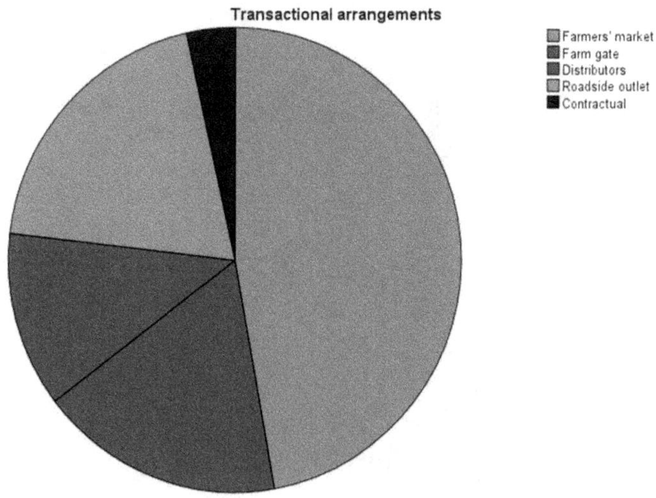

lenges were of significant concern among other challenges that farmers were facing. These included a lack of transport to deliver produce, high transport costs, and unreliable external transport arrangements. Farmers who do not own a car or any mode of transport spend more on transportation arrangements. The study results also show that besides transport challenges, farmers also face challenges with accessing the market, particularly high-value markets.

Table 9. Agricultural marketing challenges

Marketing challenges	Frequency	Percent
Transportation challenges	99	43.2
Access to storage facilities	14	6.1
Lack of market access	47	20.5
Selling price instability	17	7.4
Undesired quality of product	15	6.6
High level of competition	13	5.7
Lack of market information	14	6.1
Declined quantity sold	10	4.4
Total	**229**	**100**

Farmers' failure to access markets negatively impacts their marketability in that they suffer considerable losses, notably if they do not sell at all. At times, farmers sell their produce at the lowest possible price to avoid a complete loss of resources, ultimately impacting their sustainability. Furthermore, lack of market access leaves farmers in a predicament as more costs for storage and transportation are usually incurred in pursuit of a potential market. In a study conducted on farmers' risk perception and adaptation to climate change, it was revealed that some constraints faced by farmers included a lack of access to the

Figure 5. Agricultural marketing challenges

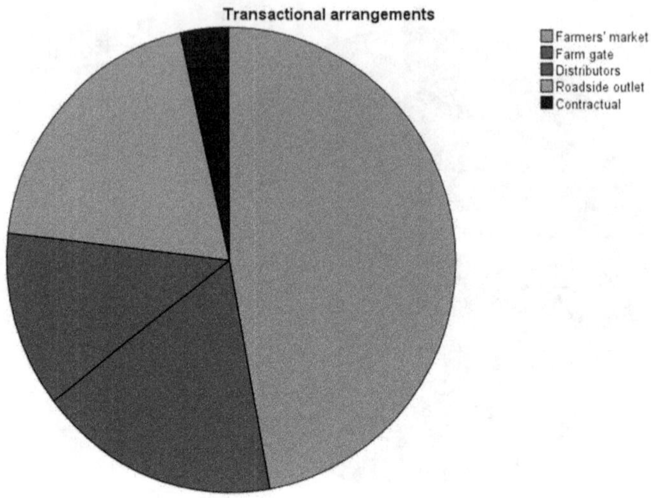

market, assets, credible sources, knowledge and information (Fahad, and Wang, 2018). In another study regarding farmers' challenges relating to marketing, it was noted that detrimental factors impacting farmers' marketing included produce quantity, price of commodities, distance to market, access to extension services, and lack of market information (Kyaw et al. 2018). The study results also show that farmers are experiencing unstable selling prices. As a result, most farmers from the study sample indicated that input prices are constantly rising while their selling prices gradually decreased. Most farmers within the study sample believed that they were selling at a loss. Their selling prices are determined by the market forces such as demand and supply, which brings considerable variations in pricing.

Regarding product quality, farmers were of the view that consumers deem their product quality poor. Furthermore, farmers indicated that their marketability was also negatively impacted by having their produce possess undesired quality. The study results also indicate that farmers experienced challenges in accessing storage facilities and lacked market information. With farmers struggling to access the market, most of them are left with no option but to maintain proper storage facilities until opportunities arose for selling their produce. Lack of complete market information impact quite negatively on farmers' market participation.

The result of the study uncovered that farmers are of the view that their marketability is challenged by a lack of access to market information. For proper planning, farmers should access adequate information timely; misallocation of resources can result from a lack of market information among farmers. Farmers have also noted high competition among themselves as one critical challenge they face as it significantly affects their marketing ability. High competition in the market leads farmers to settle for lower selling prices in an attempt to secure a share within the highly competitive market. The above finding is supported by those of Bergquist and Dinerstein (2020) who noted that agricultural markets in Africa were constrained by low farmers' revenue and high consumer food prices. The study also revealed that farmers were facing a decline in the quantity of produce sold. Farmers were of the view that, with time, they have been experiencing a gradual decline in the quantity of produce that they sell. When farmers fail to sell their products, their continuity hangs in the balance.

CONCLUSION AND RECOMMENDATIONS

The study uncovered that 82.1% of farmers within the study sample were aware of climate change, while 75.5% were familiar with climate change impacts. Regarding farmers' observation of climate change impacts on agricultural produce marketing, the study's findings revealed that the majority had noted various ways in which their marketing practices were impacted due to changing climatic conditions. Regarding farmers' perception of how climate change impacted their marketing practices, the study revealed a gradual decrease in the agricultural quantity sold, higher accumulated marketing costs, unstable selling prices, and high product perishability. Pertaining to the type of markets in which farmers within the study sample participated, the study uncovered that most farmers participated in primary markets. In contrast, most transactions took place at farmers' markets. From numerous identified marketing challenges, farmers' most prevailing included transportation of produce to the market, gradual decline in sales, unstable selling price, and lack of market information.

As regards awareness of climate change and its impact, the study recommends that various agricultural stakeholders use different forms of mechanisms to promote awareness of climate change among farmers and other critical agricultural stakeholders. The promotion of climate change information among various stakeholders will assist key role players in identifying the extent to which their agricultural practices, including the marketing of agricultural produce, are impacted by climate change.

Relating to how climate change has impacted agricultural marketing practices, the study recommends that collaborative efforts among key agricultural players be channelled towards assisting farmers in enhancing their marketing practices through various forms of support. The study also recommends a move towards some regulated markets to protect farmers from exploitations and unstable selling prices. Furthermore, the study recommends that participative agricultural development programs aimed at assisting farmers in overcoming various challenges hampering their progress be prioritized, taking into consideration the specific needs of various groups of farmers. Lastly, this study recommends that climate finance should also prioritize specialized training of different agricultural key role players to ensure effective adaptation to climate change. The latter is the intended focus of an envisaged study.

REFERENCES

Ahmad, S. I. (2019). A study on the production and marketing cost of apple fruit in Kupwara District of Jammu and Kashmir. *Advances In Management, 12*, 1.

Akano, O., Modirwa, S., Oluwasemire, K., & Oladele, O. (2022). Awareness andperception of climate change by smallholder farmers in two agroecological zones of Oyo state Southwest Nigeria. *GeoJournal, 88*(1), 1–30. doi:10.100710708-022-10590-y

Anoopkumar, M., (2021). Commodity price instability under globalisation: A study ofIndia's plantation crops.

Aydinalp, C., & Cresser, M. (2008). The effects of global climate change on agriculture.American European Journal of Agricultural Environmental Science, 3, 672–676..

Bergquist, L. F., & Dinerstein, M. (2020). Competition and entry in agricultural markets:Experimental evidence from Kenya. *The American Economic Review, 110*(12), 3705–3747.

Bozzola, M., Massetti, E., Mendelsohn, R., & Capitanio, F. (2018). A Ricardian analysis of the impact of climate change on Italian agriculture. *European Review of Agriculture Economics*, *45*(1), 57–79. doi:10.1093/erae/jbx023

Darekar, A., & Reddy, A. A. (2017). Cotton price forecasting in major producing states. *Economic Affairs*, *62*(3), 373. doi:10.5958/0976-4666.2017.00047.X

Department of Cooperative Governance (DCoG) and the Department of Traditional Affairs. (2020). *Profile and Analysis*. District Development Model. Vhembe District Municipality.

Elleby, C., Domínguez, I. P., Adenauer, M., & Genovese, G. (2020). Impacts of the COVID-19 pandemic on the global agricultural markets. *Environmental and Resource Economics*, *76*(4), 1067–1079. doi:10.100710640-020-00473-6 PMID:32836856

Fahad, S., & Wang, J. (2018). Farmers' risk perception, vulnerability, and adaptation to climate change in rural Pakistan. *Land Use Policy*, *79*, 301–309. doi:10.1016/j.landusepol.2018.08.018

Godde, C. M., Mason-D'Croz, D., Mayberry, D. E., Thornton, P. K., & Herrero, M. (2021). Impacts of climate change on the livestock food supply chain; a review of the evidence. *Global Food Security*, *28*, 100488. doi:10.1016/j.gfs.2020.100488 PMID:33738188

Gomez-Zavaglia, A., Mejuto, J. C., & Simal-Gandara, J. (2020). Mitigation of emerging implications of climate change on food production systems. *Food Research International*, *134*, 109256. doi:10.1016/j.foodres.2020.109256 PMID:32517948

Gouel, C., & Laborde, D. (2021). The crucial role of domestic and international market-mediated adaptation to climate change. *Journal of Environmental Economics and Management*, *106*, 102408. doi:10.1016/j.jeem.2020.102408

Hasegawa, T., Fujimori, S., Havlík, P., Valin, H., Bodirsky, B. L., Doelman, J. C., Fellmann, T., Kyle, P., Koopman, J. F., Lotze-Campen, H., Mason-D'Croz, D., Ochi, Y., Pérez Domínguez, I., Stehfest, E., Sulser, T. B., Tabeau, A., Takahashi, K., Takakura, J., van Meijl, H., & Witzke, P. (2018). Risk of increased food insecurity under stringent global climate change mitigation policy. *Nature Climate Change*, *8*(8), 699–703. doi:10.103841558-018-0230-x

Havlík, P., Valin, H., Herrero, M., Obersteiner, M., Schmid, E., Rufino, M.C., Mosnier, A., Thornton, P.K., Böttcher, H., Conant, R.T. and Frank, S., (2014). Climate change mitigation through livestock system transitions. *Proceedings of the National Academy of Sciences, 111*(10), pp.3709-3714.

Hristov, J., Barreiro-Hurle, J., Salputra, G., Blanco, M., & Witzke, P. (2021). Reuse of treated water in European agriculture: Potential to address water scarcity under climate change. *Agricultural Water Management*, *251*, 106872. doi:10.1016/j.agwat.2021.106872 PMID:34079159

Ifejika, S. C. (2010). Resilient adaptation to climate change in African agriculture (54). *Studies*.

Kastner, T., Rivas, M. J. I., Koch, W., & Nonhebel, S. (2012). Global changes in diets and the consequences for land requirements for food. *Proceedings of the National Academy of Sciences of the United States of America*, *109*(18), 6868–6872. doi:10.1073/pnas.1117054109 PMID:22509032

Klecka, W. R. (1980). *Series: "Quantitative Applications in the Social Sciences* (Vol. 18). SAGE Publications.

Kogo, B. K., Kumar, L., & Koech, R. (2021). Climate change and variability in Kenya: A review of impacts on agriculture and food security. *Environment, Development and Sustainability*, *23*(1), 23–43. doi:10.100710668-020-00589-1

Kumar, A., Mangla, S. K., Kumar, P., & Song, M. (2021). Mitigate risks in perishable food supply chains: Learning from COVID-19. *Technological Forecasting and Social Change*, *166*, 120643. doi:10.1016/j.techfore.2021.120643

Kyaw, N. N., Ahn, S., & Lee, S. H. (2018). Analysis of the factors influencing market participation among smallholder rice farmers in Magway region, central dry zone of Myanmar. *Sustainability*, *10*(12), 4441. doi:10.3390u10124441

Limpopo Travel Agency. (2014). *Homepage.* LTA. http://www.golimpopo.com

Liu, H., Zhang, J., Zhou, C., & Ru, Y. (2018). Optimal purchase and inventory retrieval policies for perishable seasonal agricultural products. *Omega*, *79*, 133–145. doi:10.1016/j.omega.2017.08.006

Long, S. P., Ainsworth, E. A., Leakey, A. D. B., Nosberger, J., & Ort, D. R. (2006). Food for thought: Lower-than-expected crop yield stimulation with rising CO2 concentrations. *Science*, *312*(5782), 1918–1921. doi:10.1126cience.1114722 PMID:16809532

Mahmood, N., Arshad, M., Mehmood, Y., Shahzad, M. F., & Kächele, H. (2021). Farmers' perceptions and role of institutional arrangements in climate change adaptation: Insights from rainfed Pakistan. *Climate Risk Management*, *32*, 100288. doi:10.1016/j.crm.2021.100288

Malhi, G. S., Kaur, M., & Kaushik, P. (2021). Impact of climate change on agriculture and its mitigation strategies: A review. *Sustainability*, *13*(3), 1318. doi:10.3390u13031318

Manogna, R. L., & Mishra, A. K. (2021). Forecasting spot prices of agricultural commodities in India: Application of deep-learning models. *Intelligent Systems in Accounting, Finance & Management*, *28*(1), 72–83. doi:10.1002/isaf.1487

Marwa, T., Bashir, A., Azwardi, A., Adam, M. and Thamrin, K.M., (2022). Market Integration of Agricultural Products.

Nyoro, J. K. (2019). *Agriculture and rural growth in Kenya.* Tegemeo Institute.

Pani, S.K. and Jena, D., (2020). Agricultural marketing in India: a review on challenges and prospects.

Parry, M. L., Rosenzweig, C., Iglesias, A., Livermore, M., & Fischer, G. (2004). Effects of climate change on global food production under SRES emissions and socio-economic scenarios. *Global Environmental Change*, *14*(1), 53–67. doi:10.1016/j.gloenvcha.2003.10.008

Rasul, G. (2021). Twin challenges of COVID-19 pandemic and climate change for agriculture and food security in South Asia. *Environmental Challenges*, *2*, 100027. doi:10.1016/j.envc.2021.100027

Reddy, A. R. (2018). Cost escalation in cotton cultivation: An analysis. *Economic Affairs*, *63*(4), 295480. doi:10.30954/0424-2513.4.2018.6

Ritter, G., Walkinshaw, L. P., Quinn, E. L., Ickes, S., & Johnson, D. B. (2019). An assessment of perceived barriers to farmers' market access. *Journal of Nutrition Education and Behavior*, *51*(1), 48–56. doi:10.1016/j.jneb.2018.07.020 PMID:30249521

Sabu, K. M., & Kumar, T. M. (2020). Predictive analytics in Agriculture: Forecasting prices of Arecanuts in Kerala. *Procedia Computer Science*, *171*, 699–708. doi:10.1016/j.procs.2020.04.076

Seppelt, R., Manceur, A. M., Liu, J., Fenichel, E. P., & Klotz, S. (2014). Synchronized peak-rate years of global resources use. *Ecology and Society*, *19*(4), 50. doi:10.5751/ES-07039-190450

Shin, M., Lee, H., Ryu, K., Cho, Y., & Son, Y. J. (2019). A two-phased perishable inventory model for production planning in a food industry. *Computers & Industrial Engineering*, *133*, 175–185. doi:10.1016/j.cie.2019.05.010

Tilman, D., Balzer, C., Hill, J., & Befort, B. L. (2011). Global food demand and the sustainable intensification of agriculture. *Proceedings of the National Academy of Sciences of the United States of America*, *108*(50), 20260–20264. doi:10.1073/pnas.1116437108 PMID:22106295

Ton, G., Vellema, W., Desiere, S., Weituschat, S., & D'Haese, M. (2018). Contract farming for improving smallholder incomes: What can we learn from effectiveness studies? *World Development*, *104*, 46–64. doi:10.1016/j.worlddev.2017.11.015

Tshikororo, M., Thaba, K., Nefale, T. A., & Tshikororo, M. (2021). Perception Driven from Farmers' Socio-economic Characteristics Towards Tackling Climate change. *The Journal of Agricultural Science*, *13*(6), 41–47. doi:10.5539/jas.v13n6p41

Van Meijl, H., Havlik, P., Lotze-Campen, H., Stehfest, E., Witzke, P., Domínguez, I. P., Bodirsky, B. L., van Dijk, M., Doelman, J., Fellmann, T., Humpenöder, F., Koopman, J. F. L., Müller, C., Popp, A., Tabeau, A., Valin, H., & van Zeist, W.-J. (2018). Comparing impacts of climate change and mitigation on global agriculture by 2050. *Environmental Research Letters*, *13*(6), 064021. doi:10.1088/1748-9326/aabdc4

Yakavenka, V., Mallidis, I., Vlachos, D., Iakovou, E., & Eleni, Z. (2020). Development of a multi-objective model for the design of sustainable supply chains: The case of perishable food products. *Annals of Operations Research*, *294*(1), 593–621. doi:10.100710479-019-03434-5

Yalew, A. W., Hirte, G., Lotze-Campen, H., & Tscharaktschiew, S. (2018). Climate change, agriculture, and economic development in Ethiopia. *Sustainability*, *10*(10), 3464. doi:10.3390u10103464

Yu, X., Luo, H., Wang, H., & Feil, J. H. (2020). Climate change and agricultural trade in central Asia: Evidence from Kazakhstan. *Ecosystem Health and Sustainability*, *6*(1), 1766380. doi:10.1080/20964129.2020.1766380

Zabel, F., Delzeit, R., Schneider, J. M., Seppelt, R., Mauser, W., & Václavík, T. (2019). Global impacts of future cropland expansion and intensification on agricultural markets and biodiversity. *Nature Communications*, *10*(1), 1–10. doi:10.103841467-019-10775-z PMID:31253787

Zhang, D., Chen, S., Liwen, L., & Xia, Q. (2020). Forecasting agricultural commodity prices using model selection framework with time series features and forecast horizons. *IEEE Access: Practical Innovations, Open Solutions*, *8*, 28197–28209. doi:10.1109/ACCESS.2020.2971591

Zilberman, D., Liu, X., Holst, D. R., & Sunding, D. (2004). The economics of climate change in agriculture. *Mitigation and Adaptation Strategies for Global Change*, *9*(4), 365–382. doi:10.1023/B:MITI.0000038844.72226.13

Compilation of References

Abay, K., & Hirvonen, K. (2017). Does market access mitigate the impact of seasonality on child growth? Panel data evidence from northern Ethiopia. *The Journal of Development Studies*, *53*(9), 1414–1429. doi:10.1080/00220388.2016.1251586

Abdulai, A., & Birachi, E. A. (2009). Choice of coordination mechanism in the Kenyan fresh milk supply chain. *Applied Economic Perspectives and Policy*, *31*(1), 103–121.

Abebe, G. K., Bijman, J., & Royer, A. (2016). Are middlemen facilitators or barriers to improve smallholders' welfare in rural economies? Empirical evidence from Ethiopia. *Journal of Rural Studies*, *43*, 203–213. doi:10.1016/j.jrurstud.2015.12.004

Abera, G. (2009). Commercialization of smallholder farming: Determinants and welfare outcomes. *A cross-sectional study in Enderata District, Tigari, Ethiopia.* [Unpublished master thesis, University of Agder, Kristiansand, Norway].

Abu Bakar, U. K., Pillai, V., Hashim, M., & Daud, H. M. (2005). Sharing Malaysian experience with development of biotechnology-derived food crops. *Food and Nutrition Bulletin*, *26*(4), 317–320. doi:10.1177/15648265050264S312 PMID:16465992

Aderghal, M., Lemeilleur, S., & Romagny, B. (2019). *Contribution des systèmes de distribution alimentaire à la sécurité alimentaire des villes: étude de cas sur l'agglomération de Rabat [Contribution of food distribution systems to food security in cities: case study on the agglomeration of Rabat].*

Ahmad, S. I. (2019). A study on the production and marketing cost of apple fruit in Kupwara District of Jammu and Kashmir. *Advances In Management*, *12*, 1.

Ahmed, N., Li, C., Khan, A., Qalati, S. A., Naz, S., & Rana, F. (2021). Purchase intention toward organic food among young consumers using theory of planned behavior: Role of environmental concerns and environmental awareness. *Journal of Environmental Planning and Management*, *64*(5), 796–822. doi:10.1080/09640568.2020.1785404

Aitken, R., Watkins, L., Williams, J., & Kean, A. (2020). The positive role of labelling on consumers' perceived behavioural control and intention to purchase organic food. *Journal of Cleaner Production*, *255*, 120334. doi:10.1016/j.jclepro.2020.120334

Akano, O., Modirwa, S., Oluwasemire, K., & Oladele, O. (2022). Awareness andperception of climate change by smallholder farmers in two agroecological zones of Oyo state Southwest Nigeria. *GeoJournal*, *88*(1), 1–30. doi:10.100710708-022-10590-y

Alemu, D., Guinan, A., & Hermanson, J. (2021). Contract farming, cooperatives and challenges of side selling malt barley value-chain development in Ethiopia. *Development in Practice*, *31*(4), 496–510. doi:10.1080/09614524.2020.1860194

Aliber, M., & Hall, R. (2012). Support for smallholder farmers in South Africa: Challenges of scale and strategy. *Development Southern Africa*, *29*(4), 548–562. doi:10.1080/0376835X.2012.715441

Alunga, J. U., & Jaluo, W. M. (2013). Gender roles and agribusiness in the Kenyan communities: The case of Likuyani District. *Journal of Emerging Trends in Educational Research and Policy Studies*, *4*(5), 733–738.

Amin, L., Ahmad, J., Jahi, J. M., Nor, A. R. M., Osman, M., & Mahadi, N. M. (2010). Factors influencing Malaysian public attitude to agro-biotechnology. *Public Understanding of Science (Bristol, England)*, *20*(15), 674–689. PMID:22164706

Amin, L., Azlan, N. A. A., Hamdan, M. F., Samian, A. L., & Haron, M. S. (2011c). Awareness and knowledge on modern biotechnology. *African Journal of Biotechnology*, *10*(58), 12448–12456.

Amin, L., Hashim, H., Sidik, N. M., Zainol, Z. A., & Anuar, N. (2011b). Public attitude towards modern biotechnology. *African Journal of Biotechnology*, *10*(58), 12409–12417.

Amin, L., & Ibrahim, R. (2011). How aware and knowledgeable are Malaysians of modern biotechnology? *Procedia: Social and Behavioral Sciences*, *15*, 3384–3388. doi:10.1016/j.sbspro.2011.04.305

Amin, L., Jahi, J. M., Nor, A. R. M., Osman, M., & Mahadi, N. M. (2006). Uncovering factors influencing Malaysian public attitude towards modern biotechnology. *Asia-Pacific Journal of Molecular Biology and Biotechnology*, *14*(2), 33–39.

Amin, L., Othman, J., Goh, H. L., & Jusoff, K. (2011a). Consumer information and agro-biotechnology: The experience of Malaysia. *American-Eurasian Journal of Agricultural & Environmental Sciences*, *10*(6), 1006–1017.

Anderson, J. C., Wachenheim, C. J., & Lesch, W. C. (2005). Perceptions of genetically modified and organic foods and processes: North Dakota College students. *Agribusiness and Applied Economics Report No. 571,* pp. 1-50.

Angulo, A. M., & Gil, J. M. (2007). Spanish consumers' attitudes and acceptability towards GM food products. *Agricultural Economics Review*, *8*(1), 50–63.

Anoopkumar, M., (2021). Commodity price instability under globalisation: A study ofIndia's plantation crops.

Ansah, E. O., Kaplowitz, M. D., Lupi, F., & Kerr, J. (2020). Smallholder participation and procedural compliance with sustainable cocoa certification programs. *Agroecology and Sustainable Food Systems*, *44*(1), 54–87. doi:10.1080/21683565.2019.1579776

Arumugam, S., Govindasamy, R., Vellangany, I., & Gohil, H. (2017). An analysis of organic fresh produce choice: A consumer preference study in the Mid-Atlantic USA. *Journal of Food Distribution Research*, *48*(1), 101–102. doi:10.22004/ag.econ.274576

Arya, D. (2015). *Genetically modified foods; Benefits and Risks*. Massachusetts Medical Society.

Asc consultants.co.za. (2022). *GLOBAL GAP IFA V6, What's new?* Aluzar Systems Consultants. https://ascconsultants.co.za/global-gap-ifa-v6-whats-new.

Aschemann-Witzel, J., Maroscheck, N., & Hamm, U. (2013). Are organic consumers preferring or avoiding foods with nutrition and health claims? *Food Quality and Preference*, *30*(1), 68–76. doi:10.1016/j.foodqual.2013.04.011

Aschemann-Witzel, J., & Zielke, S. (2017). Can't buy me green? A review of consumer perceptions of and behavior toward the price of organic food. *The Journal of Consumer Affairs*, *51*(1), 211–251. doi:10.1111/joca.12092

Asioli, D., Canavari, M., Pignatti, E., Obermowe, T., Sidali, K. L., Vogt, C., & Spiller, A. (2014). Sensory experiences and expectations of Italian and German organic consumers. *Journal of International Food & Agribusiness Marketing*, *26*(1), 13–27. doi:10.1080/08974438.2012.755718

Awasthi, B. D. (2007). Relevance of market information system to environment protection. *Journal of Agriculture and Environment*, *8*, 46–54. doi:10.3126/aej.v8i0.726

Aydinalp, C., & Cresser, M. (2008). The effects of global climate change on agriculture.American European Journal of Agricultural Environmental Science, 3, 672–676..

Ayinde, O. (2014). Is Marketing Board a Barrier or a Stimulant of Agricultural Production in West Africa? A Comparative Study of Ghana and Nigeria Cocoa Pricing Eras. *GJDS*, *11*(2), 50–66.

Badiye, A., Kapoor, N., & Menezes, R. (2022). *Chain of Custody*. Ncbi. nlm.nih.gov. <https://www.ncbi.nlm.nih.gov/books/NBK551677.

Baeumler, A., D'Aoust, O., Das, M. B., Gapihan, A., Goga, S., Lakovits, C., Restrepo Cavadid, P., Singh, G., & Terraza, H. (2021). *Demographic Trends and Urbanization*. W. B. Group. https://population.un.org/wup/Publications/Files/WUP2018-Report.pdf

Baker, G. A., & Burnham, T. A. (2002). The marker for genetically modified food: Consumer characteristics and policy implications. *The International Food and Agribusiness Management Review*, *4*(4), 351–360. doi:10.1016/S1096-7508(02)00063-0

Balineau, G., Bauer, A., Kessler, M., & Madariaga, N. (2021). *Food Systems in Africa: Rethinking the Role of Markets*. World Bank Publications. doi:10.1596/978-1-4648-1588-1

Barends-Jones, V. (2020). ECommerce Markets: Opportunities and Challenges for the Western Cape Agricul tural Sector. *Western Cape Department of Agriculture*. https://www.elsenburg.com/wp-content/uploads/2022/03/2020-E-commerce-Markets.pdf

Barrett, H.R., Ilbery, B.W., Angela W Browne, A.W., & Binns, T. (1998). Globalization and the changing networks of food supply: the importation of fresh horticultural produce from Kenya into the UK. *Globalization and the changing networks of food supply.*

Barrett, C. B. (2008). Smallholder market participation: Concepts and evidence from eastern and southern Africa. *Food Policy*, *33*(4), 299–317. doi:10.1016/j.foodpol.2007.10.005

Barrett, C. B. (2008). Smallholder market participation: Concepts and evidence from eastern and southern Africa. In *Food security in Africa*. Edward Elgar Publishing.,

Bateman, D. I. (1976). Agricultural marketing: A review of the literature of marketing theory and of selected applications. *Journal of Agricultural Economics*, *27*(2), 171–226. doi:10.1111/j.1477-9552.1976.tb00823.x

Battersby, J. (2020). South Africa's lockdown regulations and the reinforcement of anti-informality bias. *Agriculture and Human Values*, *37*(3), 543–544. doi:10.100710460-020-10078-w PMID:32395011

Battersby, J., & Watson, V. (2018). *Urban Food Systems Governance and Poverty in African Cities*. Routledge. doi:10.4324/9781315191195

Batt, P. J. (2003). Building trust between growers and market agents. *Supply Chain Management*, *8*(1), 65–78. doi:10.1108/13598540310463378

Baudry, J., Touvier, M., Allès, B., Péneau, S., Méjean, C., Galan, P., Hercberg, S., Lairon, D., & Kesse-Guyot, E. (2016). Typology of eaters based on conventional and organic food consumption: Results from the NutriNet-Santé cohort study. *British Journal of Nutrition*, *116*(4), 700–709. doi:10.1017/S0007114516002427 PMID:27311793

Beavon, K. (2001, 16 – 20 July 2001). *The role of transport in the rise and decline of the Johannesburg CBD, 1886-2001* 20th South African Transport Conference (SATC 2001): 'Meeting the Transport Challenges in Southern Africa, South Africa. https://repository.up.ac.za/bitstream/handle/2263/8191/5b4.pdf?sequence=1

Beinart, W., & Delius, P. (2014). The historical context and legacy of the Natives Land Act of 1913. *Journal of Southern African Studies*, *40*(4), 667–688. doi:10.1080/03057070.2014.930623

Beldad, A., Hegner, S., & Hoppen, J. (2016). The effect of virtual sales agent (VSA) gender – product gender congruence on product advice credibility, trust in VSA and online vendor, and purchase intention. *Computers in Human Behavior*, *60*, 62–72. doi:10.1016/j.chb.2016.02.046

Bembridge, T. J. (1987). Aspects of cattle production in Transkei. *South African Journal of Animal Science*, *17*(2), 74–78.

Béné, C. (2020). Resilience of local food systems and links to food security–A review of some important concepts in the context of COVID-19 and other shocks. *Food Security*, *12*(4), 805–822. doi:10.100712571-020-01076-1 PMID:32837646

Benton, T. G., & Bailey, R. (2019). The paradox of productivity: agricultural productivity promotes food system inefficiency. *Global Sustainability, 2*(e6), 1-8.

Benton, T. G., Bieg, C., Harwatt, H., Pudasaini, R., & Wellesley, L. (2021). *Food system impacts on biodiversity loss*. Three levers for food system transformation in support of nature. Chatham House. https://www.ciwf.com/media/7443948/food-system-impacts-on-biodiversity-loss-feb-2021.pdf

Bergquist, L. F., & Dinerstein, M. (2020). Competition and entry in agricultural markets:Experimental evidence from Kenya. *The American Economic Review*, *110*(12), 3705–3747.

Bernard, T., & Spielman, D. J. (2009). Reaching the rural poor through rural producer organizations? A study of agricultural marketing cooperatives in Ethiopia. *Food Policy*, *34*(1), 60–69. doi:10.1016/j.foodpol.2008.08.001

Binswanger-Mkhize, H. P. (2014). From failure to success in South African land reform. *African Journal of Agricultural and Resource Economics*, *9*(4), 253–269.

Black, R., Bellagamba, A., Botta, E., Ceesay, E., Cissokho, D., Engeler, M., Lenoël, A., Oelgemöller, C., Riccio, B., Sakho, P., Somparé, A. W., Vitturini, E., & Zingari, G. N. (2022). Migration drivers and migration choice: Interrogating responses to migration and development interventions in West Africa. *Comparative Migration Studies*, *10*(1), 1–18. doi:10.118640878-022-00283-3 PMID:35284235

Boccaletti, S., & Moro, D. (2000). Consumer willingness to pay for GM food products in Italy. *AgBioForum*, *3*, 259–267.

Bowman, A. (2019). Black economic empowerment policy and state–business relations in South Africa: The case of mining. *Review of African Political Economy*, *46*(160), 223–245. doi:10.1080/03056244.2019.1605587

Bozzola, M., Massetti, E., Mendelsohn, R., & Capitanio, F. (2018). A Ricardian analysis of the impact of climate change on Italian agriculture. *European Review of Agriculture Economics*, *45*(1), 57–79. doi:10.1093/erae/jbx023

Brandi, C., Cabani, T., Hosang, C., Schirmbeck, S., Westermann, L., & Wiese, H. (2015). Sustainability Standards for Palm Oil: Challenges for Smallholder Certification under the RSPO. *Journal of Environment & Development*, *24*(3), 292–314. doi:10.1177/1070496515593775

Brant, M., Tilley, D. S., & Mowen, J. C. (2004). *Consumer response to genetically modified foods*. Unpublished manuscript, Arkansas: Southern Agricultural Association.

Bredahl, L. (1999). Consumers' cognitions with regard to genetically modified food. Results of a qualitative study in four countries. *Appetite*, *33*(3), 343–360. doi:10.1006/appe.1999.0267 PMID:10625527

Bredahl, L. (2001). Determinants of consumer attitudes and purchase intentions with regard to genetically modified foods – results of a cross-national survey. *Journal of Consumer Policy*, *24*(1), 23–61. doi:10.1023/A:1010950406128

Bredahl, L., Grunert, K. G., & Frewer, L. J. (1998). Consumer attitude and decision-making with regard to genetically engineered food products: A review of the literature and a presentation of models for future research. *Journal of Consumer Policy*, *21*(3), 251–277. doi:10.1023/A:1006940724167

Breimyer, H. F. (1957). On price determination and aggregate price theory. *Journal of Farm Economics*, *39*(3), 676–694. doi:10.2307/1234424

Britwum, K., Bernard, J. C., & Albrecht, S. E. (2021). Does importance influence confidence in organic food attributes? *Food Quality and Preference*, *87*, 104056. doi:10.1016/j.foodqual.2020.104056

Bryman, A., & Bell, E. (2015). *Business Research Methods* (4th ed.). Oxford University Press.

Bureau Veritas. (2022). ISCC Plus: Biomaterials certification. Bureau Veritas. https://certification.bureauveritas.com/needs/iscc-plus-biomaterials-certification.

Calatayud, A., Mangan, J., & Christopher, M. (2019). The self-thinking supply chain. *Supply Chain Management*, *24*(1), 22–38. doi:10.1108/SCM-03-2018-0136

Candemir, A., Duvaleix, S., & Latruffe, L. (2021). Agricultural cooperatives and farm sustainability–A literature review. *Journal of Economic Surveys*, *35*(4), 1118–1144. doi:10.1111/joes.12417

Carelsen, C. P. R., Ncube, B., & Fanadzo, M. (2021). Classification and characterisation of smallholder farmers in South Africa: A brief review. *South African Journal of Agricultural Extension*, *49*(2), 97–106. doi:10.17159/2413-3221/2021/v49n2a12821

Carletto, C., Corral, P., & Guelfi, A. (2017). Agricultural commercialization and nutrition revisited: Empirical evidence from three African countries. *Food Policy*, *67*, 106–118. doi:10.1016/j.foodpol.2016.09.020 PMID:28413250

Carletto, C., Corral, P., & Guelfi, A. (2018). Revisiting the Gains from Agricultural Commercialization. In L. Christiaensen & L. Demery (Eds.), *Agriculture in Africa: Telling Myths from Facts* (pp. 39–46). The World Bank.

CCSA. (2017). *14 fresh produce market agents & the institute charged for price fixing* (Media release, Issue. https://www.compcom.co.za/wp-content/uploads/2017/01/fruit-and-veg-referral-draft-003.pdf

Chabalala, B. R. (2008). *An investigation of the success of comprehensive agricultural support programme projects on the farms allocated under land redistribution for agricultural development: a case study approach in the Waterberg District, Limpopo Province* [Doctoral dissertation, University of Limpopo].

Chagwiza, C., Muradian, R., & Ruben, R. (2016). Cooperative membership and dairy performance among smallholders in Ethiopia. *Food Policy*, *59*, 165–173. doi:10.1016/j.foodpol.2016.01.008

Chen, J., Lobo, A., & Rajendran, N. (2014). Drivers of organic food purchase intentions in mainland China – Evaluating potential customers' attitudes, demographics and segmentation. *International Journal of Consumer Studies*, *38*(4), 346–356. doi:10.1111/ijcs.12095

Chen, M. F. (2008). An integrated research framework to understand consumer attitudes and purchase intentions toward genetically modified foods. *British Food Journal*, *110*(6), 559–579. doi:10.1108/00070700810877889

Chen, M. F., & Li, H. L. (2007). The consumer's attitude toward genetically modified foods in Taiwan. *Food Quality and Preference*, *18*(4), 662–674. doi:10.1016/j.foodqual.2006.10.002

Chia, S. Y., Tanga, C. M., van Loon, J. J., & Dicke, M. (2019). Insects for sustainable animal feed: Inclusive business models involving smallholder farmers. *Current Opinion in Environmental Sustainability*, *41*, 23–30. doi:10.1016/j.cosust.2019.09.003

Chien, S.-H., Wu, J.-J., & Huang, C.-Y. (2018). "We made, we trust": Coproduction and image congruence in the food-tourism factories. *Asia Pacific Management Review*, *23*(4), 310–317. doi:10.1016/j.apmrv.2018.01.002

Chikazunga, D., Deall, S., Louw, A., & van Deventer, J. (2008). The role of fresh produce markets in South Africa. *Regoverning markets. Policy Brief, 4.* https://www.up.ac.za/media/shared/Legacy/sitefiles/file/48/2052/5_roleoffreshproducemarkets.pdf

Chikazunga, D., Joordan, D., Bienabe, E., & Louw, A. (2007). Patterns of restructuring food markets in South Africa: The case of fresh produce supply chains. *AAAE Second International Conference*, (pp. 53–55). AAAE.

Chipfupa, U., & Tagwi, A. (2021). Youth's participation in agriculture: A fallacy or achievable possibility? Evidence from rural South Africa. *Suid-Afrikaanse Tydskrif vir Ekonomiese en Bestuurswetenskappe*, *24*(1), 1–12. doi:10.4102ajems.v24i1.4004

Chiputwa, B., Spielman, D.J., & Qaim, R., (2015). Food Standards, Certification and Poverty among Coffee Farmers in Uganda. *World Development*, *66*, 400-412. doi:10.1016/j.worlddev.2014.09.006

Chiputwa, B., & Qaim, M. (2016). Sustainability standards, gender, and nutrition among smallholder farmers in Uganda. *The Journal of Development Studies*, *52*(9), 1241–1257. doi:10.1080/00220388.2016.1156090

Christopher, M. (2016). *Logistics & supply chain management.*

Christoph, I. B., Bruhn, M., & Roosen, J. (2008). Knowledge, attitude towards and acceptability of genetic modification in Germany. *Apetite*, *51*(1), 58–68. doi:10.1016/j.appet.2007.12.001 PMID:18243411

Cleveland, M., Kalamas, M., & Laroche, M. (2005). Shades of green: Linking environmental locus of control and pro-environmental behaviors. *Journal of Consumer Marketing*, *22*(4), 198–212. doi:10.1108/07363760510605317

COJ. (2016). *Spatial Development Framework 2040*. COJ. https://www.joburg.org.za/documents_/Pages/Key%20Documents/policies/Development%20Planning%20%EF%BC%86%20Urban%20Management/Citywide%20Spatial%20Policies/Spatial-Development-Framework-2040.aspx

COJ. (2020). *The City of Johannesburg Final Ingegrated Development Plan: 2020/21*. COJ. https://www.joburg.org.za/documents_/Documents/Intergrated%20Development%20Plan/IDP%202021-23/ITEM%2004%20ANNEXURE%20IDP.pdf

Coles, C., & Mitchell, J. (2011). Gender and agricultural value chains: A review of current knowledge. *ESA Working Paper No. 11-05.*

Colman, D., & Young, T. (1989). *Principles of agricultural economics: markets and prices in less developed countries.* Cambridge University Press. doi:10.1017/CBO9780511623509

Connor, M., & Siegrist, M. (2010). Factors influencing people's acceptance of gene technology: The role of knowledge, health expectations, naturalness and social trust. *Science Communication*, *32*(4), 514–538. doi:10.1177/1075547009358919

Conway, G. (2014). On being a smallholder. In P. Hazell & A. Rahman (Eds.), *New Directions for Smallholder Agriculture* (pp. 15–35). Oxford University Press. doi:10.1093/acprof:oso/9780199689347.003.0002

Costa-Font, J., & Mossialos, E. (2007). Are perceptions of "risks" and "benefits" of genetically modified food (in) dependent? *Food Quality and Preference*, *18*(2), 173–182. doi:10.1016/j.foodqual.2005.09.013

Costa-Font, M., & Gil, J. M. (2009). Structural equation modelling of consumer acceptance of genetically modified (GM) food in the Mediterranean Europe: A cross country study. *Food Quality and Preference*, *20*(6), 399–409. doi:10.1016/j.foodqual.2009.02.011

Costa-Font, M., Gil, J. M., & Traill, W. B. (2008). Consumer acceptance, valuation of and attitudes towards genetically modified food: Review and implications for food policy. *Food Policy*, *33*(2), 99–111. doi:10.1016/j.foodpol.2007.07.002

Cousins, B. (2015). Through a glass darkly': Towards agrarian reform in South Africa. In B. Cousins & C. Walker (Eds.), *Land Divided, Land Restored. Land Reform in South Africa for the 21st Century* (pp. 250–269). Jacana.

Cranfield, J., Deaton, B. J., & Shellikeri, S. (2009). Evaluating consumer preferences for organic food production standards. *Canadian Journal of Agricultural Economics*, *57*(1), 99–117. doi:10.1111/j.1744-7976.2008.01140.x

Crawford, I. (1997). *Agricultural and Food Marketing Management*. Food and Agriculture Organization of the United Nations.

Cripps, E. A. (2012). *Provisioning Johannesburg, 1886-1906*. University of South Africa. https://earthwormexpress.com/wp-content/uploads/2019/08/provisioning-johannesburg-1886-e28093-1906-.pdf

Cronje, F. (2015). Threats to commercial Farms. *Farmers Weekly*. https: //www.farmersweekly.co.za.

CSM. (2016). *Connecting Smallholders to Markets: An Analytical Guide*. International Civil Society Mechanism, Hands On the Land Alliance for Food Sovereignty. https://www.csm4cfs.org/connecting-smallholders-markets-analytical-guide/

Curtis, K. R., McCluskey, J. J., & Whal, T. I. (2004). Consumer acceptance of genetically modified food products in the developing world. *AgBioForum*, *7*(1-2), 70–75.

Cyranosk, D. (2005). The valley of ghosts. *Nature*, *436*(7051), 620–621. doi:10.1038/436620a PMID:16079812

DALRRD. (2009). *Report of the Ministerial Interim Committee on the restructuring of fresh produce markets in South Africa*. DALRRD. http://webapps.daff.gov.za/AmisAdmin/upload/Report%20of%20the%20Ministerial%20Interim%20Committee%20on%20restructuring%20of%20fresh%20produce%20markets%20in%20SA.pdf

DALRRD. (2020). *Statistics on Fresh Produce Markets: Report 58, 2020*. DALRRD. https://www.dalrrd.gov.za/Portals/0/Statistics%20and%20Economic%20Analysis/Statistical%20Information/Statistics%20on%20Fresh%20Produce%20Markets%202020.xlsx

DALRRD. (2021a). *Abstract of Agricultural Statistics 2021*. DALRRD. https://www.dalrrd.gov.za/Portals/0/Statistics%20and%20Economic%20Analysis/Statistical%20Information/Abstract%202021.pdf

DALRRD. (2021b). *Agriculture and Agro-processing Master Plan (AAMP). Draft version 05*. DALRRD.

Darekar, A., & Reddy, A. A. (2017). Cotton price forecasting in major producing states. *Economic Affairs*, *62*(3), 373. doi:10.5958/0976-4666.2017.00047.X

Davie, L. (2018). Joburg's market square defined the town. *The Heritage Portal*. https://www.theheritageportal.co.za/article/joburgs-market-square-defined-town

De Canio, F., & Martinelli, E. (2021). EU quality label vs organic food products: A multigroup structural equation modeling to assess consumers' intention to buy in light of sustainable motives. *Food Research International*, *139*, 109846. doi:10.1016/j.foodres.2020.109846 PMID:33509469

de Droog, S. M., Buijzen, M., & Valkenburg, P. M. (2014). Enhancing children's vegetable consumption using vegetable-promoting picture books. The impact of interactive shared reading and character–product congruence. *Appetite*, *73*, 73–80. doi:10.1016/j.appet.2013.10.018 PMID:24216486

De Jong, P. W., & de Valk, H. A. (2020). Intra-European migration decisions and welfare systems: The missing life course link. *Journal of Ethnic and Migration Studies*, *46*(9), 1773–1791. doi:10.1080/1369183X.2019.1611421

DeFries, R. S., Fanzo, J., Mondal, P., Remans, R., & Wood, S. A. (2017). Is voluntary certification of tropical agricultural commodities achieving sustainability goals for small-scale producers? A review of evidence. *Environmental Research Letters*, *12*(3), 12. doi:10.1088/1748-9326/aa625e

Department of Agriculture, Forestry, and fisheries (DAFF), (2012). National Agricultural Cooperatives Indaba, Directorate Co-operative and Enterprise Development. Pretoria.

Department of Agriculture. (2016). *Forestry and Fisheries (DAFF)*. SA-GAP Certification Programme for Smallholder Producers of Fresh Produce.

Department of Cooperative Governance (DCoG) and the Department of Traditional Affairs. (2020). *Profile and Analysis*. District Development Model. Vhembe District Municipality.

Dixie & Jaraman. (2011). Promotion of information. World Bank. http://siteresources.worldbank.org/EXTINFORMATIONANDCOMMUNICATIONANDTECHNOLOGIES/Resources/IC4D-2012-Chapter-2.pdf

Dlamini, T., Mohammed, M., & Sacolo, T. (2017). Assessing the Impact of National Marketing Boards on Consumers and Producers' Welfare. *SEPARC Working Paper 006/SZL*. Eswatini Economic Policy Analysis and Research Centre (ESEPARC).

Dompreh, E. B., Asare, R., & Gasparatos, A. (2021). Sustainable but hungry? Food security outcomes of certification for cocoa and oil palm smallholders in Ghana. *Environmental Research Letters*, *16*(5), 055001. doi:10.1088/1748-9326/abdf88

Dona, A., & Arvanitoyannis, I. S. (2009). Health Risks of Genetically Modified Foods. *Critical Reviews in Food Science and Nutrition*, *49*(2), 164–175. doi:10.1080/10408390701855993 PMID:18989835

Dorward, A., Poole, N., Morrison, J., Kydd, J., & Urey, I. (2003). Markets, institutions and technology: Missing links in livelihoods analysis. *Development Policy Review*, *21*(3), 319–332. doi:10.1111/1467-7679.00213

Dreezens, E., Martijn, C., Tenbult, P., Kok, G., & De Vries, N. K. (2005). Food and values: An examination of values underlying attitudes towards genetically modified and organic grown food products. *Appetite*, *44*(1), 115–122. doi:10.1016/j.appet.2004.07.003 PMID:15604038

DTI. (2019). Codes of good practice on Broad Based Black Economic Empowerment. DTI. https://www.gov.za/sites/default/files/gcis_document/201905/42496gen306.pdf

Du Toit, A. (2009). *Adverse incorporation and agrarian policy in South Africa, Or, How not to connect the rural poor to growth*. Institute of Poverty, Land and Agrarian Studies.

Dubb, A. (2020). Commodity Study: Small-Scale Sugar Production.

Dumais, M. (2012). *The negative consequences of agricultural marketing boards*. Montreal Economic Institute.

Duncan, D. (1991). The State Divided: Farm Labour Policy in South Africa, 1924–1948. *South African Historical Journal*, *24*(1), 67–89. doi:10.1080/02582479108671686

ECX. (2022). *Membership*. ECX. https://www.ecx.com.et/Pages/Membership.aspx#BC

Egbetokun, O. A., & Omonona, B. T. (2012). Determinants of farmers' participation in food market in Ogun State. *Global Journal of Science Frontier Research Agriculture and Veterinary Sciences*, *12*(9), 24–30.

Eicher, C. K., & Staatz, J. (1985). *Agricultural development in the third world*. The Johns Hopkins University Press.

Eisinger-Watzl, M., Wittig, F., Heuer, T., & Hoffmann, I. (2015). Customers purchasing organic food - Do they live healthier? Results of the German national nutrition survey II. *European Journal of Nutrition & Food Safety*, *5*(1), 59–71. doi:10.9734/EJNFS/2015/12734

Elleby, C., Domínguez, I. P., Adenauer, M., & Genovese, G. (2020). Impacts of the COVID-19 pandemic on the global agricultural markets. *Environmental and Resource Economics*, *76*(4), 1067–1079. doi:10.100710640-020-00473-6 PMID:32836856

Elliott, M., Elliott, L., & Sluis, E. V. D. (2018). A predictive analytics understanding of cooperative membership heterogeneity and sustainability. *Sustainability*, *10*(6), 2048. doi:10.3390u10062048

Elo, S., & Kyngäs, H. (2008). The qualitative content analysis process. *Journal of Advanced Nursing*, *62*(1), 107–115. doi:10.1111/j.1365-2648.2007.04569.x PMID:18352969

Emeana, K. N. (2017). Global value chain: the role of commodity exchange markets in linking smallholder farmers in Sub-Saharan Africa. *2nd Annual Research Conference of the Centre for African Research on Enterprise and Economic Development (CAREED)*. IEEE.

European Commission. (2010). *Eurobarometer 73.1: Biotechnology*. TNS Opinion and Social.

European Commission. (2014). *International Bioeconomy Profile Malaysia*. Joint Research Centre.

Fahad, S., & Wang, J. (2018). Farmers' risk perception, vulnerability, and adaptation to climate change in rural Pakistan. *Land Use Policy*, *79*, 301–309. doi:10.1016/j.landusepol.2018.08.018

Fan, S., & Rue, C. (2020). The role of smallholder farms in a changing world. In S. G. y Paloma, L. Riesgo, & K. Louhichi (Eds.), The role of smallholder farms in food and nutrition security (pp. 13–28). Springer, Cham. doi:10.1007/978-3-030-42148-9_2

FAO. IFAD, UNICEF, WFP, & WHO. (2021). The State of Food Security and Nutrition in the World 2021: Transforming Food Systems for Food Security, *Improved Nutrition and Affordable Healthy Diets For All*. Food and Agriculture Organisation of the United Nations.

Fatha, L., & Ayoubi, R. (2021). A revisit to the role of gender, age, subjective and objective knowledge in consumers' attitudes towards organic food. *Journal of Strategic Marketing*, 1–17. Advance online publication. doi:10.1080/0965254X.2021.1939405

Fenger, N. A., Bosselmann, A. S., Asare, R., & de Neergaard, A. (2017). The impact of certification on the natural and financial capitals of Ghanaian cocoa farmers. *Agroecology and Sustainable Food Systems*, *41*(2), 143–166. doi:10.1080/21683565.2016.1258606

Fiankor, D. D. D., Martínez-Zarzoso, I., & Brümmer, B. (2019). *Exports and governance: the role of private voluntary agrifood standards.* Agricultural Economics., doi:10.1111/agec.12488

Field, A. (2009). *Discovering statistics using SPSS* (3rd ed.). SAGE.

Firdaus-Raih, M., Senafi, S., Murad, M. A., Sidik, M. N., Lian, K. W., Daud, F., & Mohamed, R. (2005). A nationwide biotechnology outreach and awareness program for Malaysian high schools. *Electronic Journal of Biotechnology*, *8*(1), 9–16.

Fischer, E., & Qaim, M. (2012). Linking smallholders to markets: Determinants and impacts of farmer collective action in Kenya. *World Development*, *40*(6), 1255–1268. doi:10.1016/j.worlddev.2011.11.018

Fishbein, M. (1963). An investigation of the relationships between beliefs about an object and the attitude toward that object. *Human Relations*, *16*(3), 233–240. doi:10.1177/001872676301600302

Food & Drug Administration. (2017). Consumer Info About Food from Genetically Engineered Plants. FDA. https://www.fda.gov/Food/IngredientsPackagingLabeling/GEPlants/ucm461805.htm

Food and Agriculture Organization of the United Nations. (2015), The State of Food Insecurity in the World 2015. Meeting the 2015 international hunger targets: taking stock of uneven progress. FAO.

Fortin, D. R., & Renton, M. S. (2003). Consumer acceptance of genetically modified foods in New Zealand. *British Food Journal*, *105*(1/2), 42–58. doi:10.1108/00070700310467483

Francesconi, G. N., & Wouterse, F. (2019). Building the managerial capital of agricultural cooperatives in Africa. *Annals of Public and Cooperative Economics*, *90*(1), 141–159. doi:10.1111/apce.12218

Francis, J. J., Johnston, M., Robertson, C., Dlidewell, L., Entwistle, V., Eccles, M. P., & Grimshaw, J. M. (2010). What is an adequate sample size? Operationalising data saturation for theory-based interview studies. *Psychology & Health*, *25*(10), 1229–1245. doi:10.1080/08870440903194015 PMID:20204937

Freidberg, S., & Goldstein, L. (2011). Alternative food in the global south: Reflections on a direct marketing initiative in Kenya. *Journal of Rural Studies*, *27*(1), 24–34. doi:10.1016/j.jrurstud.2010.07.003

Frewer, L. J., Howard, C., Hedderley, D., & Shepherd, R. (1998). Methodological approaches to assessing risk perceptions associates with food-related hazards. *Risk Analysis*, *18*(1), 95–102. doi:10.1111/j.1539-6924.1998.tb00919.x PMID:9523447

Frewer, L. J., Howard, C., & Shepherd, R. (1996). The influence of realistic product exposure on attitudes towards genetic engineering of food. *Food Quality and Preference*, *7*(1), 61–67. doi:10.1016/0950-3293(95)00017-8

Frewer, L. J., Scholderer, J., & Bredahl, L. (2003). Communicating about the risk and benefits of genetically modified foods: The mediating role of trust. *Risk Analysis*, *23*(6), 1117–1133. doi:10.1111/j.0272-4332.2003.00385.x PMID:14641888

Frost, A., & Sullivan, W. (2009). *The Malaysian agricultural biotechnology sector*. Malaysian Biotechnology Corporation.

Fujii, L. A. (2018). *Interviewing in Social Science Research: A Relational Approach.* Routledge.

Galbraith, J. K. (1993). *American capitalism: The concept of countervailing power* (1st ed.). Routledge.

Ganiere, P., Chern, W. S., & Hahn, D. (2006). A continuum of consumer attitude toward genetically modified foods in the United States. *Journal of Agricultural and Resource Economics*, *31*(1), 129–149.

Gaskell, G., Allum, N., Wagner, W., Kronberger, N., Torgersen, H., Hampel, J., & Bardes, J. (2004). GM foods and the misperception of risk perception. *Risk Analysis*, *24*(1), 185–194. doi:10.1111/j.0272-4332.2004.00421.x PMID:15028010

Gc, R. K., & Hall, R. P. (2020). The commercialization of smallholder farming—A case study from the rural western middle hills of Nepal. *Agriculture*, *10*(5), 143. doi:10.3390/agriculture10050143

GCX. (2022). *Membership types: Ghana Commodity Exchange*. GCX. https://gcx.com.gh/membership_type/

Gebre-ab, N. (2006). Commercialization of smallholder agriculture in Ethiopia. Ethiopian Development Research Institute Notes and Papers Series No. 3. Addis Ababa, Ethiopia.

Gereffi, G., Humphrey, J., Kaplinsky, R., & Sturgeon, T. J. (2001). Introduction: Globalisation, value chains and development. *IDS Bulletin*, *32*(3), 1–8. doi:10.1111/j.1759-5436.2001.mp32003001.x

Ghosh, N. (2013). *India's agricultural marketing: Market Reforms and Emergence of New Channels*. Springer. doi:10.1007/978-81-322-1572-1

Gianluigi Guido, M. I., Peluso, A. M., Maloumby-Baka, R. C., & Buffa, C. (2010). The role of ethics and product personality in the intention to purchase organic food products: A structural equation modeling approach. *International Review of Economics*, *57*(1), 79–102. doi:10.100712232-009-0086-5

Giller, K. E., Delaune, T., Silva, J. V., Descheemaeker, K., van de Ven, G., Schut, A. G., van Wijk, M., Hammond, J., Hochman, Z., Taulya, G., Chikowo, R., Narayanan, S., Kishore, A., Bresciani, F., Teixeira, H. M., Andersson, J. A., & van Ittersum, M. K. (2021). The future of farming: Who will produce our food? *Food Security*, *13*(5), 1073–1099. doi:10.100712571-021-01184-6

Gilsenan, C., Burke, R. M., & Barry-Ryan, C. (2012). Do organic cherry vine tomatoes taste better than conventional cherry vine tomatoes? A sensory and instrumental comparative study from Ireland. *Journal of Culinary Science & Technology*, *10*(2), 154–167. doi:10.1080/15428052.2012.679232

Glaeser, E. L. (2014). A world of cities: The causes and consequences of urbanization in poorer countries. *Journal of the European Economic Association*, *12*(5), 1154–1199. doi:10.1111/jeea.12100

Glasbergen, P. (2018). Smallholders do not Eat Certiðcates. *Ecological Economics*, *147*, 243–252. doi:10.1016/j.ecolecon.2018.01.023

Glasbergen, P., & Schouten, G. (2015). Transformative capacities of global private sustain- ability standards; a reñection on scenarios in the ðeld of agricultural commodities. *Journal of Corporate Citizenship*, *58*, 11–27.

Globalgap.org. (2022a). IFA v6. What's new? *Global Gap*. https://www.globalgap.org/uk_en/for-producers/globalg.a.p./integrated-farm-assurance-ifa/IFA-V6/

Globalgap.org. (2022b). Integrated Farm Assurance (IFA) v6 at a glance. *Global Gap*. https://www.globalgap.org/uk_en/for-producers/globalg.a.p./integrated-farm-assurance-ifa/IFA-V6/

Glushkova, S., Lomakina, O., & Sakulyeva, T. (2019). The economy of developing countries in the context of globalization: Global supply chain management. *International Journal of Supply Chain Management*, *8*(1), 876–884.

Gockowski, J., Afari-Sefa, V., Sarpong, D. B., Osei-Asare, Y. B., & Dziwornu, A. K. (2011). Increasing income of Ghanaian cocoa farmers is introduction of fine flavour cocoa a viable alternative. *Zeitschrift für Ausländische Landwirtschaft*, *50*, 175–200.

Godde, C. M., Mason-D'Croz, D., Mayberry, D. E., Thornton, P. K., & Herrero, M. (2021). Impacts of climate change on the livestock food supply chain; a review of the evidence. *Global Food Security*, *28*, 100488. doi:10.1016/j.gfs.2020.100488 PMID:33738188

Gomez, G., & Vossenberg, S. (2018). Identifying ripple effects from new market institutions to household rules-Malawi's Agricultural Commodity Exchange. *NJAS Wageningen Journal of Life Sciences*, *84*(1), 41–50. doi:10.1016/j.njas.2017.10.003

Gomez-Zavaglia, A., Mejuto, J. C., & Simal-Gandara, J. (2020). Mitigation of emerging implications of climate change on food production systems. *Food Research International*, *134*, 109256. doi:10.1016/j.foodres.2020.109256 PMID:32517948

González, C., Johnson, N., & Qaim, M. (2009). Consumer acceptance of second generation GM foods: The case of biofortified cassava in the northeast of Brazil. *Journal of Agricultural Economics*, *60*(3), 604–624. doi:10.1111/j.1477-9552.2009.00219.x

Goodson, I. F. (2012). *Developing narrative theory: Life histories and personal representation*. Routledge. doi:10.4324/9780203817704

Gordon, S. (1988). *Land Reform in Post-apartheid South Africa: Some Lessons from Zimbabwe* [Doctoral dissertation, University of Cape Town].

Gouel, C., & Laborde, D. (2021). The crucial role of domestic and international market-mediated adaptation to climate change. *Journal of Environmental Economics and Management*, *106*, 102408. doi:10.1016/j.jeem.2020.102408

Granovetter, M. (1985). Economic action and social structure: The problem of embeddedness. *American Journal of Sociology*, *91*(3), 481–510. https://www.jstor.org/stable/2780199. doi:10.1086/228311

Greenberg, S. (2015). Corporate Concentration And Food Security In South Africa: Is The Commercial Agro-Food System Delivering? Guarín, A. (2013). The value of domestic supply chains: Producers, wholesalers, and urban consumers in Colombia. *Development Policy Review*, *31*(5), 511–530.

Greyling, J. C., Vink, N., & van der Merwe, E. (2018). Maize and Gold: South African Agriculture's Transition from Suppression to Support, 1886–1948. In Agricultural development in the world periphery (pp. 179-204). Palgrave Macmillan, Cham.

Greyling, J. C., Vink, N., & Mabaya, E. (2015). South Africa's agricultural sector twenty years after democracy (1994 to 2013). *Professional Agricultural Workers Journal*, *3*(1), 10.

Grubor, A., & Djokic, N. (2016). Organic food consumer profile in the Republic of Serbia. *British Food Journal*, *118*(1), 164–182. doi:10.1108/BFJ-06-2015-0225

Grunert, K., Sondergaard, H., & Scholderer, J. (2004). How can we know what we like when we don't understand it? Consumer attitude formation towards complex technical issues. From the *proceedings of the International Conference on Research in Advertising, Aarhus University*. MAPP - Centre for Research on Customer Relations in the Food Sector.

Grunert, K. G., Bredahl, L., & Scholderer, J. (2003). Four questions on European consumers' attitudes toward the use of genetic modification in food production. *Innovative Food Science & Emerging Technologies*, *4*(4), 435–445. doi:10.1016/S1466-8564(03)00035-3

Grunert, K. G., Lahteenmaki, L., Nielsen, N. A., Poulsen, J. B., Ueland, O., & Astrom, A. (2001). Consumer perceptions of food products involving genetic modification-results from a qualitative study in four Nordic countries. *Food Quality and Preference*, *12*(8), 527–542. doi:10.1016/S0950-3293(01)00049-0

Guarín, A. (2013). The value of domestic supply chains: Producers, wholesalers, and urban consumers in Colombia. *Development Policy Review*, *31*(5), 511–530. doi:10.1111/dpr.12023

Gwiriri, L. C., Bennett, J., Mapiye, C., & Burbi, S. (2019). Unpacking the 'emergent farmer' concept in agrarian reform: Evidence from livestock farmers in South Africa. *Development and Change*, *50*(6), 1664–1686. doi:10.1111/dech.12516

Gwiriri, L. C., Bennett, J., Mapiye, C., & Burbi, S. (2021). Emerging from Below? Understanding the Livelihood Trajectories of Smallholder Livestock Farmers in Eastern Cape Province, South Africa. *Land (Basel)*, *10*(2), 226. doi:10.3390/land10020226

Gyau, A., Mbugua, M., & Oduol, J. (2016). Determinants of participation and intensity of participation in collective action: Evidence from smallholder avocado farmers in Kenya. *Journal on Chain and Network Science*, *16*(2), 147–156. doi:10.3920/JCNS2015.0011

Haggblade, S. (2020). Overview. In AGRA (Ed.), Feeding Africa's Cities: Opportunities, Challenges, and Policies for Linking African Farmers with Growing Urban Food Markets (pp. 15–24). Alliance for a Green Revolution in Africa.

Hall, R., & Aliber, M. (2010). The case for re-strategising spending priorities to support small-scale farmers in South Africa. *Working Paper 17*.

Hamstra, A. (1998). *Public opinion about biotechnology. A survey of surveys*. European Federation of Biotechnology.

Han, J. H. (2006). The Effects of Perceptions on Consumer Acceptance of Genetically Modified (GM) Foods. [Unpublished Doctoral Dissertation, Department of Agricultural Economics and Agribusiness, Louisiana State University].

Hanks, L., & Mattila, A. S. (2016). Consumer response to organic food in restaurants: A serial mediation analysis. *Journal of Foodservice Business Research*, *19*(1), 109–121. doi:10.1080/15378020.2016.1129228

Hannah, C., Davies, J., Green, R., Zimmer, A., Anderson, P., Battersby, J., Baylis, K., Joshi, N., & Evans, T. P. (2022). Persistence of open-air markets in the food systems of Africa's secondary cities. *Cities (London, England)*, *124*, 103608. doi:10.1016/j.cities.2022.103608

Hao, J., Bijman, J., Gardebroek, C., Heerink, N., Heijman, W., & Huo, X. (2018). Cooperative membership and farmers' choice of marketing channels–Evidence from apple farmers in Shaanxi and Shandong Provinces, China. *Food Policy*, *74*, 53–64. doi:10.1016/j.foodpol.2017.11.004

Harper, G. C., & Makatouni, A. (2002). Consumer perception of organic food production and farm animal welfare. *British Food Journal*, *104*(3-5), 287–299. doi:10.1108/00070700210425723

Hart, K. (1985). The informal economy. *Cambridge Anthropology*, 54–58.

Hasegawa, T., Fujimori, S., Havlík, P., Valin, H., Bodirsky, B. L., Doelman, J. C., Fellmann, T., Kyle, P., Koopman, J. F., Lotze-Campen, H., Mason-D'Croz, D., Ochi, Y., Pérez Domínguez, I., Stehfest, E., Sulser, T. B., Tabeau, A., Takahashi, K., Takakura, J., van Meijl, H., & Witzke, P. (2018). Risk of increased food insecurity under stringent global climate change mitigation policy. *Nature Climate Change*, *8*(8), 699–703. doi:10.103841558-018-0230-x

Hassan, S. H., John Kua, S. B., & Harun, H. (2016). Muslim consumers' perception and purchase intention toward GM food. *International Food Research Journal*, *23*(2), 806–815.

Havlík, P., Valin, H., Herrero, M., Obersteiner, M., Schmid, E., Rufino, M.C., Mosnier, A., Thornton, P.K., Böttcher, H., Conant, R.T. and Frank, S., (2014). Climate change mitigation through livestock system transitions. *Proceedings of the National Academy of Sciences, 111*(10), pp.3709-3714.

Heijden, T., & Vink, N. (2013). Good for whom? Supermarkets and small farmers in South Africa–a critical review of current approaches to increasing access to modern markets. *Agrekon, 52*(1), 68–86. doi:10.1080/03031853.2013.778466

Hellmann, R., Griesbaum, J., & Mandl, T. (2010). Quality in blogs: How to find the best user generated content. In W. Abramowicz, & R. Tolksdorf, *Business Information Systems. BIS 2010.* [Berlin, Heidelberg: Springer.]. *Lecture Notes in Business Information Processing, 47*, 47–58. doi:10.1007/978-3-642-12814-1_5

Hernandez, M. A., Rashid, S., Lemma, S., & Kuma, T. (2017). Market institutions and price relationships: The case of coffee in the Ethiopian commodity exchange. *American Journal of Agricultural Economics, 99*(3), 683–704. doi:10.1093/ajae/aaw101

Higginson, J. (2001). Hell in small places: Agrarian elites and collective violence in the Western Transvaal, 1900-1907. *Journal of Social History, 35*(1), 95–124. doi:10.1353/jsh.2001.0084

Hilbe, J. M. (2009). *Logistic regression models.* Chapman and hall/CRC. doi:10.1201/9781420075779

Hirsch, S., Mishra, A., Möhring, N., & Finger, R. (2020). Revisiting firm flexibility and efficiency: Evidence from the EU dairy processing industry. *European Review of Agriculture Economics, 47*(3), 971–1008. doi:10.1093/erae/jbz003

Hlatshwayo, S. I., Ngidi, M., Ojo, T., Modi, A. T., Mabhaudhi, T., & Slotow, R. (2021). A typology of the level of market participation among smallholder farmers in South Africa: Limpopo and Mpumalanga Provinces. *Sustainability, 13*(14), 7699. doi:10.3390u13147699

Hlophe-Ginindza, S. N., & Mpandeli, N. S. (2021). *The Role of Small-Scale Farmers in Ensuring Food Security in Africa.* Food Secur. Afr. doi:10.5772/intechopen.91694

Honkanen, P., & Verplanken, B. (2004). Understanding attitudes towards genetically modified food: The role of values and attitudes strength. *Journal of Consumer Policy, 27*(4), 401–420. doi:10.100710603-004-2524-9

Hossain, F., & Onyango, B. (2004). Product attributes and consumer acceptance of nutritionally enhanced genetically modified foods. *International Journal of Consumer Studies, 28*(3), 255–267. doi:10.1111/j.1470-6431.2004.00352.x

Hossain, F., Onyango, B., Schilling, B., Hallman, W., & Adelaja, A. (2003). Product attributes, consumer benefits and public approval of genetically modified foods. *International Journal of Consumer Studies, 27*(5), 353–365. doi:10.1046/j.1470-6431.2003.00303.x

Hristov, J., Barreiro-Hurle, J., Salputra, G., Blanco, M., & Witzke, P. (2021). Reuse of treated water in European agriculture: Potential to address water scarcity under climate change. *Agricultural Water Management, 251*, 106872. doi:10.1016/j.agwat.2021.106872 PMID:34079159

HSPH. (2011). *Healthy Eating Plate.* Harvard University, Harvard (T.H.Chan) School of Public Health. https://www.hsph.harvard.edu/nutritionsource/healthy-eating-plate/

Huang, J., Zhi, H., Huang, Z., Jia, X., & Rozelle, S. (2009). Smallholder incomes, vegetable marketing and food safety: evidence from China. *2009 Conference of the International Association of Agricultural Economists.* AAE.

Huang, J., Qiu, H., Bai, J., & Pray, C. (2006). Awareness, acceptance of and willingness to buy genetically modified foods in urban China. *Appetite, 46*(2), 144–151. doi:10.1016/j.appet.2005.11.005 PMID:16469414

Humphrey, J., & Navas-Alemán, L. (2010). *Value chains, donor interventions and poverty reduction: A review of donor practice* (2040-0217). (IDS Research Reports, Issue. Kaplinsky, R., & Morris, M. (2001). *A handbook for value chain research* (*Vol. 113*). IDRC.

Hwang, J. (2016). Organic food as self-presentation: The role of psychological motivation in older consumers' purchase intention of organic food. *Journal of Retailing and Consumer Services*, *28*, 281–287. doi:10.1016/j.jretconser.2015.01.007

Hwang, J., & Chung, J.-E. (2019). What drives consumers to certain retailers for organic food purchase: The role of fit for consumers' retail store preference. *Journal of Retailing and Consumer Services*, *47*, 293–306. doi:10.1016/j.jretconser.2018.12.005

Ibrahim, B., Rezal, G., Mohamed, Z. A., & Sharifuddin, J. (2013). Determinants of consumer perception towards gentically modified (GM) foods: Malaysian case study. *3rd International Conference on Management Proceeding*, (pp. 488-499).

Iden, J., & Eikebrokk, T. R. (2013). Implementing IT Service Management: A systematic literature review. *International Journal of Information Management*, *33*(3), 512–523. doi:10.1016/j.ijinfomgt.2013.01.004

Ifejika, S. C. (2010). Resilient adaptation to climate change in African agriculture (54). *Studies*.

Inigo, E. A., Garst, J., Blok, V., & Pentaraki, K. M. (2020). Do voluntary standards support responsible innovation implementation and reporting in industry? The case of the European food sector. In *Assessment of Responsible Innovation* (pp. 145–168). Routledge. doi:10.4324/9780429298998-9

International Service for the Acquisition of Agri Biotech Applications. (2022) ISAAA. https://www.isaaa.org/gmapprovaldatabase/default.asp. (Accessed 14October, 2022)

International Trade Administration. (2021). *South Africa - Country Commercial Guide*. ITA. https://www.trade.gov/country-commercial-guides/south-africa-agricultural sector#:~:text=There%20are%20approximately%2032%2C000%20commercial,80%20percent%20of%20agricultural%20output

Iscc-system.org. (2022). *ISCC for Food and Feed › ISCC System*. ISCC. <https://www.iscc-system.org/process/market-applications/iscc-for-food-and-feed/ .

Ishak, S., Omar, A. R. C., Sum, S. M., Othman, A. S., & Jaafar, J. (2020). Smallholder Agriculture Cooperatives' Performance: What is in the Minds of Management? *Journal of Co-Operative Organization and Management*, *8*(2), 100110. doi:10.1016/j.jcom.2020.100110

Ismail, K., Soehod, K., Vivishna, S., Khurram, W., Jafri, S. K. A., & Ramily, M. K. B. (2012a). Genetically modified food and consumer purchase intentions: A study in Johor Bahru. *International Journal of Business and Social Science*, *3*(5), 197–207.

Ismail, K., Vivishna, S., Khurram, W., & Jafri, S. K. A. (2012b). Evaluating Consumer Purchase Intentions for Genetically Modified Food in Malaysia: A Comparative Study of Muslim and Non-Muslim Consumers. *Research Journal of Applied Sciences, Engineering and Technology*, *4*(5), 466–474.

ITC. (2022). *Standard Map app*. ITC. <https://www.standardsmap.org/en/identify?name=&products=Fruits&origin=Europe&destination=South%20Africa&themes=

ITC. (2022). Trade Map - List of importing markets from European Union (EU 28) for a product exported by South Africa. [online] *Trademap.org.* https://www.trademap.org/Country_SelProductCountry_TS.aspx?nvpm=1%7c710%7c%7c%7c14719%7c08%7c%7c%7c2%7c1%7c1%7c2%7c2%7c1%7c2%7c5%7c1%7c1>

Ito, J., Bao, Z., & Su, Q. (2012). Distributional effects of agricultural cooperatives in China: Exclusion of smallholders and potential gains on participation. *Food Policy*, *37*(6), 700–709. doi:10.1016/j.foodpol.2012.07.009

Jacoby, H. G., & Minten, B. (2009). On measuring the benefits of lower transport costs. *Journal of Development Economics*, *89*(1), 28–38. doi:10.1016/j.jdeveco.2008.06.004

Jaleta, M., Gebremedhin, B., & Hoekstra, D. (2009). Smallholder commercialization: Processes, determinants and impact. *ILRI Discussion Paper.*

Janssen, M., & Hamm, U. (2012). Product labelling in the market for organic food: Consumer preferences and willingness-to-pay for different organic certification logos. *Food Quality and Preference*, *25*(1), 9–22. doi:10.1016/j.foodqual.2011.12.004

Jari, B., & Fraser, G. (2012). Influence of institutional and technical factors on market choices of smallholder farmers in the Kat River Valley. In *Mansholt publication series-Volume 10 edited by* (p. 59). Herman D. van Schalkwyk Jan A. Groenewald. doi:10.3920/978-90-8686-168-2_3

Joiner, J., & Okeleke, K. (2019). E-commerce in agriculture: new business models for smallholders' inclusion into the formal economy. *GSMA.* https://www.gsma.com/mobilefordevelopment/resources/e-commerce-in-agriculture-new-business-models-for-smallholders-inclusion-into-the-formal-economy/

JSE. (2022). *Agricultural derivatives.* JSE. https://www.jse.co.za/trade/derivitive-market/interest-rate-derivatives-market/agricultural-derivitives

Kajale, D. B., & Becker, T. C. (2015). Factors Influencing Young Consumers' Acceptance of Genetically Modified Food in India. *Journal of Food Products Marketing*, *21*(5), 461–481. doi:10.1080/10454446.2013.845866

Kang'ethe, E. K., Grace, D., Alonso, S., Lindahl, J. F., Mutua, F., & Haggblade, S. (2020). Food safety and public health implications of growing urban food markets. In AGRA (Ed.), Africa Agriculture Status Report. Feeding Africa's Cities: Opportunities, Challenges, and Policies for Linking African Farmers with Growing Urban Food Markets (Issue 8) (pp. 101–119). Alliance for a Green Revolution in Africa (AGRA).

Kangile, R. J., Mgeni, C. P., Mpenda, Z. T., & Sieber, S. (2020). The determinants of farmers' choice of markets for staple food commodities in Dodoma and Morogoro, Tanzania. *Agriculture*, *10*(5), 142. doi:10.3390/agriculture10050142

Kanyane, M. H., & Ilorah, R. (2015). The cooperative landscape in South Africa: Analyzing critical issues. *Africa Today*, *61*(3), 3–14. doi:10.2979/africatoday.61.3.3

Karani, K. D., & Wanjoh, J. (2017). Factors Influencing Marketing Of Agricultural Produce Among Small-Scale Farmers, A Case Of Sorhgum In Giaki Location, Meru County Kenya. *International Journal of Economics, Commerce and Management, 8*, 664-678. https://ijecm.co.uk/wp-content/uploads/2017/08/5838.pdf

Kareklas, I., Carlson, J. R., & Muehling, D. D. (2014). "I eat organic for my benefit and yours": Egoistic and altruistic considerations for purchasing organic food and their implications for advertising strategists. *Journal of Advertising*, *43*(1), 18–32. doi:10.1080/00913367.2013.799450

Kassarjian, H. H. (1977). Content analysis in consumer research. *The Journal of Consumer Research*, *4*(1), 8–18. doi:10.1086/208674

Kastner, T., Rivas, M. J. I., Koch, W., & Nonhebel, S. (2012). Global changes in diets and the consequences for land requirements for food. *Proceedings of the National Academy of Sciences of the United States of America*, *109*(18), 6868–6872. doi:10.1073/pnas.1117054109 PMID:22509032

Kherallah, M., Delgado, C. L., Gabre-Madhin, E. Z., Minot, N., & Johnson, M. (2002). *Reforming agricultural markets in Africa: Achievements and challenges*. IFPRI.

Kibirige, D. (2016). Smallholder commercialization of maize and social capital in the Eastern Cape Province of South Africa. *International Journal of Economics. Commerce and Management*, *4*(9), 236–252.

Kim, R.B. (2009). Factors influencing Chinese consumer behavior when buying innovative food products. *Agricultural Economics – Czech,* 55(9), 436-445.

Kim, R.B. (2010). A multiple-attribute model of Japanese consumer`s purchase intention for GM foods. *Agricultural Economics – Czech, 56*(10), 449-459.

Kim, J. (2012). The institutionalization of YouTube: From user-generated content to professionally generated content. *Media Culture & Society*, *34*(1), 53–67. doi:10.1177/0163443711427199

Kim, R. B. (2012). Consumer attitude of risk and benefits towards genetically modified foods in South Korea: Implications for food policy. *Inzinerine Ekonomika –. The Engineering Economist*, *23*(2), 189–199.

Kirsten, J. F. (1994). *Agricultural support programmes in the developing areas of South Africa* [Doctoral dissertation, University of Pretoria].

Kirsten, J., Mapila, M., Okello, J. J., & De, S. (2013). *Managing agricultural commercialization for inclusive growth in Sub-Saharan Africa* (No. 206518). University of Pretoria, Department of Agricultural Economics, Extension and Rural Development.http://ageconsearch.umn.edu/record/206518/files/2013.pdf

Klabi, F. (2020). Self-image congruity affecting perceived quality and the moderation of brand experience: The case of local and international brands in the Kingdom of Saudi Arabia. *Journal of Global Marketing*, *33*(2), 69–83. doi:10.1080/08911762.2019.1614242

Klecka, W. R. (1980). *Series: "Quantitative Applications in the Social Sciences* (Vol. 18). SAGE Publications.

Kogo, B. K., Kumar, L., & Koech, R. (2021). Climate change and variability in Kenya: A review of impacts on agriculture and food security. *Environment, Development and Sustainability*, *23*(1), 23–43. doi:10.100710668-020-00589-1

Kohls, R. L., & Uhl, J. N. (1990). *Marketing of agricultural products*. Prentice Hall.

Konuk, F. A. (2018). The role of store image, perceived quality, trust and perceived value in predicting consumers' purchase intentions towards organic private label food. *Journal of Retailing and Consumer Services*, *43*, 304–310. doi:10.1016/j.jretconser.2018.04.011

Kumar, A., Mangla, S. K., Kumar, P., & Song, M. (2021). Mitigate risks in perishable food supply chains: Learning from COVID-19. *Technological Forecasting and Social Change*, *166*, 120643. doi:10.1016/j.techfore.2021.120643

Kumar, S. K., & Timalsina, A. K. (2016). Challenges for Adopting E-Commerce in Agriculture in Nepalese Context—A Case Study of Kathmandu Valley. *Proceedings of IOE Graduate Conference*, 305–312.

Kyaw, N. N., Ahn, S., & Lee, S. H. (2018). Analysis of the factors influencing market participation among smallholder rice farmers in magway region, central dry zone of Myanmar. *Sustainability*, *10*(12), 4441. doi:10.3390u10124441

Kyazze, L.M., Nsereko, I. and Nkote, I., (2020). Cooperative practices and non-financial performance of savings and credit cooperative societies. *International Journal of Ethics and Systems*.

Kyomugisha, H., Sebatta, C., & Mugisha, J. (2018). Potato market access, marketing efficiency and on-farm value addition in Uganda. *Scientific American, 1*, e00013.

Labarthe, P., & Laurent, C. (2013). Privatization of agricultural extension services in the EU: Towards a lack of adequate knowledge for small-scale farms? *Food Policy*, *38*, 240–252. doi:10.1016/j.foodpol.2012.10.005

Lahiff, E. P. (1997). *Agriculture and rural livelihoods in a South African 'homeland': A case study from Venda*. University of London, School of Oriental and African Studies.

Lang, T., & Heasman, M. (2004). *Food Wars: The Global Battle for Mouths, Minds and Markets: Earthscan*. James & James.

Laurent, C. (2005). Les agricultures de l'Union européenne et leur territoire. In C. Laurent & P. Thinon (Eds.), *Agricultures et territoires* (pp. 31–55). Hermès Science.

Lee, W.-C. J., Shimizu, M., Kniffin, K. M., & Wansink, B. (2013). You taste what you see: Do organic labels bias taste perceptions? *Food Quality and Preference*, *29*(1), 33–39. doi:10.1016/j.foodqual.2013.01.010

Lenta, G., & Maasdorp, G. (1984). Food production in the homelands: Constraints and remedial policies. .

Limpopo Development Plan 2020-2025 (LDP). (2022). LPG. http://www.limpopo.gov.za

Limpopo Travel Agency. (2014). *Homepage*. LTA. http://www.golimpopo.com

Liu, C., Zheng, Y., & Cao, D. (2021). An analysis of factors affecting selection of organic food: Perception of consumers in China regarding weak signal. *Appetite*, *161*, 105145. Advance online publication. doi:10.1016/j.appet.2021.105145 PMID:33515620

Liu, H., Zhang, J., Zhou, C., & Ru, Y. (2018). Optimal purchase and inventory retrieval policies for perishable seasonal agricultural products. *Omega*, *79*, 133–145. doi:10.1016/j.omega.2017.08.006

Liu, M., Min, S., Ma, W., & Liu, T. (2021). The adoption and impact of E-commerce in rural China: Application of an endogenous switching regression model. *Journal of Rural Studies*, *83*, 106–116. doi:10.1016/j.jrurstud.2021.02.021

Lobb, A. E., Mazzocchi, M., & Traill, W. B. (2007). Modelling risk perception and trust in food safety information within the theory of planned behavior. *Food Quality and Preference*, *18*(2), 384–395. doi:10.1016/j.foodqual.2006.04.004

Long, S. P., Ainsworth, E. A., Leakey, A. D. B., Nosberger, J., & Ort, D. R. (2006). Food for thought: Lower-than-expected crop yield stimulation with rising CO2 concentrations. *Science*, *312*(5782), 1918–1921. doi:10.1126cience.1114722 PMID:16809532

Loureiro, M. L., & Bugbee, M. (2005). Enhanced GM Foods: Are Consumer Ready to Pay for the Potential Benefits of Biotechnology? *The Journal of Consumer Affairs*, *39*(1), 52–70. doi:10.1111/j.1745-6606.2005.00003.x

Loureiro, M., & Hine, S. (2004). Preferences and willingness to pay for GM labelling policies. *Food Policy*, *29*(5), 467–483. doi:10.1016/j.foodpol.2004.07.001

Louw, A. (2008). The role of Fresh Produce Markets in South Africa. Regoverning Markets. [University of Pretoria.]. *Policy Brief*, 4.

Lowder, S. K., Skoet, J., & Singh, S. (2014). *What do we really know about the number and distribution of farms and family farms in the world?* Background paper for The State of Food and Agriculture 2014 (*ESA Working Paper* No. 14-02).

Lowder, S. K., Skoet, J., & Raney, T. (2016). The number, size, and distribution of farms, smallholder farms, and family farms worldwide. *World Development*, *87*, 16–29. doi:10.1016/j.worlddev.2015.10.041

Lu, L., & Gursoy, D. (2017). Does offering an organic food menu help restaurants excel in competition? An examination of diners' decision-making. *International Journal of Hospitality Management, 63*, 72–81. doi:10.1016/j.ijhm.2017.03.004

Lusk, J. L., Daniel, M. S., Mark, D. L., & Lusk, C. L. (2001). Alternative Calibration and Auction Institutions for Predicting Consumer Willingness to Pay for Nongenetically Modified Corn Chips. *Journal of Agricultural and Resource Economics, 26*(1), 40–57.

Lusk, J. L., House, L. O., Valli, C., Jaeger, S. R., Moore, M., Morrow, B., & Traill, W. B. (2004). Effect of information about benefits of biotechnology on consumer acceptance of genetically modified food: Evidence from experimental auctions in the United States, England, and France. *European Review of Agriculture Economics, 31*(2), 179–204. doi:10.1093/erae/31.2.179

Macer, D., Bazer, H., & Gough, J. (1991). *'Genetic engineering in New Zealand: Science, ethics and public policy', Information paper no.20.* Centre for Resource Management, Lincoln University.

Mafole, M. I. (2018). Memorialising the Pan-Africanist Congress of Azania1. In Social Memory as a Force for Social and Economic Transformation (pp. 257-266). Routledge.

Magesa, M. M. (2015). *Linking rural farmers to markets using ICTs.* CTA. Accessed on 12 April 2018.https://publications.cta.int/media/publications/downloads/1875_PDF.pdf

Magesa, M. M., Michael, K., & Ko, J. (2020). Access and use of agricultural market information by smallholder farmers: Measuring informational capabilities. *The Electronic Journal on Information Systems in Developing Countries, 86*(6), e12134. doi:10.1002/isd2.12134

Mahaliyanaarachchi, R. P., & Bandara, R. M. A. S. (2006). Commercialization of agriculture and role of agricultural extension. https://suslj.sljol.info/article/10.4038/suslj.v6i1.1686/galley/1419/download/

Mahmood, N., Arshad, M., Mehmood, Y., Shahzad, M. F., & Kächele, H. (2021). Farmers' perceptions and role of institutional arrangements in climate change adaptation: Insights from rainfed Pakistan. *Climate Risk Management, 32*, 100288. doi:10.1016/j.crm.2021.100288

Maiwashe, A. (2012). *Assessment of Determinants of Agricultural Commodity Marketing Channels Choice for Smallholder Farmers in Thohoyandou Service Centre, Limpopo Province, South Africa* [Master's dissertation, University of Venda].

Maiwashe, A. (2017). Towards a framework for effective performance of smallholder agricultural cooperatives in Limpopo Province, South Africa [Doctoral dissertation, University of Venda].

Malaysian Biotechnology Information Centre. (2017). *Homepage.* MBIC. http://www.bic.org.my

Malhi, G. S., Kaur, M., & Kaushik, P. (2021). Impact of climate change on agriculture and its mitigation strategies: A review. *Sustainability, 13*(3), 1318. doi:10.3390u13031318

Malodia, S., Singh, P., Goyal, V., & Sengupta, A. (2017). Measuring the impact of brand-celebrity personality congruence on purchase intention. *Journal of Marketing Communications, 23*(5), 493–512. doi:10.1080/13527266.2017.1322125

Manda, J., Khonje, M. G., Alene, A. D., Tufa, A. H., Abdoulaye, T., Mutenje, M., Setimela, P., & Manyong, V. (2020). Does cooperative membership increase and accelerate agricultural technology adoption? Empirical evidence from Zambia. *Technological Forecasting and Social Change, 158*, 120160. doi:10.1016/j.techfore.2020.120160

Mann, R. (2001). Homeland agriculture in the era of liberalisation: High-value fruit exports from parastatal projects. *The South African Geographical Journal, 83*(3), 214–220. doi:10.1080/03736245.2001.9713739

Manogna, R. L., & Mishra, A. K. (2021). Forecasting spot prices of agricultural commodities in India: Application of deep-learning models. *Intelligent Systems in Accounting, Finance & Management*, *28*(1), 72–83. doi:10.1002/isaf.1487

Maponya, P., Venter, S. L., Modise, D., Heever, E. V. D., Kekana, V., Ngqandu, A., Ntanjana, N., & Pefile, A. (2015). Determinants of Agricultural Market Participation in the Sarah Baartman District, Eastern Cape of South Africa. *Journal of Human Ecology (Delhi, India)*, *50*(1), 1–9. doi:10.1080/09709274.2015.11906854

Marian, L., Chrysochou, P., Krystallis, A., & Thøgersen, J. (2014). The role of price as a product attribute in the organic food context: An exploration based on actual purchase data. *Food Quality and Preference*, *37*, 52–60. doi:10.1016/j.foodqual.2014.05.001

Markelova, H., & Mwangi, E. (2010). Collective action for smallholder market access: Evidence and implications for Africa. *The Review of Policy Research*, *27*(5), 621–640. doi:10.1111/j.1541-1338.2010.00462.x

Martey, E., Al-Hassan, R. M., & Kuwornu, J. K. (2012). Commercialization of smallholder agriculture in Ghana: A Tobit regression analysis. *African Journal of Agricultural Research*, *7*(14), 2131–2141.

Marwa, T., Bashir, A., Azwardi, A., Adam, M. and Thamrin, K.M., (2022). Market Integration of Agricultural Products.

Marx, A., & Wouters, J. (2015). Is Everybody on Board? Voluntary Sustainability Standards and Green Restructuring. *Development*, *58*(4), 511–520. doi:10.105741301-016-0051-z

Mason, N. M., Jayne, T. S., & Myers, R. J. (2015). Smallholder supply response to marketing board activities in a dual channel marketing system: The case of Zambia. *Journal of Agricultural Economics*, *66*(1), 36–65. doi:10.1111/1477-9552.12066

Materechera, F., & Scholes, M. (2022). There's a place for big and small farms in securing South Africa's food supply. *Academic rigour, journalistic flair*. https://theconversation.com/theres-a-place-for-big-and-small-farms-in-securing-south-africas-food-supply-180826

Mather, D., & Jayne, T. S. (2019). The impact of state marketing board operations on smallholder behavior and incomes: The case of Kenya. *Gates Open Research*, *3*(993), 993.

Ma, W., & Abdulai, A. (2016a). Linking apple farmers to markets: Determinants and impacts of marketing contracts in China. *China Agricultural Economic Review*, *8*(1), 2–21. doi:10.1108/CAER-04-2015-0035

Ma, W., & Abdulai, A. (2016b). Does cooperative membership improve household welfare? Evidence from apple farmers in China. *Food Policy*, *58*, 94–102. doi:10.1016/j.foodpol.2015.12.002

Ma, W., Zheng, H., Zhu, Y., & Qi, J. (2022). Effects of cooperative membership on financial performance of banana farmers in China: A heterogeneous analysis. *Annals of Public and Cooperative Economics*, *93*(1), 5–27. doi:10.1111/apce.12326

Maylan, P. (1986). *A history of the African people of South Africa, from the early iron age to the 1970s*. David Phillips.

Ma, Z., Chen, J., Tian, G., Guo, B., & Cheng, F. (2021). Regulations on the corporate social irresponsibility in the supply chain under the multiparty game: Taking China's organic food supply chain as an example. *Journal of Cleaner Production*, *317*, 128459. doi:10.1016/j.jclepro.2021.128459

Mazibuko, N., Balarane, A., & Matsobane, M. (2016). *South Africa's statutory levies in context of same levies elsewhere-are we funding the correct area?* Research Gate. https://www.researchgate.net/publication/320164621_SOUTH_AFRICA'S_STATUTORY_LEVIES_IN_CONTEXT_OF_SAME_LEVIES_ELSEWHERE_ARE_WE_FUNDING_THE_CORRECT_AREA

McEachern, M. G., & Willock, J. (2004). Producers and consumers of organic meat: A focus on attitudes and motivations. *British Food Journal, 106*(7), 534–552. doi:10.1108/00070700410545737

McEvoy, M. (2019, March 13). *Organic 101: What the USDA organic label means*. US Department of Agriculture.

Mezui, C. A. M., Rutten, L., Sekioua, S., Zhang, J., N'Diaye, M. M., Kabanyane, N., Arvanitis, Y., Duru, U., & Nekati, B. (2013). *Guidebook on African commodity and derivatives exchanges*. African Development Bank.

Mitiku, F., de Mey, Y., Nyssen, J., & Maertens, M. (2017). Do private sustainability standards contribute to income growth and poverty alleviation? A comparison of different coffee certification schemes in Ethiopia. *Sustainability, 9*(2), 1–21. doi:10.3390u9020246

Mmbando, F. E., Wale, E. Z., & Baiyegunhi, L. J. (2015). Determinants of smallholder farmers' participation in maize and pigeonpea markets in Tanzania. *Agrekon, 54*(1), 96–119. doi:10.1080/03031853.2014.974630

Mmushi, M. P. (2020). *Towards an appropriate model for skills transfer to land reform beneficiaries in South Africa* [Doctoral dissertation, University of Pretoria].

Modise, L., & Mtshiselwa, N. (2013). The Natives Land Act of 1913 engineered the poverty of black South Africans: A historico-ecclesiastical perspective. *Studia Historiae Ecclesiasticae, 39*(2), 359–378.

Moerbeek, H., & Casimir, G. (2005). Gender differences in consumers' acceptance of genetically modified foods. *International Journal of Consumer Studies, 29*(4), 308–318. doi:10.1111/j.1470-6431.2005.00441.x

Mohr, P., Harrison, A., Wilson, C., Baghurst, K. I., & Syrette, J. (2007). Attitudes, values, and socio-demographic characteristics that predict acceptance of genetic engineering and applications of new technology in Australia. *Biotechnology Journal, 2*(9), 1169–1178. doi:10.1002/biot.200700105 PMID:17703489

Mojo, D., Fischer, C., & Degefa, T. (2017). The determinants and economic impacts of membership in coffee farmer cooperatives: Recent evidence from rural Ethiopia. *Journal of Rural Studies, 50*, 84–94. doi:10.1016/j.jrurstud.2016.12.010

Mokhtar, K. S., & Mahalingam, R. (2010). The current biotechnology outlook in Malaysia. *Economia. Seria Management, 13*(1), 11–26.

Molatlhwa, T. M. (1976). Agriculture in Bophuthatswana. In *Strategy for Development* (pp. 73–89). Palgrave Macmillan. doi:10.1007/978-1-349-02896-2_6

Montuori, A. (2005). Literature review as creative inquiry: Reframing scholarship as a creative process. *Journal of Transformative Education, 3*(4), 374–393. doi:10.1177/1541344605279381

Moon, W., & Balasubramanian, S. K. (2001). A multiple-attribute model of public acceptance of genetically modified organisms. *American Agricultural Economics Association annual meeting,* (pp. 1-19). AEA.

Moon, W., & Balasubramanian, S. K. (2004). Public attitude toward agrobiotechnology: The mediating role of risk perceptions on the impact of trust. Awareness and outrage. *Review of Agricultural Economics, 26*(2), 186–208. doi:10.1111/j.1467-9353.2004.00170.x

Mpetile, Z., & Chinyamurindi, W. (2021). Motivational factors into agriculture as a career path: narratives of Black emerging farmers as entrepreneurs in South Africa. *Journal of Enterprising Communities: People and Places in the Global Economy.*

Msibi, M.L. and Dlamini, P.Z., (2011). Water allocation reform in South Africa: History, processes and prospects for future implementation. *Water Research Commission Report, 11.*

Munuo, E. R. (2013). *Roadside vending as a business strategy and marketing option for agricultural crops: a case study of Moshi-Arusha highway*. Mzumbe University.

Muricho, G., Manda, D., Sule, F., & Kassie, M. (2017). Smallholder agricultural commercialization and poverty: empirical evidence of panel data from Kenya. *91st Annual Conference*. AgEcon. 10.22004/ag.econ.258618

Myeni, L., Moeletsi, M., Thavhana, M., Randela, M., & Mokoena, L. (2019). Barriers affecting sustainable agricultural productivity of smallholder farmers in the Eastern Free State of South Africa. *Sustainability*, *11*(11), 3003. doi:10.3390u11113003

Nagaraj, S. (2021). Role of consumer health consciousness, food safety & attitude on organic food purchase in emerging market: A serial mediation model. *Journal of Retailing and Consumer Services*, *59*, 102423. doi:10.1016/j.jretconser.2020.102423

NAMC. (1998). *Report on the investigation into fresh produce marketing: Report 1 Johannesburg Fresh Produce Market*. NAMC. https://www.namc.co.za/wp-content/uploads/2017/09/FreshProduceStudy1.pdf

NAMC. (2006). *Section 7 Investigation: National Fresh Produce Markets*. N. A. P. Council. https://www.namc.co.za/wp-content/uploads/2017/09/NationalFreshProduceMarkets.Section7InvestigationFinalReport.pdf

NAMC. (2007). *Investigation: National Fresh Produce Markets*. The National Agricultural Marketing Council.

NAMC. (2019). *Status Report On Statutory Measures Implemented In Terms Of The Marketing Of Agricultural Products Act, Act No. 47 Of 199*. NAMC. https://www.namc.co.za/wp-content/uploads/2020/03/2019-Stat-measures-REPORT.pdf

Namkung, Y., & Jang, S. (2013). Effects of restaurant green practices on brand equity formation: Do green practices really matter? *International Journal of Hospitality Management*, *33*, 85–95. doi:10.1016/j.ijhm.2012.06.006

Napier, T. L., Tucker, M., Henry, C., & Whaley, S. R. (2006). Consumer attitudes toward GMOs: The Ohio experience. *Journal of Food Science*, *69*(3), 79–76. doi:10.1111/j.1365-2621.2004.tb13344.x

Nap, J. P., Metz, P. L. J., Escaler, M., & Conner, A. J. (2003). The release of genetically modified crops into the environment. Part I. Overview of the current status and regulations. *The Plant Journal*, *33*(1), 1–18. doi:10.1046/j.0960-7412.2003.01602.x PMID:12943538

Nasir, V. A., & Karakaya, F. (2014). Consumer segments in organic foods market. *Journal of Consumer Marketing*, *31*(4), 263–277. doi:10.1108/JCM-01-2014-0845

National Agricultural Marketing Council. (2008/9). *Annual Report, 2008/09*. Agricultural Marketing Council. https://www.namc.co.za/wp-content/uploads/2017/10/ANNUALREP2009FINAL PRINTED 300909.doc.pdf

National Agricultural Marketing Council. (2018). *Annual Report, 2019*. Agricultural Marketing Council. amc.co.za/wp-content/uploads/2020/01/2018-19-NAMC-ANNUAL-REPORT-13-09-2019-Low-Reso.pdf

National Agricultural Marketing Council. (2020). *Annual Report, 2019/20*. Agricultural Marketing Council. https://www.namc.co.za/wp-content/uploads/2021/04/NAMC-ANNUAL-REPORT-2019-20.pd

National Agricultural Marketing Council. (n.d.). *Statutory Measures*. NAMC. https://www.namc.co. za/ wp-content/uploads/2017/09/Statutory-Measure-Profile.pdf

National Treasury. (2020). *Comprehensive Agriculture Support Programme*. National Treasury. https://www.gtac.gov.za/pepa/wp-content/uploads/2021/11/1.-Agriculture-Support-Programme-Spending-Review-Report.pdf

Nayga, R. M., Fisher, M. G., & Onyango, B. (2006). Acceptance of genetically modified food: Comparing consumer perspectives in the United States and South Korea. *Agricultural Economics*, *34*(3), 331–341. doi:10.1111/j.1574-0864.2006.00129.x

NEPAD. (2013). *African agriculture, transformation and outlook*. NEPAD Agency for the African Union.

Ngobo, P. V. (2011). What drives household choice of organic products in grocery stores? *Journal of Retailing*, *87*(1), 90–100. doi:10.1016/j.jretai.2010.08.001

Ngobo, P.-V., & Jean, S. (2012). Does store image influence demand for organic store brands? *Journal of Retailing and Consumer Services*, *19*(6), 621–628. doi:10.1016/j.jretconser.2012.08.003

Nicolia, A., Manzo, A., Veronesi, F., & Rosellini, D. (2014). An overview of the last 10 years of genetically engineered crop safety research. *Critical Reviews in Biotechnology*, *34*(1), 77–88. doi:10.3109/07388551.2013.823595 PMID:24041244

NPC. (2011). *National Development Plan 2030: Our future - make it work*. National Planning Commission. The Presidency, Republic Of South Africa.

Nwafor, C. (2020). Market Participation Variables in Commercializing South African Smallholder Farmers. (Preprints). doi:10.20944/preprints202003.0171.v1

Nyoro, J. K. (2019). *Agriculture and rural growth in Kenya*. Tegemeo Institute.

Obi, A., & Maya, O. (2021). Innovative Climate-Smart Agriculture (CSA) Practices in the Smallholder Farming System of South Africa. *Sustainability*, *13*(12), 6848. doi:10.3390u13126848

Ogutu, S. O., & Qaim, M. (2019). Commercialization of the small farm sector and multidimensional poverty. *World Development*, *114*, 281–293. doi:10.1016/j.worlddev.2018.10.012

Ojermark, A. (2007). Presenting Life Histories: A literature review and annotated bibliography. *Chronic Poverty Research Centre Working Paper* (101).

Okello, J. J., Lagerkvist, C.-J., Hess, S., Ngigi, M., & Karanja, N. (2012). Choice of fresh vegetable retail outlets by developing-country urban consumers: The case of kale consumers in Nairobi, Kenya. *European Journal of Development Research*, *24*(3), 434–449. doi:10.1057/ejdr.2011.58

Okem, A. E., & Lawrence, R. (2013). Exploring the opportunities and challenges of network formation for cooperatives in South Africa. *KCA Journal of Business Management*, *5*(1), 16–33.

Okou, C., Spray, J., & Unsal, D. F. (2022). Staple Food Prices in Sub-Saharan Africa: An Empirical Assessment. *IMF Working Papers, WP/22/135*

Olwande, J., Smale, M., Mathenge, M. K., Place, F., & Mithöfer, D. (2015). Agricultural marketing by smallholders in Kenya: A comparison of maize, kale and dairy. *Food Policy*, *52*, 22–32. doi:10.1016/j.foodpol.2015.02.002

Omar, A. R. C., Ishak, S., Othman, A. S., & Sum, S. M. (2022). Exploring the Catalyst and Constraint Factors of Agriculture Cooperative Performance. *International Journal of Business and Society*, *23*(1), 476–495. doi:10.33736/ijbs.4626.2022

Omiti, J.M., Otieno, D.J., Nyanamba, T.O. and McCullough, E.B., (2009). Factors influencing the intensity of market participation by smallholder farmers: A case study of rural and peri-urban areas of Kenya. *African Journal of Agricultural and Resource Economics, 3*(311-2016-5509), 57-82.

Onyango, B. (2004). Consumer acceptance of genetically modified foods: The role of product benefits and perceived risks. *Journal of Food Distribution Research, 35*(1), 154–161.

Onyango, B., Ferdaus, H., Hallman, W., Schilling, B., & Adelajan, A. (2003). Public perception of food biotechnology: Uncovering factors driving consumer acceptance of genetically modified food. *Journal of Food Distribution Research, 34*(1), 37–42.

Onyango, B., Govindasamy, R., & Nayga, J. R. M. (2004). *Measuring US consumer preferences for genetically modified foods using choice modelling experiments: the role of price, product benefits and technology*. Food Policy Institute.

Oosterveer, P., Adjei, B. E., Vellema, S., & Slingerland, M. (2014). Global sustainability standards and food security: Exploring unintended effects of voluntary certification in palm oil. *Global Food Security, 3*(3-4), 220–226. doi:10.1016/j.gfs.2014.09.006

Oppermann, M. (2000). Triangulation - A methodological discussion. *International Journal of Tourism Research, 2*(2), 141–145. doi:10.1002/(SICI)1522-1970(200003/04)2:2<141::AID-JTR217>3.0.CO;2-U

Ortmann, G. F., & King, R. P. (2007). Agricultural cooperatives II: Can they facilitate access of small-scale farmers in South Africa to input and product markets? *Agrekon, 46*(2), 219–244. doi:10.1080/03031853.2007.9523769

Ostrom, E. (2010). Beyond Markets and States: Polycentric Governance of Complex Economic Systems. *The American Economic Review, 100*(3), 408–444. doi:10.1257/aer.100.3.641

Otekunrin, O. A., Momoh, S., & Ayinde, I. A. (2019). Smallholder farmers' market participation: Concepts and methodological approach from Sub-Saharan Africa. *Current Agriculture Research Journal, 7*(2), 139–157. doi:10.12944/CARJ.7.2.02

Palinkas, L. A., Horwitz, S. M., Green, C. A., Wisdom, J. P., Duan, N., & Hoagwood, K. (2015). Purposeful sampling for qualitative data collection and analysis in mixed method implementation research. *Administration and Policy in Mental Health, 42*(5), 533–544. doi:10.100710488-013-0528-y PMID:24193818

Pallant, J. (2016). *SPSS Survival Manual: A step by step guide to data analysis using IBM SPSS*. McGraw-Hill.

Pani, S.K. and Jena, D., (2020). Agricultural marketing in India: a review on challenges and prospects.

Parry, M. L., Rosenzweig, C., Iglesias, A., Livermore, M., & Fischer, G. (2004). Effects of climate change on global food production under SRES emissions and socio-economic scenarios. *Global Environmental Change, 14*(1), 53–67. doi:10.1016/j.gloenvcha.2003.10.008

Patel, R. (2007). *Stuffed & starved*. Black Inc.

Pattison Resource Consulting Ltd. (2018). *Agri-Sustainability Standards, Market Access, and Prairie Wetland Conservation*. PRC. http://chrome-extension://efaidnbmnnnibpcajpcglclefindmkaj/viewer.html?pdfurl=https%3A%2F%2Fwww.phjv.ca%2Fwp-content%2Fuploads%2F2019%2F02%2F18-03-31-PHJV-Market-Access-Report_FINAL-.pdf&clen=2724400&chunk=true

Perrini, F., Castaldo, S., Misani, N., & Tencati, A. (2010). The impact of corporate social responsibility associations on trust in organic products marketed by mainstream retailers: A study of Italian consumers. *Business Strategy and the Environment, 19*(8), 512–526. doi:10.1002/bse.660

Pieterse, M. (2021). Anatomy of a crisis: Structural factors contributing to the collapse of urban municipal governance in Emfuleni, South Africa. Urban Forum.

Pili, O., & Ncube, B. (2022). Smallholder farmer coping and adaptation strategies for agricultural water use during drought periods in the Overberg and West Coast Districts, Western Cape, South Africa. *Water SA, 48*(1 January).

Pingali, P. L., & Rosegrant, M. W. (1995). Agricultural commercialization and diversification: Processes and policies. *Food Policy, 20*(3), 171–185. doi:10.1016/0306-9192(95)00012-4

Plaatjie, S. (1916). Native Life in South Africa before and since the European War and the Boer Rebellion. *SA History.* http://www.sahistory.orf/archives/native-life-south africa-and-european-war-and-boer-rebellion-sol-t-plaatjie

Pool, J. K., Khodadadi, M., & Asadi, A. (2018). The impact of congruence between self-concept and destination personality on behavioural intentions for visiting an Islamic-historical destination. *Tourism and Hospitality Research, 18*(3), 378–387. doi:10.1177/1467358416663820

Popke, E. J., & Ballard, R. (2004). Dislocating modernity: Identity, space and representations of street trade in Durban, South Africa. *Geoforum, 35*(1), 99–110.

Porter, G., & Phillips-Howard, K. (1997). Agricultural issues in the former homelands of South Africa: The Transkei. *Review of African Political Economy, 24*(72), 185–202. doi:10.1080/03056249708704252

Porter, M. E. (1998a). *Competitive advantage: creating and sustaining superior performance*. The Free Press. doi:10.1007/978-1-349-14865-3

Porter, M. E. (1998b). *Competitive strategy: Techniques for analyzing industries and competitors*. The Free Press.

Posel, D. (1987). The meaning of apartheid before 1948: Conflicting interests and forces within the Afrikaner nationalist alliance. *Journal of Southern African Studies, 14*(1), 123–139. doi:10.1080/03057078708708162

Pradhan, K. D. R. & MINSTEN, B. (2010). Agricultural Commercialization and Diversification in Bhutan. IFPRI (International Food Policy Research Institute), Washington, DC, USA.

Pradhan, D., Duraipandian, I., & Sethi, D. (2016). Celebrity endorsement: How celebrity–brand–user personality congruence affects brand attitude and purchase intention. *Journal of Marketing Communications, 22*(5), 456–473. doi:10.1080/13527266.2014.914561

Prati, G., Pietrantoni, L., & Zani, B. (2012). The prediction of intention to consume genetically modified food: Test of an integrated psychosocial model. *Food Quality and Preference, 25*(2), 163–170. doi:10.1016/j.foodqual.2012.02.011

Proctor, F., & Lucchesi, V. (2012). *Small-scale farming and youth in an era of rapid rural change*. IIED/HIVOS.

Rana, J., & Paul, J. (2020). Health motive and the purchase of organic food: A meta-analytic review. *International Journal of Consumer Studies, 44*(2), 162–171. doi:10.1111/ijcs.12556

Range, J. (2015). *A conceptual framework for the enhancement of trust in South Africa's fresh produce markets through the use of self-service technologies*. University of Pretoria. https://repository.up.ac.za/bitstream/handle/2263/52999/Range_Conceptual_2015.pdf?sequence=1

Range, J. J., & Leonard, A. (2015). Progressive trust formation within the South African fresh produce industry through the use of self-service technologies. *The African Journal of Information Systems, 7*(3), 1–26. https://digitalcommons.kennesaw.edu/ajis/vol7/iss3/1/

Rantlo, A. M., Tsoako, M., & Muroyiwa, B. (2020). Institutional Factors Influencing Diary Farmers Participation in Formal and Informal Milk Markets in Maseru Urban, Lesotho, South Africa. *Journal of Agricultural Extension*, *24*(2), 48–59. doi:10.4314/jae.v24i2.6

Rashid, S. (2015). Commodity Exchanges and Market Development: What Have we Learned? *Agriculture in an Interconnected World, International Conference on Agricultural Economist 8-14 August 2015*. AAE.

Rashid, S., Winter-Nelson, A., & Garcia, P. (2010). Purpose and potential for commodity exchanges in African economies. *Washington DC: International Food Policy Research Institute Discussion Paper, 1035*.

Rasul, G. (2021). Twin challenges of COVID-19 pandemic and climate change for agriculture and food security in South Asia. *Environmental Challenges*, *2*, 100027. doi:10.1016/j.envc.2021.100027

Reardon, T., Awokuse, T., Haggblade, S., Kapuya, T., Liverpool-Tasie, S., Meyer, F., Minten, B., Nyange, D., Rusike, J., & Tschirley, D. (2019). The quiet revolution in agri-food distribution (wholesale, logistics, retail) in Sub-Saharan Africa. In AGRA (Ed.), Africa Agriculture Status Report: The Hidden Middle: A Quiet Revolution in the Private Sector Driving Agricultural Transformation (pp. 13–28). Alliance for a Green Revolution in Africa.

Reardon, T., Timmer, C. P., & Minten, B. (2012). Supermarket revolution in Asia and emerging development strategies to include small-holder farmers. *Proceedings of the National Academy of Sciences*, *109*(31), 12332–12337.

Reddy, A. R. (2018). Cost escalation in cotton cultivation: An analysis. *Economic Affairs*, *63*(4), 295480. doi:10.30954/0424-2513.4.2018.6

Rees, D. J. (1979). Regulation of South African agriculture: performance of agricultural marketing boards. .

Republic of South Africa Parliament. (2020). Agricultural Produce Agents Amendment Bill. B22-2020, Parliament of the Republic of South Africa. https://www.parliament.gov.za/storage/app/media/Bills/2020/B33_2020_Agricultural_Produce_Agents_Amendment_Bill/B33_2020_Agricultural_Produce_Agents_Amendment_Bill.pdf

Resnick, D. (2017). Informal Food Markets in Africa's Cities. In Institute International Food Policy Research (Ed.), Global Food Policy Report 2017 (pp. 50–56). International Food Policy Research Institute.

Ricciardi, V., Ramankutty, N., Mehrabi, Z., Jarvis, L., & Chookolingo, B. (2018). How much of the world's food do smallholders produce? *Global Food Security*, *17*, 64–72. doi:10.1016/j.gfs.2018.05.002

Richetin, J., Mattavelli, S., & Perugini, M. (2016). Increasing implicit and explicit attitudes toward an organic food brand by referencing to oneself. *Journal of Economic Psychology*, *55*, 96–108. doi:10.1016/j.joep.2016.01.006

Riisgaard, L., Fibla, A. M. E., & Ponte, S. (2010). *Gender and value chain development: Evaluation study.*

Ritter, G., Walkinshaw, L. P., Quinn, E. L., Ickes, S., & Johnson, D. B. (2019). An assessment of perceived barriers to farmers' market access. *Journal of Nutrition Education and Behavior*, *51*(1), 48–56. doi:10.1016/j.jneb.2018.07.020 PMID:30249521

Rodriguez-Entrena, M., Salazar-Ordonez, M., & Sayadi, S. (2013). Applying partial least squares to model genetically modified food purchase intentions in southern Spain consumers. *Food Policy*, *40*, 44–53. doi:10.1016/j.foodpol.2013.02.001

Rudolph, M., Kroll, F., Muchesa, E., Paiker, M., & Fatti, P. (2021). Food Security in Urban Cities: A Case Study Conducted in Johannesburg, South Africa. *Journal of Food Security*, *9*(2), 46–55. doi:10.12691/jfs-9-2-2

Ruwanza, S., Thondhlana, G., & Falayi, M. (2022). Research progress and conceptual insights on drought impacts and responses among smallholder farmers in South Africa: A review. *Land (Basel)*, *11*(2), 159. doi:10.3390/land11020159

Sabu, K. M., & Kumar, T. M. (2020). Predictive analytics in Agriculture: Forecasting prices of Arecanuts in Kerala. *Procedia Computer Science*, *171*, 699–708. doi:10.1016/j.procs.2020.04.076

Saha, A., Sabates-Wheeler, R., & Thompson, J. (2022). Insights into smallholder capacity for agricultural commercialisation: Evidence from four African contexts. *European Journal of Development Research*, *34*(4), 1757–1802. doi:10.105741287-021-00414-z

Salazar, E., & Montenegro, G. (2009). Genetically modified crops in Chile. *Ciencia e Investigación Agraria*, *36*(3), 353–368. doi:10.4067/S0718-16202009000300003

Sarris, A., & Morrison, J. (Eds.). (2010). *Food security in Africa: Market and trade policy for staple foods in Eastern and Southern Africa*. Edward Elgar Publishing. doi:10.4337/9781849806367

Satgar, V., & Williams, M. (2008). The passion of the people: Successful cooperative experience in Africa. Johannesburg: Cooperative and Policy Alternative Centre (COPAC).

Satgar, V. (2011). Challenging the globalized agro-food complex: Farming cooperatives and the emerging solidarity economy alternative in South Africa. *The Journal of Labour and Society*, *14*(2), 177–190. doi:10.1111/j.1743-4580.2011.00331.x

SAUFM. (2022). *Comparative figure of turnover and mass of the 17 Fresh Produce Markets*. SAUFM.

Saumure, K., & Given, L. M. (2008). Data saturation. In L. M. Given (Ed.), *The SAGE Encyclopedia of Qualitative Research Methods* (pp. 195–196). SAGE.

Savadori, L., Savio, S., Nocotra, E., Rumiati, R., Finucane, M., & Slovic, P. (2004). Expert and public perception of risk from biotechnology. *Risk Analysis*, *24*(5), 1289–1299. doi:10.1111/j.0272-4332.2004.00526.x PMID:15563295

Schirmer, S. (2001). *Market regulation and agricultural development*. Econometric Research Southern Africa, University of the Witwatersrand.

Schleenbecker, R., & Hamm, U. (2013). Consumers' perception of organic product characteristics. A review. *Appetite*, *71*, 420–429. doi:10.1016/j.appet.2013.08.020 PMID:24012637

Schleifer, P. (2016). Private Governance Undermined: India the roundtable on sustainable palm oil. *Global Environmental Politics*, *16*(1), 38–58. doi:10.1162/GLEP_a_00335

Scholderer, J., & Frewer, L. J. (2003). The biotechnology communication paradox: Experimental evidence and the need for a new strategy. *Journal of Consumer Policy*, *26*(2), 125–157. doi:10.1023/A:1023695519981

Scsglobalservices.com. (2022). *ISCC PLUS Certification*. SCS Global Services. https://www.scsglobalservices.com/services/iscc-plus-certification>

Scsglobalservices.com. (n.d.). ISCC Plus Certification for the Circular Economy and Bioeconomy. SCS Global Services. https://www.scsglobalservices.com/services/iscc-plus-certification.

Sebatta, C., Mugisha, J., Katungi, E., Kashaaru, A. and Kyomugisha, H., (2014). Smallholder farmers' decision and level of participation in the potato market in Uganda.

Sebhatu, K. T., Gezahegn, T. W., Berhanu, T., Maertens, M., Van Passel, S., & D'Haese, M. (2021). Exploring variability across cooperatives: Economic performance of agricultural cooperatives in northern Ethiopia. *The International Food and Agribusiness Management Review*, *24*(3), 397–419. doi:10.22434/IFAMR2019.0215

Sebola, M. P. (2018). Financing emerging black farmers for agricultural development in South Africa: A wasteful and unworkable model for creating black farmers. *TD: The Journal for Transdisciplinary Research in Southern Africa*, *14*(1), 1–7. doi:10.4102/td.v14i1.555

Seppelt, R., Manceur, A. M., Liu, J., Fenichel, E. P., & Klotz, S. (2014). Synchronized peak-rate years of global resources use. *Ecology and Society*, *19*(4), 50. doi:10.5751/ES-07039-190450

Septianto, F., Kemper, J., & Paramita, W. (2019). The role of imagery in promoting organic food. *Journal of Business Research*, *101*, 104–115. doi:10.1016/j.jbusres.2019.04.016

Serra, R., & Davidson, K. A. (2021). Selling together: The benefits of cooperatives to women honey producers in Ethiopia. *Journal of Agricultural Economics*, *72*(1), 202–223. doi:10.1111/1477-9552.12399

Shin, M., Lee, H., Ryu, K., Cho, Y., & Son, Y. J. (2019). A two-phased perishable inventory model for production planning in a food industry. *Computers & Industrial Engineering*, *133*, 175–185. doi:10.1016/j.cie.2019.05.010

Shin, Y. H., Moon, H., Jung, S. E., & Severt, K. (2017). The effect of environmental values and attitudes on consumer willingness to pay more for organic menus: A value-attitude-behavior approach. *Journal of Hospitality and Tourism Management*, *33*, 113–121. doi:10.1016/j.jhtm.2017.10.010

Shrader, L. (2018). *Benchmarking E-Commerce Models for Africa's Smallholders: Mercy Corps Agrifin Accelerate Program*. MercyCorps. www.mercycorpsafa.org

Shuaibu, M., & Nchake, M. (2021). Impact of credit market conditions on agriculture productivity in Sub-Saharan Africa. *Agricultural Finance Review*, *81*(4), 520–534. doi:10.1108/AFR-05-2020-0063

Sibale, E. (2010). Strategies for commercialising smallholder agriculture. *How We Made it in Africa*. .https://www.howwemadeitinafrica.com/strategies-for-commercialising-smallholder-agr

Siegrist, M. (2000). The influence of trust and perceptions of risk and benefit on the acceptance of gene technology. *Risk Analysis*, *20*(2), 195–203. doi:10.1111/0272-4332.202020 PMID:10859780

Siegrist, M., & Cvetkovich, G. (2000). Perception of hazards: The role of social trust and knowledge. *Risk Analysis*, *20*(5), 713–719. doi:10.1111/0272-4332.205064 PMID:11110217

Siegrist, M., Cvetkovich, G., & Roth, C. (2000). Salient value similarity, social trust, and risk/benefit perception. *Risk Analysis*, *20*(3), 353–362. doi:10.1111/0272-4332.203034 PMID:10949414

Sihlobo, W., & Kirsten, J. (2021). Agriculture in South Africa. The Oxford Handbook of the South African Economy, 195.

Simkins, C. (1981). Agricultural production in the African reserves of South Africa, 1918–1969. *Journal of Southern African Studies*, *7*(2), 256–283. doi:10.1080/03057078108708028

Simon, H. A. (1997). *Models of bounded rationality: Empirically grounded economic reason* (Vol. 3). MIT press. doi:10.7551/mitpress/4711.001.0001

Siphesihle, Q., & Lelethu, M. (2020). Factors affecting subsistence farming in rural areas of yandeni local municipality in the Eastern Cape Province. *South African Journal of Agricultural Extension*, *48*(2), 92–105. doi:10.17159/2413-3221/2020/v48n2a540

Sirieix, L., Kledal, P. R., & Sulitang, T. (2011). Organic food consumers' trade-offs between local or imported, conventional or organic products: A qualitative study in Shanghai. *International Journal of Consumer Studies*, *35*(6), 670–678. doi:10.1111/j.1470-6431.2010.00960.x

Sitko, N. J., & Jayne, T. S. (2012). Why are African commodity exchanges languishing? A case study of the Zambian Agricultural Commodity Exchange. *Food Policy*, *37*(3), 275–282. doi:10.1016/j.foodpol.2012.02.015

Skinner, C., & Watson, V. (2020). The informal economy in urban Africa: Challenging planning theory and praxis. In M. Chen & F. Carré (Eds.), *The Informal economy revisited: Examining the past, envisioning the future* (pp. 123–131). Routledge. https://www.taylorfrancis.com/chapters/oa-edit/10.4324/9780429200724-21/informal-economy-urban-africa-caroline-skinner-vanessa-watson doi:10.4324/9780429200724-21

Smidt, H. J., & Jokonya, O. (2022). Factors affecting digital technology adoption by small-scale farmers in agriculture value chains (AVCs) in South Africa. *Information Technology for Development*, *28*(3), 558–584. doi:10.1080/02681102.2021.1975256

Smit, S., & Madubela, A. (2021, 25 October 2021). 'Terribly scary': Dysfunctional municipalities are a threat to South Africa's economy. *Mail & Guardian*. https://mg.co.za/politics/2021-10-25-terribly-scary-dysfunctional-municipalities-are-a-threat-to-south-africas-economy/

Snyder, C. M. (1996). A dynamic theory of countervailing power. *The RAND Journal of Economics*, *27*(4), 747–769. doi:10.2307/2555880

Sønderskov, K. M., & Daugbjerg, C. (2011). The state and consumer confidence in eco-labeling: Organic labeling in Denmark, Sweden, The United Kingdom and The United States. *Agriculture and Human Values*, *28*(4), 507–517. doi:10.100710460-010-9295-5

Soyez, K., Francis, J. N.-P., & Smirnova, M. M. (2012). How individual, product and situational determinants affect the intention to buy and organic food buying behavior: A cross-national comparison in five nations. *der markt, 51*, 27-35. doi:10.1007/s12642-011-0073-8

Staff, W. (2021, 24 October 2021). 8 reasons why South Africa's municipalities are completely broken. *Business Tech*. https://businesstech.co.za/news/government/528846/8-reasons-why-south-africas-municipalities-are-completely-broken/

Statista. (2021). *Organic food sales in the United States from 2005 to 2020*. Satista: https://www.statista.com/statistics/196952/organic-food-sales-in-the-us-since-2000/

Statistics South Africa. (2012). *Social profile of vulnerable groups in South Africa: 2002-2011*. SSA.

STATS SA. (2016). *Community survey Agricultural households – statistics South Africa*. SA Stats. www.statssa.gov.za>publications>prese

Stats, S. A. (2019). *Towards measuring the extent of food security in South Africa: An examination of hunger and food adequacy*. SA Stats. http://www.statssa.gov.za/publications/03-00-14/03-00-142017.pdf

Steur, D. H., Gellynck, X., Storozhenko, S., Liqun, G., Lambert, W., Van Der Straeten, D., & Viaene, J. (2010). Willingness-to-accept and purchase genetically modified rice with high folate content in Shanxi Province, China. *Appetite*, *54*(1), 118–125. doi:10.1016/j.appet.2009.09.017 PMID:19815041

Stevens, J. B. (2017). Is agricultural extension positioned to promote agripreneurship in South Africa? *South African Journal of Agricultural Extension*, *45*(2), 86–94. doi:10.17159/2413-3221/2017/v45n1a437

Stifel, D., & Minten, B. (2017). Market access, well-being, and nutrition: Evidence from Ethiopia. *World Development*, *90*, 229–241. doi:10.1016/j.worlddev.2016.09.009

Stolz, H., Stolze, M., Hamm, U., Janssen, M., & Ruto, E. (2011). Consumer attitudes towards organic versus conventional food with specific quality attributes. *NJAS Wageningen Journal of Life Sciences*, *58*(3-4), 67–72. doi:10.1016/j.njas.2010.10.002

Storstad, O., & Bjørkhaug, H. (2003). Foundations of production and consumption of organic food in Norway: Common attitudes among farmers and consumers? *Agriculture and Human Values*, *20*(2), 151–163. doi:10.1023/A:1024069627349

Sudrajat, J., Isytar, I., & Arifin, N. (2021). Farmers' perception and engagement with the role of middlemen: A case study of the vegetable farmers. *Makara Human Behavior Studies in Asia*, *25*(1), 45–54. doi:10.7454/hubs.asia.1070220

Sultan, P., Wong, H. Y., & Sigala, M. (2018). Segmenting the Australian organic food consumer market. *Asia Pacific Journal of Marketing and Logistics*, *30*(1), 163–181. doi:10.1108/APJML-10-2016-0211

Tagwi, A. (2022). The Impacts of Climate Change, Carbon Dioxide Emissions (CO2) and Renewable Energy Consumption on Agricultural Economic Growth in South Africa: ARDL Approach. *Sustainability*, *14*(24), 16468. doi:10.3390u142416468

Tagwi, A., & Chipfupa, U. (2022). Participation of Smallholder Farmers in Modern Bioenergy Value Chains in Africa: Opportunities and Constraints. *BioEnergy Research*, 1–15. doi:10.100712155-022-10451-z

Talwar, S., Jabeen, F., Tandon, A., Sakashita, M., & Dhir, A. (2021). What drives willingness to purchase and stated buying behavior toward organic food? A Stimulus–Organism–Behavior–Consequence (SOBC) perspective. *Journal of Cleaner Production*, *293*, 125882. doi:10.1016/j.jclepro.2021.125882

Tandon, A., Dhir, A., Kaur, P., Kushwah, S., & Salo, J. (2020). Why do people buy organic food? The moderating role of environmental concerns and trust. *Journal of Retailing and Consumer Services*, *57*, 102247. doi:10.1016/j.jretconser.2020.102247

Tandon, A., Jabeen, F., Talwar, S., Sakashita, M., & Dhir, A. (2021). Facilitators and inhibitors of organic food buying behavior. *Food Quality and Preference*, *88*, 104077. doi:10.1016/j.foodqual.2020.104077

Tanius, E., & Seng, S. W. (2015). Consumer awareness towards genetically modified (GM) foods. *International Journal of Business, Economics and Law*, *6*(2), 17–26.

Tarekegn, K., Haji, J., & Tegegne, B. (2017). Determinants of honey producer market outlet choice in Chena District, southern Ethiopia: A multivariate probit regression analysis. *Agricultural and Food Economics*, *5*(1), 1–14. doi:10.118640100-017-0090-0

Tavakol, M., & Dennick, R. (2011). Making sense of Cronbach's alpha. *International Journal of Medical Education*, *2*, 53–55. doi:10.5116/ijme.4dfb.8dfd PMID:28029643

Tayleur, C., Balmford, A., Buchanan, G. M., Butchart, S. H. M., Corlet Walker, C., Ducharme, H., Green, R. E., Milder, J. C., Sanderson, F. J., Thomas, D. H. L., Tracewski, L., Vickery, J., & Phalan, B. (2018). Where are commodity crops certified, and what does it mean for conservation and poverty alleviation? *Biological Conservation*, *217*, 36–46. doi:10.1016/j.biocon.2017.09.024

Thamaga-Chitja, J. M., & Morojele, P. (2014). The context of smallholder farming in South Africa: Towards a livelihood asset building framework. *Journal of Human Ecology (Delhi, India)*, *45*(2), 147–155. doi:10.1080/09709274.2014.11906688

The Limpopo Provincial Government. (2022). *Homepage.* LPG. http://www.limpopo.gov.za

Thøgersen, J., Pedersen, S., Paternoga, M., Schwendel, E., & Aschemann-Witze, J. (2017). How important is country-of-origin for organic food consumers? A review of the literature and suggestions for future research. *British Food Journal*, *119*(3), 542–557. doi:10.1108/BFJ-09-2016-0406

Tilman, D., Balzer, C., Hill, J., & Befort, B. L. (2011). Global food demand and the sustainable intensification of agriculture. *Proceedings of the National Academy of Sciences of the United States of America*, *108*(50), 20260–20264. doi:10.1073/pnas.1116437108 PMID:22106295

TMX. (2022). *How Tanzania Mercantile Exchange works.* TMX.. https://www.tmx.co.tz/trading.php

Ton, G., Vellema, W., Desiere, S., Weituschat, S., & D'Haese, M. (2018). Contract farming for improving smallholder incomes: What can we learn from effectiveness studies? *World Development*, *104*, 46–64. doi:10.1016/j.worlddev.2017.11.015

Traill, W. B., Jaeger, S. R., Yee, W. M. S., Valli, C., House, L. O., Lusk, J. L., Moore, M., & Morrow, J. L. (2004). Categories of gm risk-benefit perceptions and their antecedents. *AgBioForum*, *7*, 176–186.

Traill, W. B., Yee, W. M. S., Lusk, J. L., Jaeger, S. R., House, L. O., Morrow, J. L., Valli, C., & Moore, M. (2006). Perception of risks and benefits of GM foods and their influence on willingness to consume. *Food Economics*, *3*(1), 12–19.

Trebilcock, M., Howse, R. and Eliason, A., (2013). The Regulation of International Trade.

Truong, V. A., Lang, B., & Conroy, D. M. (2021). Are trust and consumption values important for buyers of organic food? A comparison of regular buyers, occasional buyers, and non-buyers. *Appetite*, *161*, 105123. doi:10.1016/j.appet.2021.105123 PMID:33493610

Tschirley, D. L., Ayieko, M., Mathenge, M., & Weber, M. T. (2004). *Where Do Consumers in Nairobi Purchase their Food and Why Does this Matter? The Need for Investment to Improve Kenya's "Traditional" Food Marketing System.* Tegemeo Institute for Agricultural Policy Development. https://www.tegemeo.org/

Tshikororo, M., Thaba, K., Nefale, T. A., & Tshikororo, M. (2021). Perception Driven from Farmers' Socio-economic Characteristics Towards Tackling Climate change. *The Journal of Agricultural Science*, *13*(6), 41–47. doi:10.5539/jas.v13n6p41

Tung, O. J. L. (2016). Organic food certification in South Africa: A private sector mechanism in need of state regulation? *Potchefstroom Electronic Law Journal/Potchefstroomse Elektroniese Regsblad, 19*(1).

U.S. Department of Agriculture. (2016), *Bangladesh Agricultural Biotechnology Annual 2016.* https://www.agcensus.usda.gov/Publications/

UNFSS. (2022). *Flagship Publication.* UNFSS. <https://unfss.org/home/flagship-publication/

Unsal, F., Garbers, H., Bogmans, C., & Prifiti, E. (2020). *Food (in)security in SSA under COVID-19: a macro perspective.* http://dmedms.imf.org/cyberdocs/viewdocument.asp?doc=6981213&lib=DMSDR1S

Ureña, F., Bernabéu, R., & Olmeda, M. (2008). Women, men and organic food: Differences in their attitudes and willingness to pay. A Spanish case study. *International Journal of Consumer Studies*, *32*(1), 18–26. doi:10.1111/j.1470-6431.2007.00637.x

Usman, M. A., & Callo-Concha, D. (2021). Does market access improve dietary diversity and food security? Evidence from Southwestern Ethiopian smallholder coffee producers. *Agricultural and Food Economics*, *9*(1), 1–21. doi:10.118640100-021-00190-8

Valentinov, V. (2007). Why are cooperatives important in agriculture? An organizational economics perspective. *Journal of Institutional Economics*, *3*(1), 55–69. doi:10.1017/S1744137406000555

Van Den Ban and Hawkins. (1990). *Agricultural Extension. Longman Scientific & Technical.* John Wiley & Sons, Inc.

Van den Bossche, P., & Zdouc, W. (2021). *The Law and Policy of the World Trade Organization.* Cambridge Books. doi:10.1017/9781108784542

Van der Ploeg, J. D., Huifang, W., & Congzhi, H. (2016). Markets. In J. D. Van der Ploeg & Y. Jingzhong (Eds.), *China's Peasant Agriculture and Rural Society: Changing paradigms of farming.* doi:10.4324/9781315643250

Van der Ploeg, J. D., Jingzhong, Y., & Schneider, S. (2012). Rural development through the construction of new, nested, markets: Comparative perspectives from China, Brazil and the European Union. *The Journal of Peasant Studies*, *39*(1), 133–173. doi:10.1080/03066150.2011.652619

van der Ploeg, J. D., Ye, J., & Schneider, S. (2022). Reading markets politically: On the transformativity and relevance of peasant markets. *The Journal of Peasant Studies*, 1–26. doi:10.1080/03066150.2021.2020258

van Doorn, J., & Verhoef, P. C. (2011). Willingness to pay for organic products: Differences between virtue and vice foods. *International Journal of Research in Marketing*, *28*(3), 167–180. doi:10.1016/j.ijresmar.2011.02.005

van Herpen, E., van Nierop, E., & Sloot, L. (2012). The relationship between in-store marketing and observed sales for organic versus fair trade products. *Marketing Letters*, *23*(1), 293–308. doi:10.100711002-011-9154-1

Van Meijl, H., Havlik, P., Lotze-Campen, H., Stehfest, E., Witzke, P., Domínguez, I. P., Bodirsky, B. L., van Dijk, M., Doelman, J., Fellmann, T., Humpenöder, F., Koopman, J. F. L., Müller, C., Popp, A., Tabeau, A., Valin, H., & van Zeist, W.-J. (2018). Comparing impacts of climate change and mitigation on global agriculture by 2050. *Environmental Research Letters*, *13*(6), 064021. doi:10.1088/1748-9326/aabdc4

Van Zyl, J., Kirsten, J., & Binswanger, M. P. (1996). Agricultural land reform in South Africa: Policies, markets, and mechanisms. *Agris.* https://agris.fao.org/agris-search/search.do?recordID=XF2015022733

Veeman, M., Adamowicz, W., & Hu, W. (2005). *Risk perceptions, social interactions and the influence of information on social attitudes to agricultural biotechnology.* [Unpublished manuscript, Department of Rural Economy, Faculty of Agriculture and Forestry, and Home Economics, University of Alberta, Edmonton, Canada].

Vega-Zamora, M., Torres-Ruiz, F. J., Murgado-Armenteros, E. M., & Parras-Rosa, M. (2014). Organic as a heuristic cue: What Spanish consumers mean by organic foods. *Psychology and Marketing*, *31*(5), 349–359. doi:10.1002/mar.20699

Verdier-Chouchane, A. & Karagueuzian, C. (2016). Economic development in the Middle East and North Africa. Challenges and Prospects. 8TS.

Verdurme, A. and Viaene, J. (2003). Consumer beliefs and attitude towards genetically modified food: Basis for segmentation and implications for communication. *Agribusiness, 19*(1), 91-113.

Verdurme, A., & Viaene, J. (2001). Consumer attitudes towards GM food: Literature review and recommendations for effective communication. *Journal of International Food & Agribusiness Marketing*, *13*(2-3), 77–98. doi:10.1300/J047v13n02_05

Vermeulen, W. J. V., & Kok, M. T. J. (2012). Government interventions in sustainable supply chain governance: Experience in Dutch front-running cases. *Ecological Economics*, *83*, 183–196. doi:10.1016/j.ecolecon.2012.04.006

Vilella-Vila, M., Costa-Font, J., & Mossialos, E. (2005). Consumer involvement and acceptance of biotechnology in the European Union: A specific focus on Spain and the U.K. *International Journal of Consumer Studies*, *29*(2), 108–118. doi:10.1111/j.1470-6431.2004.00425.x

Vink, N. (1999). South African agriculture in the 1970s. *South African Journal of Economic History, 14*(1_2), 90-113.

Vink, N. (2012). The long-term economic consequences of agricultural marketing legislation in South Africa. *The South African Journal of Economics*, *80*(4), 553–566. doi:10.1111/j.1813-6982.2011.01302.x

Voegtlin, C., & Scherer, A. G. (2017). Responsible innovation and the innovation of responsibility: Governing sustainable development in a globalized world. *Journal of Business Ethics*, *143*(2), 227–243. doi:10.100710551-015-2769-z

Vorley, B. (2013). *Meeting small-scale farmers in their markets: understanding and improving the institutions and governance of informal agrifood trade*. International Institute for Environment and Development/HIVOS/Mainumby.

Vorley, B., Del Pozo-Vergnes, E., & Barnett, A. (2012). *Small producer agency in the globalised market: Making choices in a changing world.* IIED/HIVOS.

Wadsworth, J.J. (2019*). Marketing Operations of Dairy Cooperatives, (No. 1502-2021-2279).*

Wale, E., & Chipfupa, U. (2021). Entrepreneurship concepts/theories and smallholder agriculture: Insights from the literature with empirical evidence from KwaZulu- Natal, South Africa. *Transactions of the Royal Society of South Africa*, *76*(1), 67–79. doi:10.1080/0035919X.2020.1861122

Wallace, E., Buil, I., & de Chernatony, L. (2017). Consumers' self-congruence with a "Liked" brand: Cognitive network influence and brand outcomes. *European Journal of Marketing*, *51*(2), 367–390. doi:10.1108/EJM-07-2015-0442

Wanyama, R., Gödecke, T., Chege, C. G. K., & Qaim, M. (2019). How important are supermarkets for the diets of the urban poor in Africa? *Food Security*, *11*(6), 1339–1353. doi:10.100712571-019-00974-3

Weatherspoon, D. D., & Reardon, T. (2003). The rise of supermarkets in Africa: Implications for agrifood systems and the rural poor. *Development Policy Review*, *21*(3), 333–355. doi:10.1111/1467-7679.00214

Wegerif, M. (2020). The Symbiotic Food System. In J. Duncan, J. S. Wiskerke, & M. S. Carolan (Eds.), *Routledge Handbook of Sustainable and Regenerative Food Systems* (pp. 188–203). Routledge. https://www.routledge.com/Routledge-Handbook-of-Sustainable-and-Regenerative-Food-Systems-1st-Edition/Duncan-Carolan-Wis
kerke/p/book/9781138608047 doi:10.4324/9780429466823-14

Wegerif, M. (2021). The impact of Covid-19 on black farmers in South Africa. *Agrekon*, 1–15.

Wegerif, M. C. A. (2018). An ethnographic exploration of food and the city. *Anthropology Today*, *34*(5), 16–19. doi:10.1111/1467-8322.12460

Wegerif, M. C. A. (2020). "Informal" food traders and food security: Experiences from the Covid-19 response in South Africa. *Food Security*, *12*(4), 797–800. doi:10.100712571-020-01078-z PMID:32837657

Wegerif, M. C. A., & Hebinck, P. (2016). The Symbiotic Food System: An 'Alternative'Agri-Food System Already Working at Scale. *Agriculture*, *6*(3), 40. doi:10.3390/agriculture6030040

Wellard, S. J., & McKenna, L. (2002). Turning tapes into text: Issues surrounding the transcription of interviews. *Contemporary Nurse*, *11*(2/3), 180–186. doi:10.5172/conu.11.2-3.180 PMID:11924614

Weng, X. (2015). *The rural informal economy: Understanding drivers and livelihood impacts in agriculture, timber and mining* (IIED Working Paper).

Willett, W., Rockström, J., Loken, B., Springmann, M., Lang, T., Vermeulen, S., Garnett, T., Tilman, D., DeClerck, F., & Wood, A. (2019). Food in the Anthropocene: the EAT–Lancet Commission on healthy diets from sustainable food systems (0140-6736). *Lancet*. https://openaccess.city.ac.uk/id/eprint/21633/

World Bank. (1994). South African agriculture: structure, performance, and options for the future. *Working Paper 12950*.

World Bank. (2008). World Development Report 2008: Agriculture for Development. Washington DC. World Bank. http://econ.worldbank.org/WBSITE/EXTERNAL/EXTDEC/EXTRESEARCH/EXTWDRS/EXTWDR2008/0,contentMDK:21410054~menuPK:3149676~pagePK:64167689~piPK:64167 673~theSitePK:2795143,00.html

World Bank. (2016). *Linking Farmers to Markets through Productive Alliances: An Assessment of the World Bank Experience in Latin America*. The World Bank.

World Health Organization. (2017), *Food, Genetically Modified*. WHO. https://www.who.int/topics/food_genetically_modified/en/ (Accessed: 17 August 2019).

Wu, X., Xiong, J., Yan, J., & Wang, Y. (2021). Perceived quality of traceability information and its effect on purchase intention towards organic food. *Journal of Marketing Management*, *37*(13/14), 1267–1286. doi:10.1080/0267257X.2021.1910328

Xie, B., Wang, L., Yang, H., Wang, Y., & Zhang, M. (2015). Consumer perceptions and attitudes of organic food products in Eastern China. *British Food Journal*, *117*(3), 1105–1121. doi:10.1108/BFJ-09-2013-0255

Xu, Z., Li, Y., Chau, S. N., Dietz, T., Li, C., Wan, L., Zhang, J., Zhang, L., Li, Y., Chung, M. G., & Liu, J. (2020). Impacts of international trade on global sustainable development. *Nature Sustainability*, *3*(11), 964–971. doi:10.103841893-020-0572-z

Yakavenka, V., Mallidis, I., Vlachos, D., Iakovou, E., & Eleni, Z. (2020). Development of a multi-objective model for the design of sustainable supply chains: The case of perishable food products. *Annals of Operations Research*, *294*(1), 593–621. doi:10.100710479-019-03434-5

Yalew, A. W., Hirte, G., Lotze-Campen, H., & Tscharaktschiew, S. (2018). Climate change, agriculture, and economic development in Ethiopia. *Sustainability*, *10*(10), 3464. doi:10.3390u10103464

Yaseen, A., Bryceson, K., & Mungai, A. N. (2018). Commercialization behaviour in production agriculture: The overlooked role of market orientation. *Journal of Agribusiness in Developing and Emerging Economies*, *8*(3), 579–602. doi:10.1108/JADEE-07-2017-0072

Yee, W. M. S., Traill, W. B., Lusk, J. L., Jaeger, S. R., House, L. O., Moore, M., Morrow, J. L. B., & Valli, C. (2008). Determinants of consumers' willingness to accept GM foods. *International Journal of Biotechnology*, *10*(2/3), 240–259. doi:10.1504/IJBT.2008.018356

Yeh, Y.-P. (2016). Market orientation and service innovation on customer perceived value: The case of supermarket retailers. *Management Research Review*, *39*(4), 449–467. doi:10.1108/MRR-08-2014-0205

Yobe, C. L., Ferrer, S. R. D., & Mudhara, M. (2020). Measuring the financial efficiency of agricultural cooperatives in South Africa: An application of the Simar–Wilson methodology. *Agrekon*, *59*(3), 269–286. doi:10.1080/03031853.2020.1761845

Yue, C., Zhao, S., Cummings, C., & Kuzma, J. (2015). Investigating factors influencing consumer willingness to buy GM food and nano-food. *Journal of Nanoparticle Research*, *17*(7), 1–19. doi:10.100711051-015-3084-4

Yu, W., Han, X., Ding, L., & He, M. (2021). Organic food corporate image and customer co-developing behavior: The mediating role of consumer trust and purchase intention. *Journal of Retailing and Consumer Services*, *59*, 102377. doi:10.1016/j.jretconser.2020.102377

Yu, X., Luo, H., Wang, H., & Feil, J. H. (2020). Climate change and agricultural trade in central Asia: Evidence from Kazakhstan. *Ecosystem Health and Sustainability*, *6*(1), 1766380. doi:10.1080/20964129.2020.1766380

Zabel, F., Delzeit, R., Schneider, J. M., Seppelt, R., Mauser, W., & Václavík, T. (2019). Global impacts of future cropland expansion and intensification on agricultural markets and biodiversity. *Nature Communications*, *10*(1), 1–10. doi:10.103841467-019-10775-z PMID:31253787

Zahonogo, P. (2018). Globalization and economic growth in developing countries: Evidence from Sub-Saharan Africa. *The International Trade Journal*, *32*(2), 189–208. doi:10.1080/08853908.2017.1333933

Zak, P. J. (2011). Moral markets. *Journal of Economic Behavior & Organization*, *77*(2), 212–233. doi:10.1016/j.jebo.2010.09.004

Zantsi, S. (2021). Unlocking the potential of the emerging smallholder farming sector in South African agriculture: An agent-based approach.

Zantsi, S., Greyling, J. C., & Vink, N. (2019). Towards a common understanding of 'emerging farmer'in a South African context using data from a survey of three district municipalities in the Eastern Cape Province. *South African Journal of Agricultural Extension*, *47*(2), 81–93. doi:10.17159/2413-3221/2019/v47n2a505

Zantsi, S., Pienaar, L. P., & Greyling, J. C. (2021). A typology of emerging farmers in three rural provinces of South Africa: What are the implications for the land redistribution policy? *International Journal of Social Economics*, *48*(5), 724–747. doi:10.1108/IJSE-10-2020-0728

Zhang, D., Chen, S., Liwen, L., & Xia, Q. (2020). Forecasting agricultural commodity prices using model selection framework with time series features and forecast horizons. *IEEE Access: Practical Innovations, Open Solutions*, *8*, 28197–28209. doi:10.1109/ACCESS.2020.2971591

Zhang, S., Sun, Z., Ma, W., & Valentinov, V. (2020). The effect of cooperative membership on agricultural technology adoption in Sichuan, China. *China Economic Review*, *62*, 101334. doi:10.1016/j.chieco.2019.101334

Zhang, X., Huang, J., Qiu, H., & Huang, Z. (2010). A consumer segmentation study with regards to genetically modified food in urban China. *Food Policy*, *35*(5), 456–462. doi:10.1016/j.foodpol.2010.04.008

Zheng, H., Ma, W., Wang, F., & Li, G. (2021). Does internet use improve technical efficiency of banana production in China? Evidence from a selectivity-corrected analysis. *Food Policy*, *102*, 102044. doi:10.1016/j.foodpol.2021.102044

Zhong, T., Si, Z., Crush, J., Scott, S., & Huang, X. (2019). Achieving urban food security through a hybrid public-private food provisioning system: The case of Nanjing, China. *Food Security*, *11*(5), 1071–1086. doi:10.100712571-019-00961-8

Zhu, X., & Xie, X. (2015). Effects of knowledge on attitude formation and change towards genetically modified foods. *Risk Analysis*, *35*(5), 790–810. doi:10.1111/risa.12319 PMID:25693867

Zika, E., Papatryfon, I., Wolf, O., Gomez-Barbero, M., Stein, A. J., & Bock, A. K. (2007). *Consequences, opportunities and challenges of modern biotechnology for Europe*. Joint Research Centre of the European Commission.

Zilberman, D., Liu, X., Holst, D. R., & Sunding, D. (2004). The economics of climate change in agriculture. *Mitigation and Adaptation Strategies for Global Change*, *9*(4), 365–382. doi:10.1023/B:MITI.0000038844.72226.13

Zondi, N. T. B., Ngidi, M. S. C., Ojo, T. O., & Hlatshwayo, S. I. (2022). Factors Influencing the Extent of the Commercialization of Indigenous Crops Among Smallholder Farmers in the Limpopo and Mpumalanga Provinces of South Africa. *Frontiers in Sustainable Food Systems*, *5*, 542. doi:10.3389/fsufs.2021.777790

Zondi, N. T. B., Ngidi, M. S. C., Ojo, T. O., & Hlatshwayo, S. I. (2022). Impact of Market Participation of Indigenous Crops on Household Food Security of Smallholder Farmers of South Africa. *Sustainability*, *14*(22), 15194. doi:10.3390u142215194

About the Contributors

Aluwani Maiwashe-Tagwi is a Senior Lecturer at the College of Agriculture and Environmental Sciences under the Department of Agriculture and Animal Health, University of South Africa. Her field of research includes agricultural marketing, cooperatives, land reform, and agricultural trade. For the past 10 years, she has worked as an Agricultural Economist, Teaching assistant, Lecturer, and senior lecturer. Her current research area is bioenergy in the agricultural sector.

Ailweli Solomon Mawela is an Associate Professor and Acting Head of Teaching and Learning at the College of Education, University of South Africa. Have obtained Doctor of Education in Didactics under college of Education. Research niche areas includes: Environmental education; Curriculum and Instructional studies; Open Distance E-Learning and Assessment in Education. Author of IGI scholarly book chapters titled "Investigating roles of school management teams in curriculum delivery."

Phineas Khazamula Chauke is a seasoned researcher and author of books who served as a school headmaster before joining the University of Venda (Univen) as lecturer in 2002. He obtained a PhD Agric) at Univen in 2006 specialising in Agricultural Policy. Both his Masters and PhD degrees focused on land reform topics, especially as regards its impact on the disenfranchised over the many years of implementing apartheid legislation especially the Natives Land Act No 27 of 1913 and other associated policies. He recently (2021) published a book entitled "Unpacking the impact of land dispossession for effective land restitution research in South Africa". His phenomenal career rise saw him achieving not only a PhD degree but also assuming the position of Head of Department: Agricultural Economics - at Univen where he served for 18 years before assuming research fellowship with the University of South Africa (Unisa) in 2021. He has supervised not less than 30 students, including more than 5 PhDs in agricultural policy, rural development, cooperatives and marketing. He also produced not less than 20 research reports for the Commission on Restitution of Land Rights.

* * *

Praveen Balakrishnan Nair is an Associate Professor attached to Edinburgh Business School at Heriot Watt University. He holds three post graduate qualifications in different functional areas of management and has earned his PhD in the area of environmental marketing. Prior to joining Heriot Watt, he was with Curtin University of Technology and Swinburne University of Technology, Australia. His current research interests lies in the area of corporate social responsibility, sustainability and business strategies and has presented and published papers in these areas.

Unity Chipfupa is a Postdoctoral Fellow in the Department of Agriculture and Animal Health, University of South Africa. His interest are agripreneurship, psychological and behavioral economics, bioenergy, youth in agriculture, sustainable agriculture, land use systems and production economics.

Makhanana Malungane is an economic researcher with an interest in the people behind entrepreneurial exploits. She tries to understand questions like: Who are they? How did they get there?. She holds an honours degree in economics specialising in international trade and finance from the University of Johannesburg. Her current obsession is in food systems and the people behind them. She is currently an MSocSci Development Studies candidate at the University of Pretoria her research focuses on the social and economic forces that shape the Johannesburg Fresh Produce Market.

Elekanyani Nekhavhambe is the Milk SA's Transformation Officer from March 2021. He graduated from the University of Venda with an MSc degree in Agriculture (Agricultural Economics) and was previously employed with the National Agricultural Marketing Council as Agricultural Economist. He gained vast experience in various agricultural industries and served on numerous industry bodies such as the SA Poultry Association Transformation Board, Macadamia South Africa NPC, and Potato SA Transformation Review Committee. His role at Milk SA is to implement the transformation programme of Milk SA in the South African dairy industry under the supervision of the Transformation Manager.

Bình Nghiêm-Phú is currently an Assistant Professor at the School of Economics and Management, University of Hyogo, Japan. The majority of his research are directed toward the understanding of consumers' perceptions and evaluations of the characteristics or images of products, services, organizations and places. He adopts the approaches of applied psychology theories to the implementation of marketing and management activities.

Ajuruchukwu Obi (Professor), is a Research Fellow at the Department of Agriculture and Animal Health, College of Agriculture and Environmental Sciences, University of South Africa, Johannesburg Campus, South Africa. He served as a Professor at the Department of Agricultural Economics & Extension, University of Fort Hare, South Africa. He is an Experienced Professor of Agricultural Economics with a demonstrated history of working in the higher education industry. Skilled in Sustainable Development, Research Design, Lecturing, Rural Development, Strong education professional with a Master's degree in Human Rights Law from University of Fort Hare, Master's degree in Economic Policy Management from McGill University in Montreal Canada, Master's degree in Agricultural Economics from the University of Nigeria and PhD in Agricultural Economics from the University of the Free State. His publications number about 100 and he has successfully supervised 19 PhDs, 65 Master's, and over 120 honours students.

Anthony Ojonimi Onoja is a Professor Extraordinarius at the Department of Agriculture and Animal Health, College of Agriculture and Environmental Sciences, University of South Africa, Johannesburg Campus, South Africa. He is a Full Professor of Resource and Environmental Economics and Director, Institute of Agricultural Research and Development at the University of Port Harcourt, Nigeria. A former Head, Department of Agricultural Economics and Extension and an adjunct Professor at the Institute of Natural Resource Environment and Sustainable Development (INRES); University of Port Harcourt, Prof. Onoja is also a Research Associate at the Resource and Environmental Policy Research Centre (REPRC),

University of Nigeria Nsukka, a Project funded by Environment for Development (EfD:). He is presently the President, Agricultural Policy Research Network (APRNet) and Executive Committee member of African Association of Agricultural Economics (AAAE). He obtained a PhD in Agricultural Economics (Resource and Environmental Economics option) from University of Nigeria, Nsukka in 2015. He was a Visiting Research Scholar (May-August 2018) at the University of South Africa (UNISA), Johannesburg in 2018 and has participated in external assessment and postgraduate supervision at University of Kwazulu Natal, South Africa, Lilongwe University of Agriculture and Natural Resources, Malawi and UNISA. Professor Onoja is a highly experienced researcher with numerous research grants/fellowships and international consultancies to his credit. He has had several research presentations at the Trade Policy Training Centre in Africa (TRAPCA) in Arusha, Tanzania. He is widely published and travelled too and his research intersects are in climate change studies especially Climate Smart Agriculture, vulnerability assessments, agricultural policy analysis, trade/macroeconomic policy, applied econometrics, sustainable agricultural livelihoods as well as agribusiness finance/project analysis. He serves in the editorial boards of four reputable journals and is an official mentor in the AWARD ICWAE Special Mentoring Orientation Program currently ongoing.

Yajeshwar Seetul holds a Bachelors Degree in Biotechnology and a Masters Degree in International Business Management with Marketing. Human behaviour is an area which stimulates his curiosity, especially how the society perceive modern scientific advancements. How beliefs and facts are wired to create a perception of reality will be constantly evolving and researchers should keep track of that evolution to ease the society into a better future.

Jillian Rae Suter is currently an Assistant Professor at the Faculty of Informatics, Shizuoka University, Japan. She obtain her Master's and Ph.D. from Keio University in Media and Governance. Her research focuses on media consumerism and intercultural interpretation of media, which recently has included elements of "Cool Japan" such as yuru-chara.

Index

I

J

L

M

N

O

P

Q

R

S

T

U

V

Y

www.ingramcontent.com/pod-product-compliance
Lightning Source LLC
LaVergne TN
LVHW080650011026
849160LV00014B/1374
* 9 7 8 1 6 6 8 4 4 7 8 1 9 *